ELECTROCHEMISTRY: THEORETICAL FOUNDATIONS

ELECTROCHEMISTRY: THEORETICAL FOUNDATIONS
QUANTUM AND STATISTICAL MECHANICS, THERMODYNAMICS, THE SOLID STATE

Jerry Goodisman
Department of Chemistry
Syracuse University
Syracuse, New York

A Wiley-Interscience Publication
JOHN WILEY & SONS
New York • Chichester • Brisbane • Toronto • Singapore

Copyright © 1987 by John Wiley & Sons, Inc.

All rights reserved. Published simultaneously in Canada.

Reproduction or translation of any part of this work beyond that permitted by Section 107 or 108 of the 1976 United States Copyright Act without the permission of the copyright owner is unlawful. Requests for permission or further information should be addressed to the Permissions Department, John Wiley & Sons, Inc.

Library of Congress Cataloging in Publication Data:

Goodisman, Jerry.
   Electrochemistry: theoretical foundations, quantum and statistical mechanics, thermodynamics, the solid state.

   "A Wiley-Interscience publication."
   Bibliography: p.
   Includes index.
   1. Electrochemistry. I. Title.
QD553.G68   1987   541.3'7   86-32495
ISBN 0-471-82850-5

Printed in the United States of America
10 9 8 7 6 5 4 3 2 1

PREFACE

Matter is composed of charged particles. The chemical properties of matter are determined by this fact, because the Coulombic interaction is the only interparticle potential in the Schrödinger equation. The Coulombic interaction thus determines electronic structure and hence chemistry. However, this fact need not be dealt with explicitly—except in electrochemistry. Electrochemical phenomena reflect directly the electrostatic interaction between the particles that comprise matter. Yet, most of electrochemical theory was built on a macroscopic level, without explicit consideration of the fundamental charged particles and their properties. Progress in quantum chemistry, statistical mechanics, and so on, has made it possible to provide a description on the molecular level of some of the macroscopic electrochemical properties and laws. In this book, we describe some of what has been developed in this domain and try to relate the molecular description to the macroscopic description and to the observed phenomena of electrochemistry.

In the traditional (thermodynamic) description of electrochemistry, the cell potential is introduced as a consequence of the existence of a cell reaction that is not at equilibrium. The cell potential is subsequently divided into contributions of the two half-cells making up the cell and eventually into individual electrode potentials, related to the two half-cell reactions. The more modern approach, as exemplified in the books by Bockris, begins from a physical description on the molecular level of the electrochemical interfaces and of conducting media. The electrochemical cell potential is then a sum of actual potential differences across the interfaces and (when current is flowing) across homogeneous regions, which, when linked together in series, make up the cell.

Since understanding what goes into a cell potential involves knowledge of solid-state and liquid-state physics and chemistry, statistical mechanics, thermodynamics, quantum mechanics, and other areas of science, study of electrochemistry ought to be central to physical chemistry. This is not at all what one would conclude from the way the subject is discussed in many physical chemistry texts. Many authors, far from relating the subject to the other areas, discuss electrochemistry as a "special topic" with its own strange and not very well founded lore. As a result, "distaste and disgust are the predominant emotions that the normal student feels for electrochemistry" (Albery 1975). In this book we try to discuss basic material in electrostatics, thermodynamics, and statistical mechanics, as well as the electrochemistry itself, in the hope of helping to alleviate this problem.

We hope that our presentation may be useful to chemists and physicists desiring to learn about the subject, and to those who are involved in teaching it to others, as well as to electrochemists whose knowledge of quantum mechanics and molecular science lags behind their knowledge of electrochemistry. The book is, in fact, the attempt of the author (a quantum chemist by training) to get it all straight for himself. He hopes, therefore, that others may find it useful and that it may help interest scientists in the problems at the electrochemical interface.

It is convenient to separate electrochemistry into phenomena of conduction, which occur in homogeneous systems, and phenomena occurring at the electrochemical interface, which is the macroscopically negligible region between two different homogeneous media. We are concerned mainly with the latter part. In the interface, charge density, particle density, electrical potential, and other intensive properties vary in space, sometimes extremely rapidly. The extreme inhomogeneity constitutes the main difficulty in generating theories and descriptions of the interface.

Chapter 1 gives a general description and overview of electrochemistry, with only minimal theoretical framework. Chapter 2 discusses thermodynamics and electrostatics, the concepts of which are used for macroscopic descriptions of the interface. Chapter 3 turns to statistical mechanics, whose province it is to relate molecular-level concepts to macroscopic quantities. Succeeding chapters (4, 5, and 6) deal with the structure of surfaces and interfaces, in the framework of the first few chapters. The remainder of the book discusses dynamical processes: diffusion (Chapter 7) and electrochemical kinetics. In Chapter 8, the processes of electrode kinetics are discussed. Chapter 9 considers the quantum mechanical aspects of electron transfer, which are incorporated in the theories of electrode reactions discussed in Chapter 10.

JERRY GOODISMAN

Syracuse, New York
April, 1987

CONTENTS

1. ELECTROCHEMISTRY   1
    A. Introduction: The Electrochemical Cell / 1
    B. Electromotive Force and Half-Cells / 4
    C. Electrochemical Potentials and Electrical Potentials / 8
    D. Kinds of Electrodes / 13
    E. Redox Electrodes / 15
    F. Polarizable Interfaces / 18
    G. Liquid Junction Potentials and Concentration Cells / 20
    H. More Complicated Cells—Applications / 23
    I. Overvoltages / 27
    J. Conduction Through Homogeneous Phases / 29
    K. Transport Numbers / 33
    L. Activation Overpotential / 35
    M. Tafel Law / 39
    N. Concentration Overpotential / 43
    O. Diffusion Plus Reaction / 46
    P. Discharge Potentials / 49
    Q. Polarography / 50

2. THERMODYNAMICS AND ELECTROSTATICS   54
    A. Surface Quantities / 54
    B. Surface Tension / 57
    C. Adsorption / 61
    D. Lippmann Equation / 64

E. Adsorption of Neutrals / 70
  F. Solid Surfaces / 75
  G. Electrostatics / 78
  H. Double Layers / 83
  I. Electrochemical Potentials / 88

3. **STATISTICAL MECHANICS**   91
  A. Introduction / 91
  B. Pressure and Surface Tension / 97
  C. Mechanical Equilibrium / 102
  D. Dielectric Constants / 108
  E. Evaluating the Partition Function / 113
  F. Electrochemical Potentials / 116
  G. Surface Charge / 121
  H. Electrode Potentials / 123

4. **STRUCTURE OF SURFACES**   130
  A. Liquid Water Surface / 130
  B. Diffuse-Layer Theory / 136
  C. Compact Layer / 144
  D. Ions in the Compact Layer / 147
  E. Metals / 153
  F. Metal Surfaces / 158
  G. Semiconductors / 165
  H. Semiconductor Surfaces / 168
  I. Surface States / 176

5. **INTERFACES**   183
  A. Metal–Metal Interface / 183
  B. Metal–Polar-Liquid Interface / 188
  C. Metal–Molten-Electrolyte Interface / 196
  D. Water Layer on Metal / 198
  E. Adsorbed Water Structure / 206
  F. Specific Adsorption / 219
  G. Semiconductor–Electrolyte Interface / 220

6. **FURTHER DEVELOPMENTS**   225
  A. Improvements on Gouy–Chapman Theory / 225
  B. Improved Models for the Interface / 232

C. Hydrated Electron / 239
D. Theories of the Hydrated Electron / 243

7. **DIFFUSION** 252
   A. Laws of Diffusion / 252
   B. Liquid Junction Potential / 255
   C. Diffusion of Electrolytes / 257
   D. Integration of Diffusion Equation / 259

8. **ELECTRODE KINETICS** 266
   A. Introduction / 266
   B. Reaction and Mass Transport / 269
   C. Multistep Reactions / 275
   D. Parallel Electrode Processes / 284
   E. Reaction Overvoltage / 286
   F. Electrocatalysis / 288
   G. Electronic Aspects / 292
   H. Semiconductor Electrodes / 294

9. **QUANTUM THEORY OF ELECTRON TRANSFER** 300
   A. Electronic Transitions / 300
   B. Role of the Solvent / 305
   C. Quantum Mechanical Tunneling / 313
   D. Wentzel–Kramers–Brillouin Tunneling Formula / 317
   E. Electron Emission / 324
   F. Other Applications / 331

10. **QUANTUM KINETICS** 336
    A. Current Density / 336
    B. Activation Process / 343
    C. Solvent Fluctuations / 350
    D. Implications / 357

REFERENCES 365

INDEX 371

1

ELECTROCHEMISTRY

A. INTRODUCTION: THE ELECTROCHEMICAL CELL

An electrochemical cell is a collection of chemical reactants connected to the surroundings by two terminals (electronic conductors) of identical composition. The arrangement of the reactants inside the cell is such that, for a chemical reaction between some of them to occur, electrical current must flow in the surroundings from one terminal to another. If, when the circuit is closed by making electrical contact between the terminals, the chemical reaction occurs spontaneously with a current flow resulting, we refer to the cell as a *galvanic cell*. If an external source of electrical potential is introduced between the terminals, with the current produced being accompanied by the chemical reaction, we refer to the cell as an *electrolytic cell*. We may suppose in either case that a voltmeter and an ammeter are connected to the cell so that one can measure the difference in electrical potential between the terminals (this is why we require them to be of identical composition) and the current, if any, flowing between them.

Inside the cell, there must be a physical separation between reactants to prevent direct reaction from occurring. This requires that several phases be present, with mixing being hindered by some barrier or by the nature of the phases. It is conventional to specify the composition of the cell by listing the phases from the left-hand terminal to the right-hand terminal, with a vertical line indicating a phase boundary. Thus the Daniell cell, the cell reaction of which is the oxidation of Zn to Zn^{2+} by Cu^{2+} ion, is represented as

$$Cu \mid Zn \mid ZnSO_4(a_1) \parallel CuSO_4(a_2) \mid Cu \qquad [1]$$

if the terminals are copper (a_i is an activity). The double vertical line indicates a liquid–liquid junction across which the electrical potential difference has been eliminated, or is assumed to be negligible. So long as Cu^{2+} ions do not reach the Zn electrode, direct chemical reaction does not occur. The liquid–liquid junction prevents such mixing while allowing electrical contact between the solutions. This is sometimes accomplished by inserting a concentrated solution or gel containing an inert electrolyte between the two solutions of the cell.

The electromotive force (emf) of a cell, in the most commonly used convention, is the electrical potential of the terminal on the right minus the electrical potential of the terminal on the left, when no current is flowing (open circuit). The value and sign of the electromotive force $\mathcal{E}$ indicate the spontaneity of the chemical reaction that will occur on closing the circuit. If $\mathcal{E}$ is positive, as is the case for the Daniell cell as given in equation [1], the tendency will be for positive electrical current to flow in the surroundings (external circuit) from the right-hand terminal to the left-hand terminal, that is, electrons will enter the cell from the surroundings through the right-hand terminal and leave the cell through the left-hand terminal. Thus, within the cell, reduction will occur at the right-hand electrode (e.g., $Cu^{2+} + 2e^- \rightarrow Cu$) and oxidation will occur at the left-hand electrode (e.g., $Zn \rightarrow Zn^{2+} + 2e^-$). The value of the emf depends, of course, on the composition and arrangement of the phases constituting the cell.

Since there are no electrical fields within any phase when no current is flowing, each phase is an equipotential at open circuit, so that the emf involves a sum of potential differences across phase boundaries. For the Daniell cell, one may write

$$\begin{aligned}\mathcal{E} &= \phi^{(R)} - \phi^{(L)} \\ &= [\phi^{(R)} - \phi^{(2)}] + [\phi^{(2)} - \phi^{(1)}] \\ &\quad + [\phi^{(1)} - \phi^{(Zn)}] + [\phi^{(Zn)} - \phi^{(L)}]\end{aligned} \quad [2]$$

where R and L identify the right-hand and left-hand copper terminals, respectively, and 1 and 2 identify the $ZnSO_4$ and $CuSO_4$ solution phases, respectively. The assumption of no liquid junction potential means $\phi^{(2)} - \phi^{(1)} = 0$. The potential difference between zinc and copper in contact, for example, $\phi^{(Zn)} - \phi^{(L)}$, is a property of this combination of metals, whereas each of the two metal–solution potential differences should depend on the solution concentration as well as the chemical natures of metal and solution. The fact that it is experimentally impossible to measure the difference of electrical potential between two chemically nonidentical phases will not prevent us from discussing it.

Indeed, one of our major concerns is to describe the progress that has been made in understanding the interphase potential differences and in explaining them in terms of the properties of the two phases in contact. On a molecular level, of course, there is no sharp boundary between phases, but rather a region across which intensive properties vary in values from one phase to another. The structure of this region depends upon the same kinds of intermolecular interactions that determine the structure of the adjacent homogeneous phases. We will also be concerned with understanding how the potential difference between phases may be altered by changing the electrical conditions, that is, by introducing electrical charge, or, where possible, by changing the composition of the phases. The electrical charge, too, is located at interphase boundaries and may be described as a surface charge density, because there cannot be a nonzero volume charge density through a homogeneous phase. If there are no electric fields in the homogeneous phases, the total surface charge density at each interface must be zero, so that there must be compensating layers of positive and negative charge density ("double layer"), thus giving rise to a difference of electrical potential but not to any electrical field outside the interface. Understanding double layers will be another important concern.

When current is flowing through an electrochemical cell, there are electrical fields within the homogeneous phases, which contribute to the overall potential drop across the cell. At the same time, of course, the electrical structure of the interphase regions may be altered. Within the homogeneous phases, we are concerned with the phenomena of conductivity. At each junction between two phases, there is a charge transfer, involving a heterogeneous chemical reaction, to be analyzed. Writing the overall cell reaction for the Daniell cell,

$$Cu^{2+}(aq) + Zn(s) \rightarrow Cu(s) + Zn^{2+}(aq)$$

does not present any picture of the complicated processes involved, whose overall effect is the transfer of electrical charge through the cell and the conversion of reactants to products. It is these processes that must be understood if one is to understand the way in which the current flowing through a galvanic or electrolytic cell depends on the difference in electrical potential between the terminals. Of course, the current is simply related to the rate of the overall cell reaction by Faraday's laws, which state that the rate at which a product is formed or a reactant consumed is given by the current through the cell divided by the number of elemental electrical charges per molecule of substance. Here, 1 elemental charge = charge of an electron = 1 faraday $\div$ Avogadro's number, where 1 faraday ($\mathcal{F}$) = 96,493 coulombs/mole (C/mol) and $N_{\text{Avogadro}} = 6.0225 \times 10^{23}$ mole^{-1}. These laws, like the cell reaction itself, give no information about what is really happening.

In the remainder of this chapter, we review the phenomena of electrochemistry, which the rest of the book attempts to explain or at least discuss. We first consider the emf, discussing the Nernst equation and introducing electrochemical potentials and their chemical and electrical parts. Not much more than elementary ideas will be presented regarding the structure of the interfacial regions. Then we will consider, still from a macroscopic point of view, something of what happens when the external circuit of a cell is closed. Subsequent chapters will deal with the above-mentioned phenomena, after the introduction of a molecular description of the electrochemical interface, and consider the connection between the molecular description and the macroscopic laws.

B. ELECTROMOTIVE FORCE AND HALF-CELLS

If we limit ourselves to open circuit, so that no current is passing through our electrochemical cell, we can measure only the difference of electrical potential between the terminals, that is, the emf, defined by

$$\mathcal{E} = \phi^{(R)} - \phi^{(L)}$$

where R and L indicate the right-hand and left-hand terminals, respectively. If $\mathcal{E}$ is nonzero, short-circuit of the cell will lead to a current flow and a chemical reaction within the cell: Oxidation will take place at one electrode and reduction will take place at the other. This would be an irreversible process: The occurrence of reaction indicates the absence of equilibrium. However, we may consider that, before closing the circuit, one has equilibrium everywhere within the cell and that this equilibrium may be maintained at closed circuit if an external voltage source equal and opposite to $\mathcal{E}$ is inserted between the terminals, stopping the current that would otherwise flow. If the external voltage differs infinitesimally from $\mathcal{E}$, an infinitesimal current will flow. Under these circumstances, the transfer of charge around the circuit, with the accompanying chemical reactions within the cell, will be occurring reversibly.

The work that the cell performs on the surroundings may then be equated to the decrease in the cell's free energy. Assuming the process occurs at constant pressure, we have

$$(-\Delta G_M)\, dx = \mathcal{E}\, dq \qquad [3]$$

where ΔG_M is the free energy change when the amount of reaction, x, changes by 1 mole at the given values of intensive parameters; dx represents an infinitesimal amount of reaction; and dq is the amount of charge transferred from the terminal on the right to the terminal on the left, through the external circuit, when dx degree of reaction occurs. The changes in the numbers of moles of reactants and products are related to dx by the stoichiometric coefficients, and the relation of dq to dx is obtained by writing the reaction as a sum of oxidation and reduction half-reactions. For the Daniell cell we have

$$\text{Oxidation:} \quad \text{Zn} \to \text{Zn}^{2+} + 2e^- \quad \text{(left)}$$
$$\text{Reduction:} \quad \text{Cu}^{2+} + 2e^- \to \text{Cu} \quad \text{(right)}$$

so 1 mole of reaction involves 2 faradays ($\mathcal{F}$) of electricity in the form of electrons. Then $dq = 2\mathcal{F}\,dx$ and

$$\Delta G_M = -n\mathcal{F}\mathcal{E} \qquad [4]$$

where $n = 2$ for the Daniell cell as written. Inserting this into the Gibbs–Helmholtz equation, $\Delta G = \Delta H + T(\partial \Delta G/\partial T)$ we find

B. ELECTROMOTIVE FORCE AND HALF-CELLS

$$\Delta H = -n\mathfrak{F}\mathcal{E} + n\mathfrak{F}\frac{\partial \mathcal{E}}{\partial T} \quad [5]$$

so that from the cell emf, as a function of temperature, one can calculate the heat of reaction.

The free energy change for a reaction is given by

$$\Delta G_M = \sum_i \mu_i \nu_i$$

where the chemical potential μ_i of each reactant or product is multiplied by the stoichiometric coefficient ν_i, taken from the cell reaction, with ν_i positive for products and negative for reactants. Thus, for the Daniell cell,

$$2\mathfrak{F}\mathcal{E} = \mu(Zn) - \mu(Zn^{2+}) - \mu(Cu) + \mu(Cu^{2+})$$
$$= -\mu(ZnSO_4) + \mu(CuSO_4) + \mu(Zn) - \mu(Cu)$$

where the reaction is considered as if it occurred directly.

If the chemical potential for each substance is written as its value for the standard state of that substance plus a term involving its activity a_i, we have

$$\Delta G_M = \sum \nu_i \mu_i^0 + RT \sum \nu_i \ln a_i$$

Since $\Delta G_M = -n\mathfrak{F}\mathcal{E}$, it is natural to represent the first summation on the right as $-n\mathfrak{F}\mathcal{E}^0$, where $\mathcal{E}^0$ is the emf the cell would exhibit if all reactants and products were in their standard states. Then

$$\mathcal{E} = \mathcal{E}^0 - \frac{RT}{n\mathfrak{F}} \sum \nu_i \ln a_i \quad [6]$$

which is a form of the Nernst equation. If the reaction were at equilibrium, the value of the summation would be the logarithm of the equilibrium constant K:

$$\sum \nu_i \ln a_i = \ln \left(\prod a_i^{\nu_i} \right) = K \quad [7]$$

In the electrochemical cell, the existence of this condition is revealed by an emf of zero: There is no tendency for electron flow or chemical reaction on connecting the terminals. Thus

$$\mathcal{E}^0 = \frac{RT}{n\mathfrak{F}} \ln K \quad [8]$$

which relates the standard cell potential to the equilibrium constant.

It is also possible to write $\mathcal{E}$ as a sum of two contributions, one for each half-reaction. This is useful as a "bookkeeping device," with *no* implication that the actual potential difference across the cell is the sum of terms representing potential differences across the half-cells (we discuss such a separation later). From here on, we follow the convention of writing all half-cell reactions as reductions (electrons appear on the left-hand side of the reaction equation), so that the cell reaction is obtained by subtracting the reaction for the left-hand half-cell from the reaction for the right-hand half-cell, just as $\mathcal{E}$ is obtained by subtracting the potential of the left-hand terminal from the potential of the right-hand terminal. Thus, if the reaction takes place in the direction it is written, reduction will occur at the right-hand electrode and oxidation will occur at the left-hand electrode. For the general cell reaction

$$O_R + R_L \rightarrow R_R + O_L$$

(where O and R refer to oxidized species and reduced species, respectively, and subscripts R and L refer to right-hand and left-hand half-cells, respectively), the half-cell reactions are written as

$$O_R + ne^- \rightarrow R_R \quad \text{on the right}$$

and

$$O_L + ne^- \rightarrow R_L \quad \text{on the left}$$

their difference being the cell reaction. Separating chemical potentials for the two half-cell reactions, we have

$$-n\mathcal{F}\mathcal{E} = \overset{R}{\sum} \nu_i \mu_i + \overset{L}{\sum} \nu_i \mu_i$$
$$= -n\mathcal{F}(\mathcal{E}^R_{1/2} - \mathcal{E}^L_{1/2}) \qquad [9]$$

where the electrons do not appear. The emf of the cell has thus been written as the difference of terms referring to the half-cells.

Now consider several half-cells that may be combined into cells:

$$O_I + ne^- \rightarrow R_I$$
$$O_{II} + ne^- \rightarrow R_{II}$$
$$O_{III} + ne^- \rightarrow R_{III}$$

We may suppose that each half-cell reaction has been multiplied by some integer to make the number of electrons the same for each, so that any one may be subtracted from another to give a cell reaction. If I is used as the left-hand half-cell and II is used as the right-hand half-cell, the cell reaction will be the sum of reaction II and reaction I in reverse, with molar free energy change

$$\Delta G_{\text{I-II}} = \overset{\text{II}}{\sum} \nu_i \mu_i - \overset{\text{I}}{\sum} \nu_i \mu_i = -n\mathcal{F}\mathcal{E}_{\text{I-II}}$$

If III is used on the right instead of II, we will have

$$\Delta G_{\text{I-III}} = \overset{\text{III}}{\sum} \nu_i \mu_i - \overset{\text{I}}{\sum} \nu_i \mu_i = -n\mathcal{F}\mathcal{E}_{\text{I-III}}$$

and if II is used on the left and III is used on the right,

$$\Delta G_{\text{II-III}} = \overset{\text{III}}{\sum} \nu_i \mu_i - \overset{\text{II}}{\sum} \nu_i \mu_i = -n\mathcal{F}\mathcal{E}_{\text{II-III}}$$

Obviously, the free energy changes obey

$$\Delta G_{\text{II-III}} = \Delta G_{\text{I-III}} - \Delta G_{\text{I-II}}$$

so that the emf's of the cells obey

$$\mathcal{E}_{\text{II-III}} = \mathcal{E}_{\text{I-III}} - \mathcal{E}_{\text{I-II}} \qquad [10]$$

This allows the possibility of assigning "half-cell potentials" to the half-cell reactions, such that the emf of any cell will be equal to the difference of the potentials for the half-cells of which it is composed: $\mathcal{E}^R_{1/2} - \mathcal{E}^L_{1/2}$.

Writing chemical potentials in terms of activities according to

$$\mu_i = \mu_i^0 + RT \ln a_i$$

and defining $\mathcal{E}^0_{1/2}$ as the value of the half-cell potential when all participants in the half-cell reaction (electrons not included) are at unit activity means that

$$\mathcal{E}_{1/2} = \mathcal{E}^0_{1/2} - \frac{RT}{n\mathcal{F}} \sum \nu_i \ln a_i \qquad [11]$$

Here the stoichiometric coefficients correspond to the participants in a half-cell reaction, written as a reduction. When we combine two relations like equation [11] to obtain the emf of a full electrochemical cell, we recover

$$\mathcal{E} = \mathcal{E}^0 - \frac{RT}{n\mathcal{F}} \sum \nu_i \ln a_i \qquad [12]$$

where the sum now includes all participants in the cell reaction, that is, both half-cells, with those of the left-hand half-cell having ν_i of sign opposite to what appears in the half-cell reaction. It may also be noted that multiplying a half-cell reaction by a constant c does not change $\mathcal{E}^0_{1/2}$, because n and all ν_i are multiplied by c. The standard half-cell potentials $\mathcal{E}^0_{1/2}$ are given in tables as standard half-cell reduction potentials (or as standard oxidation potentials if half-cell reactions are written as oxidations instead of reductions). Since only differences in the values of $\mathcal{E}_{1/2}$ are meaningful, the value of one may be arbitrarily assigned; that corresponding to the standard hydrogen electrode (SHE), with half-cell reaction

$$\tfrac{1}{2}H_2(g,\ 1\ \text{atm}) + e^- \rightarrow H^+(a=1)$$

is by convention given the value zero. Clearly, the values given to the other half-cell reduction potentials depend on this convention and have no absolute meaning.

C. ELECTROCHEMICAL POTENTIALS AND ELECTRICAL POTENTIALS

The chemical potentials in the preceding equations include those for charged species. If the chemical potential of a substance is defined as the work necessary to introduce 1 mole of the substance into a large system, we note the following: (a) There is a problem of measurability for chemical potentials of charged species; (b) the chemical potential is most likely dependent on the electrical potential of the system. With respect to (b), we note that a cell reaction is written as if all the substances were in the same medium, with the reaction occurring outside of the electrochemical cell. Since the total charge of the reactants must equal the total charge of the products, changing the electrical potential of the medium would affect reactants and products equally, leaving ΔG_M unaffected. Furthermore, one can always add oppositely charged species to both sides of the equation for the cell reaction to produce one involving neutral species. For example, the Daniell cell reaction,

$$Zn + Cu^{2+}(aq) \rightarrow Zn^{2+}(aq) + Cu$$

is quite equivalent to

$$Zn + Cu^{2+}(aq) + SO_4^{2-}(aq) \rightarrow Zn^{2+}(aq) + SO_4^{2-}(aq) + Cu$$

or just

$$Zn + CuSO_4(aq) \rightarrow ZnSO_4(aq) + Cu$$

with the electrical state of the medium being irrelevant.

With respect to (a), we may say that, although individual chemical potentials for charged species cannot be measured, they may still be considered theoretically.

C. ELECTROCHEMICAL POTENTIALS AND ELECTRICAL POTENTIALS

However, the work needed to introduce a charged species into a phase must depend on the electrical state of the phase, which leads to the concept of electrochemical potential. The electrochemical potential $\tilde{\mu}_i$ is defined as the work necessary to transfer 1 mole of species i into a phase from an infinitely dilute gaseous state at an infinite distance from the phase, or the free-energy change accompanying such a transfer. Thus, if two phases are in contact so species i can move from one to another, the equilibrium condition is that $\tilde{\mu}_i$ be the same for both.

The value of the electrochemical potential obviously depends on the phase: In addition to the work against the short-range (chemical) forces within the phase, there will be electrical work against the long-range Coulombic forces, if species i is charged. We write

$$\tilde{\mu}_i^{(P)} = \mu_i^{(P)} + q_i \mathcal{F} \phi^{(P)} \qquad [13]$$

where P refers to the phase, $\mu_i^{(P)}$ is the chemical potential of substance i in phase P, and $\phi^{(P)}$ is the electrostatic potential inside phase P. This is the potential referred to in the definition of the emf. Its value is relative to zero for a vacuum an infinite distance away.

Since the same constituents of matter are responsible for its electrical charge and its chemical properties, the separation implied by equation [13] cannot be completely rigorous, but simple calculations show that the amount of electrons required to give even a small, but macroscopic, amount of matter a potential of the order of volts is far too small to affect the chemical properties. The potential $\phi^{(P)}$ is called the *inner electrical potential* or *Galvani potential* of phase P and is an average over a region large compared to molecular size. The chemical potential $\mu_i^{(P)}$ represents the work required to transfer a mole of i to P in the case $\phi^{(P)} = 0$.

It may not be possible in a particular case to realize the situation $\phi^{(P)} = 0$. This is because an electrical potential difference is likely to exist between a point just outside the surface of a phase and a point just inside the surface. This surface potential results from the structure of the inhomogeneous surface region, in which large electric fields may exist; this structure, which is governed by the interactions between constituents of the phase, is inseparable from the bulk properties of the phase. It is usual to express this by

$$\phi^{(P)} = \chi^{(P)} + \psi^{(P)} \qquad [14]$$

where $\chi^{(P)}$ is the surface potential and $\psi^{(P)}$ is the outer potential or Volta potential. The electrical work required to transfer a charged species j from infinity to the interior of P is thus the sum of the work $q_j \psi^{(P)}$ required to bring it to a point just outside P (macroscopic distance of zero) and the work associated with the surface potential $\chi^{(P)}$. The value of $\psi^{(P)}$ is associated with the charge of P.

The difference in outer or Volta potentials between two chemically different phases is measurable (since it involves the work required to move a test charge through vacuum from one phase to the other), and hence the value of the Volta potential of a phase is measurable if the zero is taken as the vacuum at infinite distance. The difference in Galvani potentials between two chemically different phases is not measurable (since it involves passing a test charge from the interior of one phase to the interior of another, which implicates the chemical forces). If the phases are chemically identical, the Galvani potential difference is measurable, since the surface potentials are the same. Then the Galvani potential difference is the same as the Volta potential difference, and the chemical potential of any species j will be the same in the two phases, so that

$$\tilde{\mu}_j^{(P)} - \tilde{\mu}_j^{(Q)} = q_j \mathcal{F}(\psi^{(P)} - \psi^{(Q)}) \qquad [15]$$

This is the situation for the emf of a cell with chemically identical terminals.

In some cases, the work required to emit a charged particle from the interior of a phase P (to vacuum at infinity) can be measured. Per mole, this is the negative of

$$\tilde{\mu}_i^{(P)} = \mu_i^{(P)} + q_i \mathcal{F} \chi^{(P)} + q_i \mathcal{F} \psi^{(P)} \qquad [16]$$

where the zero for electrochemical potentials, as well as for electrostatic potentials, is taken as vacuum at infinity. One can now say that

$$\alpha_i^{(P)} = \mu_i^{(P)} + q_i \mathcal{F} \chi^{(P)} \qquad [17]$$

which is called the *real potential*, is measurable, although $\mu_i^{(P)}$ and $\chi^{(P)}$ are not measurable separately. For electrons in a metal, the negative of $\alpha_i^{(P)}$ is the *work function*. The work included in $\alpha_i^{(P)}$ for an electron in a metal or semiconductor involves image forces (as the electron gets very close to the surface), dispersion forces, and chemical forces.

The difference in the real potentials α_i of two metals can be measured in the arrangement shown in Figure 1, where the space between the metals contains a low-pressure gas. Here, G is a galvanometer and P is a potentiometer that can

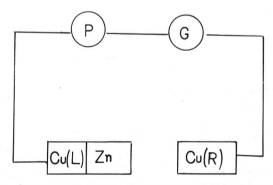

FIGURE 1. Schematic diagram for measurement of real-potential differences.

C. ELECTROCHEMICAL POTENTIALS AND ELECTRICAL POTENTIALS

apply an electrical potential difference between copper phases R and L. This potential difference V is equal to the difference of inner potentials of the phases R and L. Since the chemical potentials of the electron are equal in the two chemically identical copper phases, V is also $-\mathcal{F}^{-1}$ times the difference of electrochemical potentials of the electrons. Since the electrochemical potentials of the electrons are equal in Cu(L) and Zn because they are in contact and equilibrium is established, we can write

$$V = \phi^{(R)} - \phi^{(L)} = -\mathcal{F}^{-1}(\tilde{\mu}_{e^-}^{(R)} - \tilde{\mu}_{e^-}^{(L)})$$
$$= -\mathcal{F}^{-1}(\tilde{\mu}_{e^-}^{(R)} - \tilde{\mu}_{e^-}^{(Zn)})$$

Introducing the real potentials, we have

$$-\mathcal{F}V = \alpha_{e^-}^{(R)} - \mathcal{F}\psi^{(R)} - \alpha_{e^-}^{(Zn)} + \mathcal{F}\psi^{(Zn)}$$

and, assuming that V has been adjusted to equalize $\psi^{(R)}$ and $\psi^{(Zn)}$, we can measure $\alpha_{e^-}^{(Cu)} - \alpha_{e^-}^{(Zn)}$. This difference in real potentials is just V. Without the compensating potential supplied by P, there would be a difference in outer potentials between Zn and Cu(R), implying the existence of surface charges on these metals. If the distance between Zn and Cu(R) were varied, a current would flow through G, because the surface charge density must change to maintain the potential difference.

The equilibrium condition for transfer of a charged species between two phases is now the equality of electrochemical potentials:

$$\tilde{\mu}_i^{(P)} = \tilde{\mu}_i^{(Q)} \qquad [18]$$

This represents a balance between chemical and electrical forces:

$$\mu_i^{(P)} - \mu_i^{(Q)} = -z_i\mathcal{F}(\phi^{(P)} - \phi^{(Q)}) \qquad [19]$$

The situation is exemplified by the contact of two phases with the common component i, which can move from one phase to the other, when no transport of i is occurring. For example, it can be applied to an electrochemical cell at open circuit. The equilibrium condition for any charged species that can transfer between two phases thus fixes the inner potential difference between the phases, which is a contribution to the cell emf.

Within a phase, any equilibrium condition can be written in terms of the $\tilde{\mu}_i$, with the same inner potential for all species. Since the total electrical charge will be the same for reactants and products in a reaction, the electrical terms cancel out, leaving an equality between sums of chemical potentials. Thus, for Zn^{2+}, e^-, and Zn in Zn metal, we have

$$\tilde{\mu}_{Zn^{2+}}^{(Zn)} + 2\tilde{\mu}_{e^-}^{(Zn)} = \tilde{\mu}_{Zn}^{(Zn)} = \mu_{Zn}^{(Zn)}$$

and the left-hand side is

$$\mu_{Zn^{2+}}^{(Zn)} + 2\mathcal{F}\phi^{(Zn)} + 2(\mu_{e^-}^{(Zn)} - \mathcal{F}\phi^{(Zn)})$$

Thus, for the Daniell cell, we consider that (a) the Cu and Zn wires consist of Cu^{2+} and Zn^{2+} ions, the corresponding atoms, and electrons, (b) the solutions consist of solvent and ions, (c) transport of electrons between metals in contact is possible; and (d) ions may be transferred from a metal to solution. If we denote the two Cu leads by (L) and (R) and denote the $ZnSO_4$ and $CuSO_4$ solutions by (1) and (2), we have the equilibrium conditions at open circuit:

$$\tilde{\mu}_{e^-}^{(L)} = \tilde{\mu}_{e^-}^{(Zn)} \qquad [20]$$

$$\tilde{\mu}_{Zn^{2+}}^{(Zn)} = \tilde{\mu}_{Zn^{2+}}^{(1)} \qquad [21]$$

$$\tilde{\mu}_{Cu^{2+}}^{(2)} = \tilde{\mu}_{Cu^{2+}}^{(R)} \qquad [22]$$

Now the emf of the cell is

$$\mathcal{E} = \psi^{(R)} - \psi^{(L)} = \phi^{(R)} - \phi^{(L)}$$

because both leads are copper; we have

$$-\mathcal{F}(\phi^{(R)} - \phi^{(L)}) = \tilde{\mu}_{e^-}^{(R)} - \tilde{\mu}_{e^-}^{(L)}$$

for the same reason. Therefore, with equation [20],

$$-\mathcal{F}\mathcal{E} = \tilde{\mu}_{e^-}^{(R)} - \tilde{\mu}_{e^-}^{(Zn)} \qquad [23]$$

Because the combination of two electrons and one Cu^{2+} ion is electrically neutral, the equation

$$2\tilde{\mu}_{e^-}^{(R)} + \tilde{\mu}_{Cu^{2+}}^{(R)} = \tilde{\mu}_{Cu}^{(Cu)} = \mu_{Cu}^{(Cu)}$$

is independent of the electrical potential of the phase R, and we consider that it represents the electrochemical potential (or the chemical potential) of a neutral copper atom in copper. Similarly, the equation

$$2\tilde{\mu}_{e^-}^{(Zn)} + \tilde{\mu}_{Zn^{2+}}^{(Zn)} = \mu_{Zn}^{(Zn)}$$

corresponds to the chemical potential of a zinc atom in metallic zinc. Combining these two definitions with [23], we have

$$-2\mathcal{F}\mathcal{E} = (\mu_{Cu}^{(Cu)} - \tilde{\mu}_{Cu^{2+}}^{(2)}) - (\mu_{Zn}^{(Zn)} - \tilde{\mu}_{Zn^{2+}}^{(1)}) \qquad [24]$$

Formally, we have written $\mathcal{E}$ as the difference of quantities referring to the right-hand (Cu) and left-hand (Zn) half-cells that make up the cell.

Now the assumption of no liquid junction potential means $\phi^{(1)} = \phi^{(2)}$, so the difference of electrochemical potentials of the ions may be replaced by the difference of chemical potentials. Each chemical potential may then be written in terms of ionic activities to get a Nernst equation:

$$-2\mathcal{F}\mathcal{E} = (\mu_{Cu}^{(Cu)} - \mu_{Cu^{2+}}^{(2)0} - RT \ln a_{Cu^{2+}}^{(2)})$$
$$- (\mu_{Zn}^{(Zn)} - \mu_{Zn^{2+}}^{(1)0} - RT \ln a_{Zn^{2+}}^{(1)})$$
$$= -2\mathcal{F}(\mathcal{E}_{1/2}^{(R)} - \mathcal{E}_{1/2}^{(L)})$$

The calculation of a half-cell potential would require calculation of ionic chemical potentials or, if real potentials for ions were known, of surface potentials.

D. KINDS OF ELECTRODES

Often, an electrochemical cell consists of a "working electrode" on the right-hand side and a standard electrode on the left-hand side. Varying the potential drop $\phi^{(R)} - \phi^{(L)}$ by means of a source of potential in the external circuit is equivalent to varying the relative electrochemical potentials and hence the relative energies of electrons in the two terminals. If the value of $\phi^{(R)} - \phi^{(L)}$ is high enough, the energy of electrons on the right-hand side will be low enough so electrons will transfer into the right-hand electrode from the adjacent phase, giving rise to an oxidation current at the working electrode. For low enough values of $\phi^{(R)} - \phi^{(L)}$, the energy of electrons on the right-hand side will be high enough for the reverse to occur, with a reduction current at the working electrode.

The development of the Nernst equation for half-cell and full-cell potentials requires that equilibrium be established between reactants and products in the corresponding half-reactions or full reactions, so that potential differences have well-defined values. This would require in turn that the rates of oxidation and reduction processes, the forward and reverse reactions that define the equilibria, be infinite. In practice, a half-cell or electrode is said to be reversible if the rates of its forward and reverse reactions are rapid compared to the rate at which conditions are changed in a given experiment. Thus reversibility of a half-cell is a practical, not an absolute, concept (reversibility of an electrochemical cell has a somewhat different definition, as discussed later). In the ideal case of a perfectly reversible electrode, the oxidation and reduction reactions are infinitely fast, so equilibrium is always maintained and the potential differences are always well-defined. The ideal case at the other extreme is the *ideally polarizable electrode,* for which the oxidation and reduction processes occur infinitely slowly, so that the potential difference across the half-cell can be changed without inducing reactions that would tend to restore the original value. We discuss this in Section E.

Reversible electrodes are classified according to the arrangement of phases between which electron transfer takes place. In *electrodes of the first kind,* the electron transfer is between an atom-containing phase and a solution containing cations or anions of the atoms of the first phase. This includes metal or amalgam electrodes, as well as gas electrodes. In the former, equilibrium is between metal atoms (reduced form) in the metal and metal cations (oxidized form) in solution. The copper and zinc electrodes of the Daniell cell, which we have been using in our discussion, are examples. In a gas electrode, atoms or molecules of the gas replace the atoms of the metal; the ions in solution may be cations (e.g., $2H^+ + 2e^- = H_2$) or anions (e.g., $Cl_2 + 2e^- = 2Cl^-$). The gas in a gas electrode, as well as the solution, must be in contact with a metallic conductor that transports electrons into the half-cell.

An *electrode of the second kind* involves equilibrium between a metal, a sparingly soluble salt of the metal, and a solution containing the same anion as the salt. Thus, the silver–silver chloride electrode has a half-cell reaction

$$AgCl(s) + e^- \rightarrow Ag(s) + Cl^-(aq)$$

and a Nernst equation

$$\mathcal{E}_{1/2} = \mathcal{E}^0_{1/2} - \frac{RT}{\mathcal{F}} \ln a_{Cl^-}$$

It may be thought of as a silver electrode (electrode of the first kind) in which the half-cell reaction is $Ag^+ + e^- \rightarrow Ag$ and the activity of the silver ion is controlled by the solubility equilibrium of the metal salt, that is,

$$\mathcal{E}_{1/2, AgCl} = \mathcal{E}^0_{1/2, AgCl} - \frac{RT}{\mathcal{F}} \ln a_{Cl^-}$$

$$= \mathcal{E}^0_{1/2, AgCl} - \frac{RT}{\mathcal{F}} \ln K^{AgCl}_{sp} + \frac{RT}{\mathcal{F}} \ln a_{Ag^+}$$

$$= \mathcal{E}^0_{1/2, Ag} - \frac{RT}{\mathcal{F}} \ln (a_{Ag^+})^{-1}$$

where $K^{AgCl}_{sp} = a_{Ag^+} a_{Cl^-}$ is the solubility product of AgCl. Other important electrodes of the second kind are the calomel electrode, with the reaction

$$Hg_2Cl_2(s) + 2e^- \rightarrow 2Hg + 2Cl^-$$

and others involving insoluble salts of Ag, Hg, and Pb. The saturated calomel electrode (SCE), represented by $Hg\,|\,Hg_2Cl_2\,|\,KCl$ (saturated, aqueous), has a potential of 0.242 V vs. the SHE; it is a convenient and often-used reference electrode.

In an *electrode of the third kind*, the step from first kind to second kind (by means of an insoluble salt controlling concentration) is taken twice. Thus, with Pb, Pb(COO)$_2$(s), and Ca(COO)$_2$(s) in contact with CaCl$_2$, one could consider a lead electrode, with $a_{Pb^{2+}}$ controlled by the solubility of lead oxalate, and $a_{C_2O_4^{2-}}$ controlled by Ca^{2+} via the calcium oxalate solubility product, so that

$$\mathcal{E}_{1/2,\,PbOx} = \mathcal{E}^0_{1/2,\,Pb} - \frac{RT}{2\mathcal{F}} \ln (a_{Pb^{2+}})^{-1}$$

$$= \mathcal{E}^0_{1/2,\,Pb} + \frac{RT}{2\mathcal{F}} (\ln K_{Sp}^{Pb(COO)_2} - \ln a_{(COO)_2^{2-}})$$

$$= \mathcal{E}^0_{1/2,\,Pb} + \frac{RT}{2\mathcal{F}} (\ln K_{Sp}^{Pb(COO)_2} - \ln K_{Sp}^{Ca(COO)_2} + \ln a_{Ca^{2+}})$$

The half-cell reaction, overall, is

$$Pb(COO)_2(s) + 2e^- + Ca^{2+}(aq) \rightarrow Pb + Ca(COO)_2(s)$$

Finally, in oxidation–reduction (redox) electrodes, both oxidized and reduced species are present in the solution, namely, as two ions of the same metal with different oxidation numbers. The electrode, made of a different metal, is not a reactant, serving merely as a conduit of electrons. Correspondingly, the half-cell reaction does not involve the electrode and the half-cell potential is independent of the electrode material. For the Fe^{2+}/Fe^{3+} electrode, we expect, because the reduction is Fe^{3+} + e$^-$ → Fe^{2+}, the Nernst equation

$$\mathcal{E}_{1/2} = \mathcal{E}^0_{1/2} - \frac{RT}{\mathcal{F}} \ln \frac{a_{Fe^{2+}}}{a_{Fe^{3+}}}$$

In general, the logarithmic term will be $-(RT/n\mathcal{F}) \ln (a_{red}/a_{ox})$, where n electrons are involved per ion in converting "ox" to "red," unless one ion or the other is in equilibrium with other substances present in the solution, in which case manipulations like that of the previous paragraph are appropriate.

E. REDOX ELECTRODES

To discuss Fe^{2+}/Fe^{3+} from the point of view of electrochemical potentials, consider the cell

$$Pt(L)\,|\,H_2(g)\,|\,H^+(aq)\,\|\,Fe^{2+}(aq),\,Fe^{3+}(aq)\,|\,Pt(R)$$

The emf is equal to

$$\phi^{(R)} - \phi^{(L)} = \phi^{(R)} - \phi^{(sol)} - (\phi^{(L)} - \phi^{(sol)}) \qquad [25]$$

assuming that there is no liquid junction potential, so the two solutions have the same Galvani potential. Equilibrium for the reduction $Fe^{3+} + e^- \to Fe^{2+}$ implies

$$\tilde{\mu}(Fe^{3+}, aq) + \tilde{\mu}(e^-, R) = \tilde{\mu}(Fe^{2+}, aq)$$

which, when writing each electrochemical potential as chemical and electrical parts, yields

$$\mu(Fe^{3+}, aq) + \mu(e^-, Pt) - \mu(Fe^{2+}, aq)$$
$$= 2\mathcal{F}\phi^{(sol)} + \mathcal{F}\phi^{(R)} - 3\mathcal{F}\phi^{(sol)} \qquad [26]$$
$$= \mathcal{F}(\phi^{(R)} - \phi^{(sol)})$$

Now the equality between $\tilde{\mu}(H^+, aq) + \tilde{\mu}(e^-, L)$ and $\tfrac{1}{2}\mu(H_2)$ is

$$\mu(H^+, aq) + \mu(e^-, Pt) - \tfrac{1}{2}\mu(H_2) = \mathcal{F}(\phi^{(L)} - \phi^{(sol)}) \qquad [27]$$

Thus the emf, which is the difference between equations [26] and [27], does not involve $\mu(e^-, Pt)$, and so is independent of the nature of the electrode. In fact, the emf is just the difference of half-cell potentials

$$\mathcal{E}_{1/2}^{(R)} = \mathcal{F}^{-1}[\mu(Fe^{3+}, aq) - \mu(Fe^{2+}, aq)]$$

and

$$\mathcal{E}_{1/2}^{(L)} = \mathcal{F}^{-1}[\mu(H^+, aq) - \tfrac{1}{2}\mu(H_2)]$$

The emf of a cell with a redox electrode on the right-hand side and some reference electrode as the left-hand half-cell would be

$$\mathcal{E} = \mathcal{E}' - \frac{RT}{n\mathcal{F}} \ln \frac{a_{Red}}{a_{ox}}$$

where $\mathcal{E}'$ includes the electrode potential of the reference electrode and the standard potential of the redox electrode. If the ratio of activity coefficients γ_{Red}/γ_{ox} is independent of the degree of reduction of the redox species, we can write

$$\mathcal{E} = \mathcal{E}'' - \frac{RT}{n\mathcal{F}} \ln(\alpha^{-1} - 1)$$

where $\alpha = c_{ox}/(c_{Red} + c_{ox})$ = fraction oxidized and $\mathcal{E}''$ is a constant including activity coefficients. As a function of α, we note that $\mathcal{E}$ has an inflection point ($d^2\mathcal{E}/d\alpha^2 = 0$) at $\alpha = \frac{1}{2}$, and its slope at this point is

$$\left(\frac{d\mathcal{E}}{d\alpha}\right)_{1/2} = \frac{4RT}{n\mathfrak{F}}$$

Thus one can determine the number of electrons involved in the reaction.

The standard potential for a redox electrode equals the emf of a cell with the redox electrode on the right-hand side and a standard hydrogen electrode on the left-hand side. Thus, a higher standard potential for a redox system means the reduction half-reaction occurs with a larger decrease in free energy or that the redox system acts as a better oxidizing agent. A redox system with a higher standard half-cell potential will oxidize one with a lower one: for example, Cu^{2+}/Cu^+ has $\mathcal{E}^0_{1/2} = 0.15$ V, Cr^{3+}/Cr^{2+} has $\mathcal{E}^0_{1/2} = -0.41$ V, so Cu^+ will oxidize Cr^{2+} to Cr^{3+}.

The examples given so far show how the cell potential relates to half-cell potentials and electrode–solution potential differences. One also has to consider the inner potential difference between two solutions (liquid junction potential) as well as between two phases that exchange charged particles (contact potential). The latter is simple and has already been considered for electrons, since equating electrochemical potentials in the two phases yields the electrical potential difference as the difference of chemical potentials. The liquid junction potential between two electrolyte solutions, on the other hand, is not an equilibrium phenomenon. This potential is a result of concentration gradients in the transition layer between the solutions which are associated with diffusion of charged particles (see Section G). The other potential differences are associated with equilibrium for the process of charge transfer between phases.

An electrochemical *cell* is said to be reversible if all of the change in Gibbs free energy associated with the cell reaction appears in the surroundings as electrical work. The extreme of total irreversibility is achieved when the reactants contact and react directly, so that no electrical work appears. To have reversibility, it must be possible to make the cell reaction proceed in one direction or another by making the opposing voltage in the external circuit slightly greater, or slightly less, than the cell emf. This is sometimes called *material reversibility*. The electrical work expended when the cell reaction occurs in one direction should be the same as that obtained when the reaction occurs in the other; this is sometimes known as *energetic reversibility*. This requires that the only processes that occur in the cell should be those connected with the current flow in the external circuit, so that when the cell is not operating (open circuit), there should be no diffusion or dissolution of electrodes. When the cell operates, some electrical work is lost to irreversible processes such as resistance heating; thus reversible operation requires negligible current flow, implying that the potential in the external circuit must differ only infinitesimally from the cell emf. Some of the phenomena associated with finite current flows (overpotentials) are discussed in Section H and thereafter.

F. POLARIZABLE INTERFACES

A purely polarizable interface is the opposite of an ideal reversible interface. In an ideal reversible electrode, the equilibrium between oxidized and reduced species (including electrons) maintains a particular value for the potential difference across the interface (usually between metal and solution) at given temperature and pressure. If an attempt is made to change this potential difference by charging the interface (with equal and opposite surface charge densities in the two phases on either side), the charge passes through the interface, with the accompanying electrode reactions restoring the potential difference. The term *polarization* is used to refer to changes in a potential difference relative to its equilibrium value, so the reversible electrode is nonpolarizable. A polarizable electrode is one whose potential difference can be changed by charging. This charging does not result in electrochemical reaction, but does result in a change in the double layer charge density and in the way in which it is distributed and hence leads to a change in the potential difference across the interface.

It is implied that there exists no process capable of exchanging charged particles between the two phases to respond to the changed potential. In reality, it suffices that any such processes that do exist should be slow on the time scale of an experiment. Thus, a mercury electrode in contact with KCl solution acts as a polarizable electrode over a 1.5-V range of potential. The oxidation of Hg to Hg_2Cl_2,

$$2Hg + 2Cl^- \rightarrow Hg_2Cl_2 + 2e^-$$

may be neglected for electrode potentials less than 0.15 V (relative to the SHE): The standard reduction potential for the reverse reaction to the above is 0.27 V, so that the Nernst equation is

$$0.27 - \mathcal{E} = 0.0257 \ln a_{Cl^-}$$

and oxidation at $\mathcal{E} < 0.15$ V would require $a_{Cl^-} > 10^2$. The reduction of $K^+(aq)$ to $K(Hg)$ may be neglected except at potentials much below -2 V because the equilibrium concentration of K in mercury is otherwise vanishingly small. The oxidation of $Cl^-(aq)$ to Cl_2 may be neglected for potentials below 0.15 V since the standard reduction potential for $Cl_2 + 2e^- \rightarrow 2Cl^-$ is 1.36 V, so that

$$0.0257 \ln \frac{p_{Cl_2}}{(a_{Cl^-})^2} = \mathcal{E} - 1.36$$

and at $\mathcal{E} = 0.15$ V, $a_{Cl^-} = 1$, the partial pressure of Cl_2 is 10^{-21} atm. The reduction of water to $\frac{1}{2}H_2 + OH^-(aq)$ is thermodynamically possible but does not occur because there is a high overvoltage, that is, the rate of the reaction is extremely slow unless the electrode potential is far below the equilibrium value. An alternative way of looking at the problem (Parsons 1980, Section 1.1) is to calculate the change in the equilibrium concentration of a species in one of the above reactions when the electrode potential is changed within the polarization range and then to calculate the charge associated with carrying out this change. Except for the H_2/H_2O reaction, the charge required is immeasurably small.

An ideally polarizable interface acts as a capacitor: The differential capacitance is the ratio between a small change in surface charge density and the corresponding infinitesimal change in potential difference produced by it. Such an electrode has an additional degree of freedom as compared with a reversible electrode, because the specification of the equilibrium state requires specification of the potential difference across the interface. If q^M is the surface charge density (charge per unit area) on the metal side of the interface, $q^S = -q^M$ is the surface charge density on the solution side of the interface, and V_{m-sol} is the potential difference across the interface (metal − solution), the differential capacitance is $C = dq^M/dV_{m-sol}$. Values for C are typically 10–40 $\mu F/cm^2$ (1 F = 1 farad = 1 C/V); they depend on V_{m-sol}. A cell consisting of a polarizable electrode and a reversible electrode can be represented as far as its electrical behavior is concerned as a resistor, representing the latter, in series with a capacitor, representing the former. The capacitance of the reversible electrode, which in principle should be included, is generally much larger than C, so it can be neglected.

The charge density q^M in a simple metal is probably in the form of an excess or deficiency of conduction electrons (see Chapter 4). The density of conduction electrons spreads outward from the positively charged ion cores of the metal, giving rise to a double layer of thickness less than 1 Å (10^{-10} m) and a potential difference of the order of volts even for $q^M = 0$. The charge density q^S is in the form of excess ions of one species or another, thought to be either adsorbed on the metal surface or solvated and distributed in a "diffuse layer" that may extend hundreds or thousands of angstroms into the solution, with lower concentration yielding larger diffuse-layer thickness (see Chapter 4). The contribution of the solution phase to ϕ_{m-sol} is, like that of the metal phase, nonzero even for $q^S = 0$; preferential adsorption of one kind of ion over another and preferential orientation of dipolar solvent molecules at the metal surface produces a potential drop across the solution part of the interface.

For the mercury–solution interface, the surface tension can be measured as a function of the solution composition at fixed potential V_{m-sol} and as a function of V_{m-sol} at fixed solution composition. From the slope of the latter curve, values of q^M are obtained (see Chapter 2, Section D). From the surface tension as a function of solution composition, one obtains surface concentrations, as discussed in Chapter 2. For consideration of ϕ_{m-sol}, a convenient reference point is $q^M = q^S = 0$ (not corresponding to zero surface potential, as mentioned above); the surface tension as a function of potential has a maximum at this point ("electrocapillary maximum").

G. LIQUID JUNCTION POTENTIALS AND CONCENTRATION CELLS

In most electrochemical cells, the Daniell cell included, the two electrodes have different solution phases. The two solutions are separated by a membrane, porous glass plate, or other device that allows electrical contact, but no gross mixing. If the solutions are not identical, there will be diffusion of ions whose concentrations are different in the two solutions. Thus, supposing that equal concentrations of $ZnSO_4$ and $CuSO_4$ are used in the Daniell cell, Cu^{2+} ions will diffuse into the $ZnSO_4$ and Zn^{2+} ions from the $ZnSO_4$ to the $CuSO_4$ solution. Unless the diffusion rates are identical, there will be a net transfer of charge across the interface between the solutions, with formation of a double layer and establishment of an electrical potential difference. The solution with the less rapidly diffusing cation (Zn^{2+} in the present example) will have a net positive charge and will be at higher potential. This will tend to accelerate the Zn^{2+} and decelerate the Cu^{2+} ions until the diffusion rates become equal. It will also lead to diffusion of SO_4^{2-} from the $CuSO_4$ solution to the $ZnSO_4$ solution. Eventually, a steady state (not an equilibrium) will be reached and the potential difference between the solutions will take on a constant value, namely, the liquid junction potential or diffusion potential.

The magnitude and sign of a liquid junction potential of course depend on the two solutions between which diffusion occurs. The size is not large, usually less than 0.1 V, but, especially in precise measurements, an attempt is made to make it as close to zero as possible. For this purpose, an auxiliary solution, with a high concentration of a salt such as KCl whose ions have closely equal diffusion coefficients, is introduced between the two solutions of the cell. At each boundary between this salt bridge and a cell solution, the diffusion potential is mainly due to the ions of the salt bridge. This potential is small because the cations and anions of the salt bridge diffuse at the same rate. Furthermore, essentially the same potential appears, in opposite directions, at the two boundaries (in our example, from $ZnSO_4$ to the salt bridge and from the salt bridge to $CuSO_4$ solution), giving further cancellation.

The liquid junction potential can be measured in a concentration cell, constructed from chemically identical half-cells differing in the activity of solute. First we consider such a cell with elimination of the liquid junction potential by a salt bridge or other device:

$$Ag \mid AgNO_3(a_+^L) \parallel AgNO_3(a_+^R) \mid Ag$$

Both half-cell reactions are

$$Ag^+(aq) + e^- \rightarrow Ag$$

so that the cell reaction is

$$Ag^+(R) \rightarrow Ag^+(L)$$

G. LIQUID JUNCTION POTENTIALS AND CONCENTRATION CELLS

and the emf is

$$\mathcal{E} = \mathcal{E}^0 - \frac{RT}{\mathcal{F}} \ln \frac{a_+^L}{a_+^R} \qquad [28]$$

with $\mathcal{E}^0 = 0$. Of course, diffusion of ions into and out of the salt bridge accompanies the cell reaction and maintains electroneutrality in all the solutions. The cation activities a_+ are not measurable and should be replaced in equation [28] by the mean ionic activities $a_\pm$ of AgNO$_3$. In general, for an electrolyte each molecule of which dissociates into ν_+ cations and ν_- anions, the mean ionic activity is defined by

$$(a_\pm)^\nu = (a_+)^{\nu_+} (a_-)^{\nu_-}$$

where $\nu = \nu_+ + \nu_-$. In the case of AgNO$_3$,

$$\frac{(a_\pm^R)^2}{(a_\pm^L)^2} = \frac{(a_+^R a_-^R)}{(a_+^L a_-^L)}$$

Assuming $a_-^R/a_-^L = a_+^R/a_+^L$,

$$\mathcal{E} = -\frac{RT}{\mathcal{F}} \ln \frac{a_\pm^L}{a_\pm^R} \qquad [29]$$

for our cell.

For the same cell, but without suppression of the liquid junction potential, the transport of ions must be considered. For each faraday passing through the cell, 1 mole of Ag$^+$ is reduced to Ag at the right-hand electrode (acting as cathode), 1 mole of Ag dissolves as Ag$^+$ at the left-hand electrode (acting as anode), and 1 faraday of positive electricity is carried by cations and anions from left to right through the cell. Of this, let t_+ faradays be transported by Ag$^+$ cations moving left to right and let $t_- = 1 - t_+$ faradays be transported by NO$_3^-$ anions moving right to left (t_+ and t_- are transport numbers, discussed in more detail in Section H). Then the right-hand solution gains $t_+ - 1 = -t_-$ moles of Ag$^+$ and loses t_- moles of NO$_3^-$, while the left-hand solution gains $1 - t_+$ moles of Ag$^+$ and t_- moles of NO$_3^-$. Overall, the cell reaction is

$$t_- \text{Ag}^+(a_+^R) + t_- \text{NO}_3^-(a_-^R) \rightarrow t_- \text{Ag}^+(a_+^L) + t_- \text{NO}_3^-(a_-^L)$$

The free energy change is

$$\Delta G = RT \ln \frac{(a_+^L a_-^L)^{t_-}}{(a_+^R a_-^R)^{t_-}} = 2t_- RT \ln \frac{a_\pm^L}{a_\pm^R}$$

Since ΔG is equal to $-\mathcal{F}\mathcal{E}$, the emf of the cell with transport is

$$\mathcal{E}^T = -2t_- \frac{RT}{\mathcal{F}} \ln \frac{a_\pm^L}{a_\pm^R} \qquad [30]$$

This is identical to the emf of the cell with the salt bridge only if $t_- = \frac{1}{2}$ (equal currents carried by cations and anions).

The cell potential, $\mathcal{E}^T = \phi_{Ag}^R - \phi_{Ag}^L$, is a sum of electrode–solution potential differences and a liquid junction potential. The electrode–solution potential differences are the same in the cell without transference, so the liquid junction potential is

$$\mathcal{E}^T - \mathcal{E} = \phi_{sol}^R - \phi_{sol}^L = (2t_- - 1) \frac{RT}{\mathcal{F}} \ln \frac{a_\pm^R}{a_\pm^L} \qquad [31]$$

The factor $2t_- - 1$ may also be written as $t_- - t_+$. The origin of the liquid junction potential appears clearly as the difference in current-carrying abilities of cation and anion. As will be discussed in Section H, the values of t_+ and t_- reflect the relative speeds attained by the ions in the presence of an electric field, which in turn depend on their sizes and interactions with each other and the solvent. We note that equation [31] is not exact because of the assumption about a_-^R/a_-^L used to obtain $\mathcal{E}$ in terms of mean ionic activities (equation [29]).

A similar analysis applies to the concentration cell

$$\text{Ag}|\text{AgCl}(s)|\text{KCl}(a_L)|\text{KCl}(a_R)|\text{AgCl}(s)|\text{Ag}$$

which is made from two identical electrodes of the second kind. If the junction potential between the two KCl solutions were eliminated, the emf would be

$$\mathcal{E} = \left(\mathcal{E}_{1/2,\text{AgCl}}^0 - \frac{RT}{\mathcal{F}} \ln a_-^R \right) - \left(\mathcal{E}_{1/2,\text{AgCl}}^0 - \frac{RT}{\mathcal{F}} \ln a_-^L \right)$$

$$= \frac{RT}{\mathcal{F}} \ln \frac{a_-^L}{a_-^R}$$

If we assume that the ratio of activities for K^+ for the two solutions is the same as the ratio of activities for Cl^-, we can write

$$\mathcal{E} = \frac{RT}{\mathcal{F}} \ln \frac{a_\pm^L}{a_\pm^R}$$

With transport across the liquid junction, for each mole of Cl^- created on the right-hand side and consumed on the left-hand side, t_- moles of Cl^- flow from right to left and t_+ moles of K^+ flow from left to right. The net result is

$$t_+ K^+(a_+^L) + (1 - t_-) Cl^-(a_-^L) \to t_+ K^+(a_+^R) + (1 - t_-) Cl^-(a_-^R)$$

so the emf is

$$\mathcal{E}^T = 2t_+ \frac{RT}{\mathcal{F}} \ln \frac{a_\pm^L}{a_\pm^R}$$

and the liquid junction potential is

$$\mathcal{E}^T - \mathcal{E} = (t_+ - t_-) \ln \frac{a_\pm^L}{a_\pm^R}$$

A concentration cell may also be constructed with identical electrodes but with *metals* at different activities, such as

$$Zn(Hg, L) | ZnSO_4(aq) | Zn(Hg, R)$$

for which the half-cell reactions are

$$Zn(Hg) \to Zn^{2+}(aq) + 2e^-$$

The cell reaction is the removal of Zn from the amalgam on the left-hand side (by oxidation to Zn^{2+}) and the insertion of Zn into the amalgam on the right-hand side. If the Zn activities in the amalgams on the left-hand side and right-hand side are a^L and a^R, respectively, we can write

$$\mathcal{E} = \frac{RT}{2\mathcal{F}} \ln \left(\frac{a^L}{a^R} \right) \qquad [32]$$

A final kind of concentration cell involves coupling two identical gas electrodes with the gas at different pressures, using a single solution. It is clearly similar to the cell we presented earlier in this paragraph.

H. MORE COMPLICATED CELLS—APPLICATIONS

Double Cell

The following cell is a concentration cell without transport:

$$Zn | ZnSO_4(a_\pm^L) | Hg_2SO_4(s) | Hg | Hg_2SO_4(s) | ZnSO_4(a_\pm^R) | Zn$$

In reality, this cell consists of two cells in series, with the central mercury serving as right-hand electrode in the first cell and as left-hand electrode in the second.

We write the emf as

$$\mathcal{E} = \phi^{(R)} - \phi^{(L)} = (\phi^{(R)} - \phi^{(Hg)}) + (\phi^{(Hg)} - \phi^{(R)})$$
$$= \mathcal{E}^{II} - \mathcal{E}^{I} \qquad [33]$$

The half-cell reactions involved are

$$Hg_2SO_4 + 2e^- \rightarrow 2Hg + SO_4^{2-}(aq)$$

and

$$Zn^{2+}(aq) + 2e^- \rightarrow Zn$$

In the first cell the cell reaction is

$$Hg_2SO_4 + Zn \rightarrow 2Hg + Zn^{2+} + SO_4^{2-}$$

so

$$\mathcal{E}^{I} = \mathcal{E}^{0} - \frac{RT}{2\mathcal{F}} \ln a_-^L - \frac{RT}{2\mathcal{F}} \ln a_+^L$$

where $\mathcal{E}^0$ is the Hg_2SO_4/Hg standard half-cell potential minus the Zn^{2+}/Zn standard half-cell potential. For the second cell the cell reaction is the reverse of that for the first and thus

$$\mathcal{E}^{II} = -\mathcal{E}^{0} + \left(\frac{RT}{2\mathcal{F}}\right) \ln a_-^R + \frac{RT}{2\mathcal{F}} \ln a_+^R$$

Therefore $\mathcal{E}^{II} - \mathcal{E}^{I}$ gives

$$\mathcal{E} = \frac{RT}{2\mathcal{F}}(-\ln a_-^L a_+^L + \ln a_-^R a_+^R) = \frac{RT}{\mathcal{F}} \ln \frac{a_\pm^R}{a_\pm^L}$$

The overall cell reaction is

$$Zn^{2+}(a_+^R) + SO_4^{2-}(a_-^R) \rightarrow Zn^{2+}(a_+^L) + SO_4^{2-}(a_-^L)$$

or the transfer of $ZnSO_4$ from the right-hand cell to the left-hand cell.

Glass Electrode

The glass electrode is a pH-measuring device, at the center of which is a blown glass membrane separating a solution of known and constant pH from a solution whose pH is to be measured. An identical indicator electrode is connected to each solution, giving a double cell like the following:

$$\text{Hg} \mid \text{Hg}_2\text{Cl}_2 \mid \text{KCl(sat)} \parallel \text{H}^+(a_1) \mid \text{H}^+(a_2) \parallel \text{KCl(sat)} \mid \text{Hg}_2\text{Cl}_2 \mid \text{Hg}$$

Here we have used saturated calomel electrodes as indicator electrodes; $-\log_{10} a_1$ is the known constant pH and a_2 is to be measured; the glass membrane is represented by $\mid$. The emf of this cell, which is just the potential difference across the glass membrane, is found to obey

$$\mathcal{E} = \mathcal{E}' + \frac{RT}{\mathcal{F}} \ln a_2 \qquad [34]$$

except for highly alkaline solutions, where $\mathcal{E}$ is lower than predicted by equation [34], with different compositions of glass giving different behavior. The value of the constant $\mathcal{E}'$ is obtained by calibration with a standard solution. Writing $\mathcal{E}'$ as $\mathcal{E}_a - (RT/\mathcal{F}) \ln a_1$, one finds that $\mathcal{E}_a$ (called the asymmetry potential) depends on the glass used and on how the membrane is prepared; it is not zero. This means that $\mathcal{E}$ does not vanish when $a_2 = a_1$, so that the glass electrode is not acting as a concentration cell with a membrane impermeable to H^+.

One theory on the dynamics of the glass electrode invokes exchange reactions on the glass surface between hydrogen ions and alkali cations of the glass. For a sodium glass,

$$\text{Na}^+(\text{glass}) + \text{H}^+(\text{solution}) \leftrightarrow \text{H}^+(\text{glass}) + \text{Na}^+(\text{solution})$$

Let the equilibrium constant for this be K and suppose that there are N sites per unit area which can be occupied by H^+ or Na^+. Then

$$\frac{a^s_{\text{H}^+} a_{\text{Na}^+}}{a_{\text{H}^+} a^s_{\text{Na}^+}} = K \qquad [35]$$

and

$$m^s_{\text{H}^+} + m^s_{\text{Na}^+} = N \qquad [36]$$

where the superscript s refers to the glass surface and m^s is a surface concentration in moles per square decimeter. When we eliminate $m^s_{\text{Na}^+}$ from equation [35] and equation [36], we get

$$\frac{a^s_{\text{H}^+} a_{\text{Na}^+}}{a_{\text{H}^+}} = K \gamma^s_{\text{Na}^+}\left(N - \frac{a^s_{\text{H}^+}}{\gamma^s_{\text{H}^+}}\right)$$

$$a^s_{\text{H}^+}(a_{\text{Na}^+} + K a_{\text{H}^+}) = KN \gamma^s_{\text{Na}^+} a_{\text{H}^+}$$

where the γ_i are activity coefficients; $\gamma^s_{\text{Na}^+}/\gamma^s_{\text{H}^+}$ has been assumed to be unity.

Equilibrium between H^+ ions in solution and H^+ on surface sites requires

$$\mu^0_{H^+} + RT \ln a_{H^+} + \mathcal{F}\phi^{sol} = \mu^0_{H^+,s} + RT \ln a^s_{H^+} + \mathcal{F}\phi^{glass}$$

so that

$$\phi^{glass} - \phi^{sol} = C' + \frac{RT}{\mathcal{F}} \ln \left(\frac{a_{H^+}}{a^s_{H^+}}\right)$$

$$= C + \frac{RT}{\mathcal{F}} \ln (a_{Na^+} + K a_{H^+})$$

where C' and C are constants ($\gamma^s_{Na^+}$ has been incorporated in C as a constant). For an acidic or neutral medium, one can suppose that $K a_{H^+} \gg a_{Na^+}$ and that the glass–solution potential drop is a constant plus $(RT/\mathcal{F}) \ln a_{H^+}$ as in equation [34]. However, if a_{Na^+} is comparable to $K a_{H^+}$, as in alkaline solutions or solutions with high concentrations of sodium salt, this will no longer hold. In fact, if $a_{Na^+} \gg K a_{H^+}$, $\phi^{glass} - \phi^{sol}$ will be $C + (RT/\mathcal{F}) \ln a_{Na^+}$, so that the emf is independent of a_{H^+} and the electrode functions as a sodium-measuring electrode. The "alkaline error" is, according to this theory,

$$\frac{RT}{\mathcal{F}} \ln (a_{Na^+} + K a_{H^+}) - \frac{RT}{\mathcal{F}} \ln K a_{H^+}$$

Activity Coefficients

Mean ionic activity coefficients can be determined by measuring the emf's of suitably chosen cells. For instance, consider the double cell

$$Ag | AgCl(s) | KCl(m_1) | K(Hg) | KCl(m_2) | AgCl(s) | Ag$$

The emf is the emf of the right-hand cell minus that of the left-hand cell, which gives

$$\mathcal{E} = -\frac{2RT}{\mathcal{F}} \ln \frac{a^{(2)}_\pm}{a^{(1)}_\pm} \qquad [35]$$

which is rearranged to

$$\frac{2RT}{\mathcal{F}} \ln m^{(2)}_\pm + \frac{2RT}{\mathcal{F}} \ln \gamma^{(2)}_\pm + \mathcal{E} = \frac{2RT}{\mathcal{F}} \ln a^{(1)}_\pm \qquad [37]$$

If the emf is measured for a series of concentrations of solution 2 (with solution 1 always the same), a plot of $(2RT/\mathcal{F}) \ln m^{(2)}_\pm + \mathcal{E}$ vs. $m^{(2)}_\pm$, extrapolated to $m^{(2)}_\pm = 0$, gives $(2RT/\mathcal{F}) \ln a^{(1)}_\pm$ (according to the Debye–Hückel theory, plotting against $m^{1/2}$ should give a linear plot). Given the value of $a_\pm^{(1)}$, measurement of $\mathcal{E}$ for any $m_\pm^{(2)}$ then yields the mean ionic activity coefficient $\gamma_\pm^{(2)}$.

As another example, consider a concentration cell with transport built from cationic electrodes, with emf

$$\mathcal{E} = \frac{zRT}{\mathcal{F}} t_- \ln \frac{a_\pm^{(2)}}{a_\pm^{(1)}}$$

where z is the charge on the cation and t_- is the transport number for the anion. When we bring the activity coefficients to the left-hand side, we have

$$\ln \frac{\gamma_\pm^{(2)}}{\gamma_\pm^{(1)}} = -\ln \frac{m_\pm^{(2)}}{m_\pm^{(1)}} + \frac{\mathcal{E}\mathcal{F}}{zRTt_-}$$

If one knows emf's and transport numbers for two concentrations, plus the activity coefficient at one concentration (perhaps by the Debye–Hückel law), one can obtain the activity coefficient at the other concentration. Measurement of transport numbers is discussed in Section K.

I. OVERVOLTAGES

Except for the liquid junction potentials, we have so far considered only equilibrium properties. We have considered the potential difference across an electrochemical cell at open circuit, expressing it in terms of potential differences across phase boundaries, which in turn depend on equilibria for charge transfer between phases (except for the polarizable interface). The potential difference measures the decrease in free energy associated with the cell reaction and gives the external potential that, if inserted in the external circuit, would yield an equilibrium situation such that no current would flow at closed circuit. For a reversible cell, an external potential infinitesimally higher or infinitesimally lower than this value would lead to an infinitesimal current and accompanying cell reaction in one direction or the other. If the external potential differed by a finite amount from the cell emf, a finite current would flow. Then one has to consider the charge-transfer process at phase boundaries and the transport of charge across phases that are homogeneous in the absence of current flow. New potential differences appear in the circuit, one of which would be the resistance drop across the electrolyte associated with passage of current. The potential difference at each interface would differ from its equilibrium (no current) value. One refers to an electrode operating at a potential other than its reversible potential as *polarized*.

The potential of a working electrode is determined by coupling it to a standard electrode with a salt bridge and measuring the potential of the resulting cell with a potentiometer. The potentiometer is balanced so no current flows through the salt bridge. At the same time, the working electrode is connected to another electrode, forming a circuit through which current flows. The other electrode is called the *auxiliary electrode* or *counter electrode*. A problem is that the potential measured by the potentiometer includes an *IR* (resistance) drop between the working electrode and the salt bridge, since current is flowing through the solution around the electrode, which constitutes part of the working circuit. To minimize this extra potential, the end of the salt bridge is brought very close to the working electrode by means of a capillary (Luggin tip). If current is high or solution resistance is large, this may not suffice to make the *IR* drop negligibly small.

A remedy is to measure potential for a series of distances between the capillary tip and the working electrode and extrapolate to zero distance. Another method of separating the *IR* drop from the change in potential that one is trying to measure involves interrupting the current and measuring the electrode potential at several later times (of the order of tenths of milliseconds). The *IR* drop becomes zero essentially instantly whereas the overpotential decays to zero less rapidly. Then the potential can be extrapolated back to time zero to obtain the overpotential at the time the current was interrupted.

Suppose the electrode of the first kind whose half-cell reaction is

$$M^+(\text{aq}) + e^- \to M(s)$$

is part of an electrochemical cell with emf $\mathcal{E}$. The value of $\mathcal{E}$ depends on the other half-cell, on the nature of M, and on the concentration of M^+. With the potential difference $\phi^{(R)} - \phi^{(L)}$ between right-hand electrode and left-hand electrode maintained equal to $\mathcal{E}$ by a source of potential in the external circuit (no current flowing), the potential difference between the metal M and the adjacent solution has a value defined by M and the concentration of M^+, which we refer to as the *reversible potential*. If the value of $\phi^{(R)} - \phi^{(L)}$ were increased, the potential difference between M and the solution would increase above the equilibrium value, decreasing the sum of electrochemical potentials,

$$\tilde{\mu}_{M^+} + \tilde{\mu}_{e^-}^{(M)} = \mu_{M^+} + \mathcal{F}\phi^{(\text{sol})} + \mu_{e^-}^{(M)} - \mathcal{F}\phi^{(M)} \qquad [38]$$

The chemical potential of the neutral species M would be unaffected, so that the dissolution reaction

$$M(M) \to e^-(M) + M^+(\text{aq})$$

would be thermodynamically favored by the increase in $\mathcal{E}$. Oxidation would occur at M and current would flow through the cell from right to left. If the value of $\phi^{(R)} - \phi^{(L)}$ were decreased below $\mathcal{E}$, then $\phi^{(M)} - \phi^{(\text{sol})}$ would become less than its equilibrium value, increasing $\tilde{\mu}_{m^+} + \tilde{\mu}_{e^-}^{(M)}$ and favoring the reduction of M^+ at M, which would then function as a cathode.

More correctly, one should consider that, on a molecular level, processes corresponding to oxidation and reduction occur simultaneously. At equilibrium, their rates are equal, but an increase of $\phi^{(M)} - \phi^{(\text{sol})}$ favors oxidation processes over the reverse, leading to a net oxidation current, and vice versa. When oxidation is taking place, the electrode is "polarized positively," that is, the potential of metal relative to solution is higher than its reversible potential, and the electrode functions as an anode. The electrode is polarized negatively when it functions as a cathode, with reduction taking place. The difference between the potential of a working electrode and its reversible potential is called the overpotential or overvoltage and is denoted by η. The overpotential is positive for anodes and negative for cathodes. The overall potential difference across the operating cell is the cell emf (open circuit potential) plus the overvoltages at the two electrodes plus the potential differences (IR drops) due to electrical resistances in the electrolyte and possibly other phases.

The process of reduction of M^+(aq) to M must involve transport of ions to the electrode surface, discharge or combination of ions with electrons of the metal to form M atoms, and incorporation of these atoms into the metal lattice. If all these occurred as fast as electrons could be supplied to M, $\phi^{(M)} - \phi^{(sol)}$ would always remain at the reversible potential and there would be no overpotential. In fact, the rate of reduction is finite, being limited by the slowest step, which may be anywhere in the scheme outlined.

Of course, any electrode process involves a series of steps that limit the rate of the overall process and determine the overpotential. In addition to processes associated with transport and charge transfer, there are others whose kinetics may affect the electrode reaction indirectly. These include chemical reactions between components of the electrolyte, adsorption of reactants on the electrode (sometimes, but not always, necessary for reaction), adsorption of products and electro-inactive substances (this may retard the electrode reaction), chemical reactions between adsorbed species, and phase changes such as crystallization. It is necessary to distinguish between the overall reaction (e.g., $Cd(CN)_4^{2-} + 2e^- \rightarrow Cd + 4CN^-$) and the mechanism, which is a series of partial processes (e.g., dissociation of $Cd(CN)_4^{2-}$ to $Cd(CN)_3^- + CN^-$ and of $Cd(CN)_3^-$ to $Cd(CN)_2 + CN^-$, electron transfer to $Cd(CN)_2$ to release $2CN^-$, and dissolution of Cd in the electrode). The overall electrode process gives little indication of the actual mechanism.

J. CONDUCTION THROUGH HOMOGENEOUS PHASES

We first consider the process of conduction in an electrolyte. An ionic solution, like a metal, behaves as an electrical resistor. This means that Ohm's law is obeyed, that is, the current flowing across a sample is proportional to the potential difference across it. Writing this as

$$I = \frac{V}{R} \qquad [39]$$

where I is the current, V is the potential, and R is the resistance, one finds that, for a rectangular sample, R is inversely proportional to the cross-sectional area a and directly proportional to the length l along the current direction. Thus the resistivity

$$\rho = \frac{Ra}{l}$$

and the conductivity

$$\kappa = \frac{1}{\rho}$$

are intensive properties of the conducting medium. When we introduce the conductivity into Ohm's law, we have that $I = Va\kappa/l$ or

$$j = \kappa E \qquad [40]$$

where $j = I/a$ is the current density and $E = V/l$ is the electric field. For silver at 25°C, κ is $6.3 \times 10^5 \, \Omega^{-1} \cdot \text{cm}^{-1}$, where $1 \, \Omega = 1 \text{ ohm} = 1 \, V/A$, $1 \, A = 1$ ampere $= 1 \, C/s$. For 0.1 M KCl, κ is $1.3 \times 10^2 \, \Omega^{-1} \cdot \text{cm}^{-1}$, whereas κ is 4.0×10^{-8} for pure water.

The mechanism of current flow is quite different for metals and solutions, since the current in a metal is carried by the electrons, which are quantum mechanical particles, and the current in a solution is carried by the ions, heavy particles whose motion is describable by classical mechanics. The basic fact is that the current-carrying species, under the influence of an electric field E, are accelerated to a terminal velocity whose value is proportional to E. This follows when we write j as Σj_i, where

$$j_i = q_i n_i v_i \qquad [41]$$

where n_i is the number density of particles of species i, q_i is their charge, and v_i is their velocity. Because the number densities and charges are fixed, $j = \kappa E$ implies that all v_i are proportional to E.

Because of the proportionality of j_i to n_i, the contribution of each ionic species to κ is proportional to its concentration or number density. Therefore one often defines the molar conductivity of an electrolyte as

$$\Lambda = \frac{\kappa}{c}$$

The molar conductivity becomes independent of concentration at high dilution, since ions then migrate independently under the influence of the electric field. The limiting molar conductivity Λ^∞, which is the limit of Λ as $c \to 0$, is then a sum of contributions of individual ions, so that, for example, $\Lambda^\infty_{\text{KCl}} + \Lambda^\infty_{\text{NaI}} = \Lambda^\infty_{\text{KI}} + \Lambda^\infty_{\text{NaCl}}$, and $\Lambda^\infty_{\text{NaX}} - \Lambda^\infty_{\text{KX}}$ is independent of the anion X. The individual ionic contributions are measurable and calculable, and tables of ionic conductivities at infinite dilution exist.

It is clear from the formula for j_i (equation [41]) that, in comparing ions, it is appropriate to divide the contribution of an ion to Λ by its charge, obtaining the *equivalent ionic conductivity at infinite dilution*. From a table of equivalent ionic conductivities at infinite dilution, one can calculate the conductivity of any strong electrolyte at concentrations that are low enough so that interferences between ions are unimportant. Thus, from the equivalent conductivities of Na^+ (50.11 $\Omega^{-1} \cdot cm^2 \cdot equiv^{-1}$) and SO_4^{2-} (80.00 $\Omega^{-1} \cdot cm^2 \cdot equiv^{-1}$) we find $\Lambda_{Na_2SO_4}^\infty = 2(50.11) + 2(80.00)$. The behavior of Λ at high concentration is more complicated.

Some so-called strong electrolytes exhibit fairly high values of Λ for all concentrations (e.g., $\lambda_{KCl} = 147 \ \Omega^{-1} \cdot cm^2 \cdot mol^{-1}$ at 0.001 M, decreasing to 129 $\Omega^{-1} \cdot cm^2 \cdot mol^{-1}$ at 0.1 M), with the values decreasing linearly with $c^{1/2}$ for c below about 0.1 M. These electrolytes dissociate completely to ions; the linear decrease of Λ with $c^{1/2}$ is a result of interionic interactions. Because of electrostatic forces, the region around each ion will have an excess of ions of opposite charge, and this ion atmosphere affects the motion of the central ion when an electric field is applied; the atmosphere becomes less important at higher ionic dilution. Weak electrolytes, which usually have lower molar conductivities (Λ for acetic acid at 25°C is 49 $\Omega^{-1} \cdot cm^2 \cdot mol^{-1}$ at 0.001 M and 5 $\Omega^{-1} \cdot cm^2 \cdot mol^{-1}$ at 0.10 M, show a more complicated behavior as a funtion of c. This is due to incomplete dissociation to ions, for example, for acetic acid

$$\frac{a_{H^+} a_{Ac^-}}{a_{HAc}} = 1.86 \times 10^{-5} \frac{mol}{dm^3} \quad [42]$$

If the actual ionic concentration is used in Λ instead of the total electrolyte concentration, values of Λ are quite in line with those of other electrolytes: From equation [42], for 0.10 M acetic acid we find that, equating activities with concentrations, $c_{H^+} = c_{Ac^-} = 0.00136$, which would give $\Lambda = 381 \ \Omega^{-1} \cdot cm^2 \cdot mol^{-1}$.

It may be noted that the limiting molar conductivities of H^+ and OH^- in water are much higher than those of any other singly charged ions. This indicates that the mechanism of conduction for these ions is different from that of other ions. Thus, apparent proton transfer in water can occur very rapidly, since it can be accomplished by moving electron pairs. This is one example of information about phenomena on a molecular scale which can be inferred from measurement of conductivities.

The limiting molar conductivities of Li^+, Na^+, and K^+ in aqueous solution are in the order $K^+ > Na^+ > Li^+$, opposite to what one would expect from the relative masses and sizes of the ions. The explanation is that the ions are hydrated: In the presence of an electric field, it is the hydrated ion, with its attached water, that moves. For a smaller ion, the electrostatic forces that hold the dipolar water molecules are stronger so that, effectively, more water molecules are attached, leading to a larger species that moves more slowly. At high ionic concentrations, the water transported by the ions must be taken into account in measuring transport numbers with a Hittorf cell (see Section K).

If the conductivity of a saturated solution of a sparingly soluble, but completely dissociated, salt is measured, the solubility can be determined with the aid of a table of ionic conductivities. The sum of conductivities of the ions of the salt, multiplied by ionic concentrations, should give the conductance κ_S of the salt, obtained by subtracting the conductance of the solvent from that measured for the saturated solution. Thus if ν_+ cations and ν_- anions result from dissociation of the salt, the concentration c of the saturated solution is

$$c = \frac{\kappa_S}{\nu_+ \Lambda_+^\infty + \nu_- \Lambda_-^\infty}$$

Similarly, for a solution of a weak electrolyte, measurement of the conductance, with use of a table of ionic conductivities, allows calculation of the concentration of the ions and hence, with the nominal electrolyte concentration, calculation of the equilibrium constant for dissociation.

In this case, it is often necessary to take into account the variation of ionic activity coefficients with concentration. Thus to determine

$$K_a = \frac{a_{H^+} a_{A^-}}{a_{HA}}$$

for the dissociation $HA \to H^+ + A^-$ one assumes that γ_{HA} is approximately unity and that the mean ionic activity coefficient $\gamma_\pm$ obeys the Debye–Hückel law

$$\ln \gamma_\pm = -AI^{1/2}$$

where A is a constant and the ionic strength I is given by $(c_{H^+} + c_{A^-})/2$, so that

$$\ln K_a = \ln \frac{c_{H^+} c_{A^-}}{c_{HA}} - 2AI^{1/2} \qquad [43]$$

From the conductivity measurement one obtains $c_{H^+} = c_{A^-} = \alpha c$ (c = nominal concentration of electrolyte), so equation [43] becomes

$$\ln \frac{\alpha^2 c}{1 - \alpha} = \ln K_a + 2A(\alpha c)^{1/2}$$

and a plot of the left-hand side vs. $(\alpha c)^{1/2}$ should give a straight line with $\ln K_a$ as the intercept.

Conductivity measurements are also used to follow the course of titrations involving ions. A plot of κV (κ = conductance, V = total solution volume) vs. volume of added titrant usually is linear with an abrupt change of slope at the equivalence point. For example, if a strong acid is titrated with a strong base, the replacement of hydrogen ions (which have very high molar conductivity) by cations of the base leads to a decrease in conductivity: after the equivalence point, the addition of base increases the conductivity because cations and hydroxide ions are being added.

K. TRANSPORT NUMBERS

For a binary electrolyte that dissociates to ν_+ cations and ν_- anions, we may write the current density as

$$j = j_+ + j_- = \alpha c N_A (\nu_+ q_+ v_+ + \nu_- q_- v_-)$$

where c is the concentration and α is the fraction dissociated (N_A = Avogadro's number, introduced because c is normally expressed in moles per liter). Expressing the charge on an ion as $q_i = z_i e$ and introducing the faraday $\mathfrak{F} = N_A e$, we have

$$\frac{j}{E} = \alpha c \mathfrak{F} \left(\nu_+ z_+ \frac{v_+}{E} + \nu_- |z_-| \left| \frac{v_-}{E} \right| \right) \qquad [44]$$

The positive quantities $v_+/E = u_+$ and $|v_-/E| = u_-$, called *mobilities*, are the speeds attained by cation and anion under unit electric fields. The molar conductivity is given in terms of mobilities by

$$\Lambda = \mathfrak{F} (\nu_+ z_+ u_+ + \nu_- |z_-| u_-) \qquad [45]$$

If Λ is 100 $\Omega^{-1} \cdot$ cm$^2 \cdot$ mol^{-1} for a uni-univalent electrolyte, the average mobility of the ions is 5.2×10^{-8} m$^2 \cdot$ V$^{-1} \cdot$ s^{-1}, so under a field of 1 V/m the ions would move at 5.2×10^{-8} m/s.

The transport number or transference number of an ion is the fraction of the current carried by that ion. For a binary electrolyte, the transport number of the cation is

$$t_+ = \frac{\nu_+ z_+ u_+}{\nu_+ z_+ u_+ + \nu_- |z_-| u_-}$$

$$= \frac{c_+ z_+ u_+}{c_+ z_+ u_+ + c_- |z_-| u_-}$$

where $c_+ = \nu_+ c$ gives the cationic concentration. Obviously $t_+ + t_- = 1$. For a general electrolyte the transport number of ion i is

$$t_i = \frac{c |z_i| u_i}{\sum_j c_j |z_j| u_j} \qquad [46]$$

A knowledge of transport numbers and conductivities of an electrolyte suffices to establish individual ionic conductivities. Transport numbers may be determined experimentally by examining the changes in the solutions at the cathode and anode after electrolysis.

In a Hittorf cell there are cathode and anode compartments (separated by a thin tube), from which samples may be withdrawn for analysis. The electrolysis, which causes reduction at the cathode and oxidation at the anode, is accompanied by conduction of anions from cathode to anode and cations from anode to cathode through the cell. The electrolysis also contributes to the change in ion concentrations.

Consider the simple case in which the cation C^{n+} is reduced to C at the cathode and C is oxidized to C^{n+} at the anode, with the solution containing only the electrolyte CX_n. If f faradays of electricity pass through the cell, the electrolysis introduces f/n moles of C^{n+} at the anode and removes f/n moles of C^{n+} at the cathode. The transport process removes $t_+ f/n$ moles of C^{n+} from the anode (since each carries n charges) and brings $t_+ f/n$ moles of C^{n+} to the cathode, while $t_- f$ moles of X^- move in the opposite direction. Thus the anode compartment gains $(1 - t_+) f/n$ moles of C^{n+} and $t_- f$ moles of X^-, which together constitute $t_- f/n$ moles of CX_n. The cathode loses the same amount. From measurements of electrolyte concentration before and after electrolysis one can then derive t_+ and t_-.

The moving boundary methods for determining transport numbers deal with ionic mobilities directly. In one kind of moving boundary cell, a solution of the electrolyte CA is placed on top of a solution of CA' (same cation, different anion) and a solution of C'A (different cation from CA) is placed on top of the CA solution (see Figure 2). If the densities are in the order $d_{CA'} > d_{CA} > d_{C'A}$, there will be

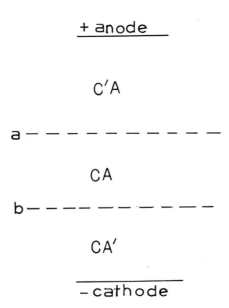

FIGURE 2. Arrangement for determining transport numbers by moving-boundary method.

no gross mixing of solutions, but at each of the boundaries between solutions the difference in diffusion rates between anions or cations gives rise to a steady-state potential difference, as in the case of a liquid junction potential (Section G). Now suppose electrolysis is carried out with the anode at the top, so that anions move upward and cations move downward. Boundary a, which is a boundary between solutions differing in cations, will move down, toward the cathode; boundary b will move toward the anode, since it is defined by solutions having the same cations but different anions. If C moves faster than C′ and A moves faster than A′, there will still be no mixing of solutions. Suppose that in time t boundary a moves a distance l_a and boundary b moves a distance l_b. The ratio l_a/l_b must equal the ratio v_c/v_A. Thus $t_+ = v_+/(v_+ + v_-) = l_a/(l_a + l_b)$.

It is possible to find negative values for transport numbers, and correspondingly (since the sum of transport numbers for an electrolyte is unity) it is possible to find transport numbers greater than unity. This is explained by formation of complex ions, as in the case of CdI_2, for which $t_+ < 0$ for concentrations exceeding 0.5 mol/dm^3. Equilibrium for the reaction

$$Cd^{2+} + 4I^- \rightarrow [CdI_4]^{2-}$$

means that the concentration of the complex ion $[CdI_4]^{2-}$ increases as the solution concentration increases. This ion moves to the anode, transporting Cd. At concentrations above 0.5 mol/dm^3, there is enough of this ion present so that more Cd is transported to the anode than to the cathode, making t_+ negative.

We now turn to the overpotential at an electrode, associated with the electrode process or with the movement of ions to and from the electrode by diffusion. The former will be briefly discussed in Sections L and M and the latter will be discussed in Sections N and thereafter. More detailed discussions appear in later chapters.

L. ACTIVATION OVERPOTENTIAL

In general, an electrode process may be represented as the result of the reduction

$$O + ze^- \rightarrow R$$

and the reverse reaction, the oxidation

$$R \rightarrow O + ze^-$$

Here, O and R each represent a combination of reactants. The rate of the reduction reaction will be $k_r a_O$, where a_O is the product of activities of the reactants O, each raised to the proper power, and k_r is the rate constant for reduction. The rate of the oxidation reaction may correspondingly be written as $k_o a_R$, where k_o is a rate constant and a_R is the proper combination of activities of the species in R. At equilibrium, $k_r a_O = k_o a_R$. The rates of both processes will be proportional to the area of the electrode. Thus it is customary to define k_r and k_o so that the rates $k_r a_O$ and $k_o a_R$ are in moles per unit area. (Although the true surface area of an electrode, considered on a microscopic scale, may be considerably bigger than the geometric surface area, especially for a solid electrode, it is the geometric surface area that is used.) Then a rate may be transformed to a current density by multiplication by $z\mathcal{F}$.

The dependence of rates on electrode potential is contained in the rate constants. According to absolute reaction rate theory, any rate constant may be written as $(kT/h)\exp(-\Delta G^{\ddagger}/RT)$, where k = Boltzmann's constant, h = Planck's constant, and $\Delta G^{\ddagger}$ is the free energy of activation, which apparently must depend on the electrode potential. It is customary to imagine the free energy as a function of a reaction coordinate that leads from reactants to products. In such a scheme, reactants and products are located at minima, separated by a maximum associated with the activated complex. If the free energies of O, R, and activated complex are G_O, G_R, and G_C, respectively, the free energy of activation for the oxidation is

$$\Delta G_o^{\ddagger} = G_C - G_R \qquad [46]$$

and the free energy of activation for the reduction is

$$\Delta G_r^{\ddagger} = G_C - G_O \qquad [47]$$

Clearly,

$$\Delta G_o^{\ddagger} - \Delta G_r^{\ddagger} = G_O - G_R \qquad [48]$$

which represents the overall free energy change for the oxidation process.

The electrode–solution potential difference will determine the electrical potential throughout the region of the electrode, thus affecting the free energies G_R and G_O and, if one considers the activated complex as intermediate between oxidized species O and reduced species R, G_C as well. Changing $\phi^{(M)} - \phi^{(\text{sol})}$ will then change $\Delta G_r^{\ddagger}$ and $\Delta G_o^{\ddagger}$ and the rate constants k_r and k_o. In the M^+/M half-cell discussed in Section I, changing $\phi^{(M)} - \phi^{(\text{sol})}$ by E would change $G_O - G_R = \tilde{\mu}_{M^+} + \tilde{\mu}_{e^-}^{(M)} - \mu_M$ by $-\mathcal{F}E$ (see equation [38]). One may suppose that the change in the activation free energy is proportional to the change in the free energy of reaction. Such a proportionality, implying a linear relation between the logarithm of a rate constant and the logarithm of an equilibrium constant, is often observed.

For reduction of a metal ion M^+ to M, one can imagine (see Figure 3) $V_{\text{sol}}(z)$, the potential energy of solvated M^+ as a function of distance from the electrode; it has a minimum at some $z = z_2$, several atomic diameters from the metal. The potential energy of an ion in the electrode, $V_{\text{met}}(z)$, has a minimum at $z = z_1$ ($z_1 < z_2$). The two potential curves intersect at z_m ($z_1 < z_m < z_2$), and $V_{\text{met}}(z_m) - V_{\text{met}}(z_1)$ is the activation energy for oxidation. An increase in the electrode–solution potential difference by ϕ shifts V_{met} upward compared to V_{sol} by $q\phi$, and the

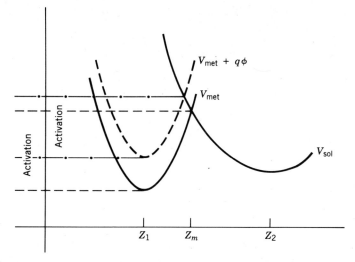

FIGURE 3. Intersection of potential energy curves, showing activation energy and effect on activation energy of a change in the metal–solution potential difference.

intersection with V_{sol} occurs at a value of z smaller than z_m. The activation energy decreases by a fraction of $q\phi$, which is the change in the free energy of reaction.

In general, suppose the change in $G_R - G_O$ is $z\mathcal{F}E$ and that $G_C - G_O$ changes by a fraction α of $z\mathcal{F}E$. This α is called the *transfer coefficient*. Then according to equation [48], $\Delta G_O^\ddagger$ must change by $(\alpha - 1)\,z\mathcal{F}E$. The rate constants may be written

$$k_r = k_r^0 \exp\left(\frac{-\alpha z \mathcal{F} E}{RT}\right)$$

and

$$k_o = k_o^0 \exp\left[\frac{(1-\alpha)\,z\mathcal{F}E}{RT}\right]$$

where the superscript zero refers to the situation $E = 0$, which corresponds to no potential difference between electrode and solution. If there is equilibrium and no net current flow (as at open circuit), we have

$$k_r a_O = k_o a_R$$

and substitution of the expressions for k_r and k_o gives

$$\frac{a_R}{a_O} = \frac{k_r^0}{k_o^0} \exp\left(\frac{-z\mathcal{F}E}{RT}\right) \qquad [49]$$

When all activities are unity, the value of E is $E^0_{1/2}$, so that

$$1 = \frac{k^0_r}{k^0_o} \exp\left(\frac{-z\mathfrak{F} E^0_{1/2}}{RT}\right) \qquad [50]$$

and, dividing equation [50] into equation [49] and taking logarithms, we get

$$\ln \frac{a_R}{a_O} = \frac{-z\mathfrak{F}(E - E^0_{1/2})}{RT}$$

Thus one has a Nernst-type equation

$$E = E^0_{1/2} - \frac{RT}{z\mathfrak{F}} \ln \frac{a_R}{a_O}$$

It should be remembered that $E^0_{1/2}$, an actual electrode–solution potential difference, differs from the standard half-cell potential $\mathcal{E}^0_{1/2}$ by a constant whose value relates to the choice of $\mathcal{E}^0_{1/2}(H^+/H_2)$ as O.

The cathodic current density is

$$j_c = z\mathfrak{F} k^0_r a_O \exp\left(\frac{-\alpha z \mathfrak{F} E}{RT}\right) \qquad [51]$$

and the anodic current density, associated with oxidation, is

$$j_a = z\mathfrak{F} k^0_o a_R \exp\left[\frac{(1-\alpha) z \mathfrak{F} E}{RT}\right] \qquad [52]$$

Since reduction corresponds to (1) a flow of electrons from the surroundings into the cell through the electrode and (2) a positive current flow through the cell from left to right when the electrode of interest is the right-hand electrode of a cell, the cathodic current density is taken as positive and the net current density is $j_c - j_a$.

Let us define k^0 by $k^0_r \exp(-\alpha z \mathfrak{F} E^0_{1/2}/RT)$, so that, from equation [50], k^0_o is $k^0 \exp[-(1-\alpha) z \mathfrak{F} E^0_{1/2}/RT]$ and the net current density is

$$j_c - j_a = z\mathfrak{F} k^0 \Big\{ a_O \exp\left[\frac{-\alpha z \mathfrak{F}(E - E^0_{1/2})}{RT}\right]$$

$$- a_R \exp\left[\frac{(1-\alpha) z \mathfrak{F}(E - E^0_{1/2})}{RT}\right] \Big\} \qquad [53]$$

The kinetic parameters α and k^0 characterize the electrode reaction. They are often considered to be independent of potential. Values of α are between 0 and 1, and k^0 is generally between 10^{-5} and 10^{-1} cm/sec.

The value of either j_c or j_a at open circuit (equilibrium) is denoted by j_0 and called the *exchange current density*. The equations suggest that, when current is flowing through the cell ($j_c - j_a \neq 0$), j_0 does not change, so that

$$j_c = j_0 \exp\left[\frac{-\alpha z \mathcal{F}(E - E')}{RT}\right]$$

and correspondingly for j_a, where E is the electrode–solution potential difference and E' is its equilibrium value. Obviously, j_c and j_a are no longer equal. The current density is

$$j = j_0 \left\{ \exp\left[\frac{-\alpha z \mathcal{F}(E - E')}{RT}\right] - \exp\left[\frac{(1-\alpha) z \mathcal{F}(E - E')}{RT}\right] \right\} \quad [54]$$

where

$$j_0 = z\mathcal{F}k_r^0 a_O \exp\left[\frac{-\alpha z \mathcal{F}E'}{RT}\right] = z\mathcal{F}k_0^0 a_R \exp\left[\frac{(1-\alpha) z \mathcal{F}E'}{RT}\right] \quad [55]$$

If $E - E' < 0$, then $j_c > j_a$ so that the net current is cathodic, which is considered positive.

M. TAFEL LAW

Equation [54] gives the dependence of the current density on the overpotential or overvoltage $\eta = E - E'$. If the electrode is part of a cell, the other electrode of which is reversible, η is also the deviation of the cell potential from its equilibrium or open-circuit value. This overpotential is associated with the activation energy of the reaction and is referred to as *activation overpotential*. We see that j is monotonic in η with $j = 0$ for $\eta = 0$, j is positive (cathodic current) for $\eta < 0$ with $j \to \infty$ for $\eta \to -\infty$, and j is negative (anodic current) for $\eta > 0$ with $j \to -\infty$ and $\eta \to \infty$. A plot of current density vs. overpotential is shown in Figure 4, for $\alpha = \frac{1}{2}$.

When η is small, one may use the approximation $e^x \sim 1 + x$ for each exponential, and we have

$$j = j_0 \left[1 - \frac{\alpha z \mathcal{F}\eta}{RT} - 1 - \frac{(1-\alpha) z \mathcal{F}\eta}{RT}\right] = \frac{j_0 z \mathcal{F}\eta}{RT} \quad [56]$$

Thus for small η, we note that j is directly proportional to η with the exchange current density giving the constant of proportionality. When the magnitude of η is

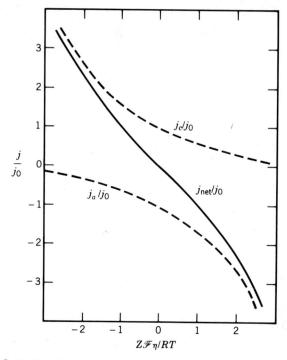

FIGURE 4. Cathodic, anodic, and net currents as a function of overpotential ($\alpha = \frac{1}{2}$).

large, one term in j becomes exponentially small and negligible. For η large and negative, we have

$$j \simeq j_0 \exp\left(\frac{-\alpha z \mathscr{F} \eta}{RT}\right)$$

which rearranges to

$$\frac{\alpha z \mathscr{F} \eta}{RT} = \ln j_0 - \ln j \qquad [57]$$

so η is linear in $\ln j$ (j positive). The equation

$$\eta = a - b \log_{10} j \qquad [58]$$

is known as the Tafel equation for cathodic currents. For η large and positive, we have

$$-j = j_0 \exp\left[\frac{(1-\alpha)\mathscr{F}\eta}{RT}\right]$$

which rearranges to

$$\frac{(1-\alpha)z\mathcal{F}\eta}{RT} = -\ln j_0 + \ln(-j) \qquad [59]$$

where j is negative. The equation

$$\eta = a' + b'\log_{10}(-j) \qquad [60]$$

is the Tafel equation for anodic behavior. By fitting experimental current–overpotential data to either Tafel equation, one can determine the exchange current density and, if z is known, the transfer coefficient.

A plot of η vs. $\log_{10}|j|$ should be a straight line except for small $|\eta|$. For cathodic currents, the intercept is $(RT/\alpha z\mathcal{F})\ln j_0$ and the slope is $-2.303 RT/\alpha z\mathcal{F}$; for anodic currents the intercept is $[RT/(\alpha-1)z\mathcal{F}]\ln j_0$ and the slope is $2.303 RT/(1-\alpha)z\mathcal{F}$. A plot of η against $\ln|j|$ will involve two curves, each having a horizontal asymptote for small $\ln|j|$ and each approaching a linear asymptote for large $\ln|j|$. The two linear asymptotes intersect at the point $\eta = 0$, their common values being $|j| = j_0$.

The exchange current density j_0 of an electrode can be anything from 10^{-2} to 10^{-16} A/cm^2. The magnitude of j_0 indicates the speeds of the anodic and cathodic electrode reactions, so that, the higher the exchange current density, the more difficult the electrode is to polarize. The equation $j = j_0 z\mathcal{F}\eta/RT$, valid for small η, shows that the ratio of overpotential to net current is inversely proportional to j_0. Differentiating equation [54] shows that $-(\partial j/\partial \eta)_0 = j_0 z\mathcal{F}/RT$. The quantity $RT/j_0 z\mathcal{F}$ has dimensions of resistance per unit area and is called the *polarization resistance at the equilibrium electrode potential*. For large $|\eta|$, we have

$$\ln\frac{|j|}{j_0} = \begin{cases} \dfrac{-\alpha z\mathcal{F}\eta}{RT}, & \text{cathodic} \\ \dfrac{(1-\alpha)z\mathcal{F}\eta}{RT}, & \text{anodic} \end{cases}$$

so for a given current j a larger j_0 means the magnitude of the overpotential is smaller. Smaller overpotentials mean faster electrode reactions.

For very small values of j_0, such as for very dilute solutions, establishment of equilibrium is slow. Indeed, impurities in an electrochemical system often provide a current density of 10^{-7} A/cm^2. If j_0 is lower than the current density due to impurities, the potential of the electrode will be governed by the reactions of the impurities rather than by the electrode reactions.

Although a reduction involving several electrons probably involves several one-electron steps, the expression for the current may still resemble that for a single multi-electron step. Suppose the reduction

$$O + 2e^- \rightarrow R$$

actually occurs in two steps:

$$O + e^- \rightarrow I$$

with forward and reverse rate constants k_{r1} and k_{o1}, followed by

$$I + e^- \rightarrow R$$

with forward and reverse rate constants k_{r2} and k_{o2}. Let the standard potentials for the two steps be E_1^0 and E_2^0 and the transfer coefficients be α_1 and α_2, and let the concentrations of O, I, and R be denoted by c_O, c_I, and c_R, respectively. Neglecting the concentration differences, if any, across the double layer, we have the following for the current density:

$$j = \mathcal{F}(k_{r1}c_O + k_{r2}c_I - k_{o1}c_I - k_{o2}c_R) \qquad [61]$$

From the steady-state condition

$$\frac{dc_I}{dt} = k_{r1}c_O - k_{r2}c_I + k_{o1}c_I + k_{o2}c_R = 0$$

we can solve for the concentration of the intermediate I. Substituting for c_I into equation [61], we have

$$j = \frac{2\mathcal{F}(k_{r1}k_{r2}c_O - k_{o1}k_{o2}c_R)}{k_{r2} + k_{o1}} \qquad [62]$$

Now suppose the first reduction step is rate-determining so that I is reduced whenever it is formed, that is, $k_{o1} \ll k_{r2}$ so we can neglect k_{o1} in the denominator of equation [62]. Then substituting for the remaining rate constants according to (see equations [51]–[53])

$$k_{ri} = k_i^0 \exp\left[\frac{-\alpha_i \mathcal{F}(E - E_i^0)}{RT}\right], \qquad i = 1, 2$$

$$k_{oi} = k_i^0 \exp\left[\frac{(1 - \alpha_i)\mathcal{F}(E - E_i^0)}{RT}\right], \qquad i = 1, 2$$

we have the following for the current density of equation [62]:

$$j = 2\mathcal{F}k_i^0 \left\{ c_O \exp\left[\frac{-\alpha_1 \mathcal{F}(E - E_1^0)}{RT}\right] - c_R \exp\left[\frac{(1 - \alpha_1)\mathcal{F}(E - E_1^0)}{RT}\right] \right.$$
$$\left. \cdot \exp\left[\frac{\mathcal{F}(E - E_2^0)}{RT}\right] \right\} \qquad [63]$$

With the substitutions

$$E^0 = \frac{E_1^0 + E_2^0}{2}, \quad k^0 = k_i^0 \exp\left[\frac{\alpha \mathcal{F}(E_1^0 - E_2^0)}{RT}\right], \quad \alpha = \frac{\alpha_1}{2}$$

equation [63] becomes

$$j = 2\mathcal{F}k^0 \left\{ c_O \exp\left[\frac{-2\alpha \mathcal{F}(E - E^0)}{RT}\right] - c_R \exp\left[\frac{2(1-\alpha)\mathcal{F}(E - E^0)}{RT}\right] \right\}$$

This is just the current-density expression for a two-electron reduction, as given in equation [53]. One can similarly consider $k_{o1} \gg k_{r2}$, making the reduction of I the rate-determining step, and obtain a similar expression. If the rate constants k_{o1} and k_{r2} are comparable in size, however, the current-density expression can be more complicated.

N. CONCENTRATION OVERPOTENTIAL

Electron transfer at an electrode is both preceded and followed by diffusion. When diffusion is much faster than electron transfer, or vice versa, it is possible to consider only one of the steps in discussing the rate of the overall process. The rate of the overall process is determined by parameters relating to the slower step, which is the rate-determining step. The other step is either instantaneous, if it follows the rate-determining step, or comes to equilibrium, if it precedes the rate-determining step.

Associated with the diffusion is another overpotential, *concentration overpotential*, because of the necessity for transport of reactant ions to, and product ions away from, the electrode surface of a working electrode, if a current is to be maintained. This requires a concentration gradient of these ions in the adjoining solution, so that the ionic concentration at the electrode surface must differ from the concentration in bulk solution. The potential calculated according to a Nernst equation will likewise differ from what one would get by using the nominal (bulk) solution concentration.

For example, for the M^{z+}/M electrode, a cathodic current will reduce M^{z+} ions to M, diminishing their concentration at the electrode surface, which will lead to a flow of M^{z+} ions toward the electrode. At steady state, the concentration of M^{z+} will vary from its value in bulk solution, c_s, to a smaller value c_c at the cathode surface. Similarly, if an anodic current is imposed, the concentration of M^{z+} at the electrode surface when steady state is reached will be larger than c_s, with the concentration varying from c_a at the surface to c_s far from the surface. In a cell formed from identical M^{z+}/M electrodes, the imposition of a potential difference between the electrodes to produce electrolysis will effectively lead to a concentration cell since c_a (at the anode surface) will exceed c_c. Part of the potential difference across the cell will be due to the two concentration overpotentials at the

1. ELECTROCHEMISTRY

electrodes, and part will be due to the activation overpotential. If the exchange current density is high enough, the electrode reactions will be rapid enough to maintain equilibrium at the electrode surfaces and make the activation overpotential negligible.

At an M^+/M electrode acting as a cathode, the potential difference between metal and solution will be

$$E' = E^0 + \frac{RT}{z\mathcal{F}} \ln c_c$$

where E^0 is a standard potential corresponding to unit activity of the ion in solution; we have assumed equilibrium is maintained with respect to the reduction and deposition of M^{z+} and approximated the activity of M^{z+} by its concentration. With no net current flowing, $c_c = c_s$ and $E' = E$. When reduction is occurring, $c_c < c_s$ so $E' < E$. If there were no stirring of the solution, the electrode reaction would, as time passed, deplete solution further and further away from the electrode surface in M^{z+}, leading to a concentration profile that would decrease from c_c to c_s more and more gradually. Since the gradient of concentration at the electrode would decrease in size, the diffusion process would become slower and slower and the reduction current would become smaller and smaller. If stirring took place, ions would be brought to the electrode by the movement of the solution and a steady state could be attained.

The situation is often pictured as involving a *diffusion layer* of finite thickness adjacent to the electrode, with the concentration of ions maintained equal to its bulk value outside by the stirring. The concentration then increases monotonically from c_c (at the electrode) to c_s across the diffusion layer. There is a constant concentration overpotential

$$\eta = E' - E = \frac{RT}{z\mathcal{F}} \ln \frac{c_c}{c_s} \qquad [64]$$

and a constant current flow. A further approximation, originally due to Nernst, is to assume a linear variation of c across the diffusion layer. If the thickness of the layer is δ, the gradient of concentration is $(c_s - c_c)/\delta$, in the direction from electrode to solution. According to Fick's first law of diffusion, the amount of substance diffusing through unit area per unit time in the x-direction, at a point where its concentration is c, is given by

$$J = -D \frac{\partial c}{\partial x} \qquad [65]$$

where D is the diffusion coefficient. Then the flow of M^{z+} ions is

$$J = \frac{-D(c_s - c_c)}{\delta} \qquad [66]$$

where the minus sign indicates flow in the negative x-direction, that is, toward the electrode. Usually, the thickness δ turns out to be much less than 1 mm. According to the Nernst approximation, J is constant through the diffusion layer. The value of J at the electrode surface gives the cell current I on multiplication by the area of the electrode and by $z\mathcal{F}$, the charge carried by 1 mole of M^{z+}.

As discussed in Section J, the current I is carried through the bulk solution by all ions, and the fraction carried by the ion i is given by its transport number t_i. Let us assume that only M^{z+} and an anion A^- are available, with transport numbers t_+ and $t_- = 1 - t_+$. For every faraday of electricity transported through the cell, t_+/z moles of M^{z+} pass in the direction from anode to cathode and t_- moles of A^- pass from cathode to anode. The reduction of M^{z+} at the cathode consumes $1/z$ moles of M^{z+} which must be transported to the electrode through the diffusion layer. This means that $(1 - t_+)/z$ moles of M^{z+} must be transported by diffusion, in addition to that transported in connection with electrical conduction through the system (migration).

The amount of A^+ transported from cathode to anode through bulk solution is t_-. The solution near the cathode becomes depleted in A^- and the solution near the anode becomes concentrated in A^- until, at steady state, a gradient of A^- concentration exists, which causes a diffusion current of A^- in the direction opposite to migration, so that the net transport of A^- through the diffusion layer is zero. If this electrode were functioning as an anode, each faraday of electricity would produce $1/z$ moles of M^{z+}, of which t_+/z would flow away as a consequence of migration, so that $(1 - t_+)/z$ must be removed by diffusion. Then t_- moles of A would be transported toward the M^{z+}/M electrode, building up the A^- concentration until the diffusion current of A^- exactly canceled its migration current.

These arguments show that within the diffusion layer the entire current is carried by the electroactive ion (M^{z+}), whether the electrode is cathodic or anodic. For this ion, the current in the diffusion layer is the migration current $t_+ I$ and the diffusion current $(1 - t_+)I$ since the total current is I; for the other ion the diffusion current is $t_- I$ since it exactly cancels its migration current $-t_- I$ (opposite in sign to the migration current of the electroactive ion). If there is another cation N^{y+} present in excess, the transport number of M^{z+} will be small and most of the current through the bulk solution will be carried by N^{y+} and by the anion. However, the current through the diffusion layer will still be carried wholly by M^{z+} because only M^{z+} is electroactive, transferring charge to and from the electrode. Thus within the diffusion layer the diffusion currents for the anion and for the other cation are equal and opposite to their migration currents. Using the Nernst approximation for the diffusion current and taking current into the electrode as positive, we get

$$I = \frac{z\mathcal{F}Da(c_s - c_c)}{\delta} + t_+ I \qquad [67]$$

where a is the electrode area, and the total current is the same everywhere in the circuit.

When we solve equation [67] for I we find I proportional to D, to $(c_s - c_c)/\delta$, and to $(1 - t_+)^{-1}$. As t_+ decreases, with other factors remaining unchanged, I decreases. If $t_+ \to 0$, I becomes identical with the diffusion current. As the current increases, so does the value of $c_s - c_c$, since this determines the gradient and the diffusion rate. The overpotential at this cathode becomes more negative; with $t_+ = 0$, equation [67] gives

$$c_c = c_s - \frac{I\delta}{z\mathcal{F}Da}$$

so that equation [64] becomes

$$\eta = \frac{RT}{z\mathcal{F}} \ln \frac{1 - I\delta}{z\mathcal{F}Dac_s}$$

The overpotential decreases if D increases or I decreases. It of course vanishes when $I = 0$, like the activation overpotential.

Since c_s is fixed and c_c cannot be less than zero there is a limiting diffusion current, given by

$$I_1 = \frac{z\mathcal{F}Dac_s}{\delta} \qquad [68]$$

The overpotential may be written in terms of I_1 as

$$\eta = \frac{RT}{z\mathcal{F}} \ln 1 - \frac{I}{I_1}$$

Thus

$$\frac{I}{I_1} = 1 - \exp\left(\frac{z\mathcal{F}\eta}{RT}\right) \qquad [69]$$

and $I \to I_1$ exponentially as $\eta \to -\infty$. It will be noted that I_1 is proportional to the bulk solution concentration of electroactive species. This is the basis of polarography (Section Q).

O. DIFFUSION PLUS REACTION

In addition to reactions at the electrode, chemical reactions in the bulk phases may also be important in determining the rate of current flow. Although it is possible for diffusion, chemical reaction, and electrode reaction all to be important, here we suppose the electrode reaction is fast enough so that only the first two need to be considered explicitly.

The electrode reaction is supposed to be fast, but the concentration of the species which is reduced at the electrode is controlled by a chemical equilibrium. For example, one may study the reduction of H^+ to H_2 with the H^+ produced by the dissociation of a weak acid (Albery 1975, pp. 125–130). The diffusion of the acid to the electrode must be considered together with the dissociation kinetics. The equilibrium constant K_A, where

$$K_A = \frac{c_{H^+} c_{A^-}}{c_{HA}} \quad [70]$$

is supposed to be small, so $c_{HA} \gg c_{H^+}$, and a salt of A^- is supposed to be present, so $c_{A^-} \gg K_A$. If the rate of reduction is fast enough to reduce c_{H^+} to 0 at the electrode surface, the rate of the overall process depends on the interplay between the diffusion of HA and its dissociation. As a result of the reduction, the equilibrium [70] is disturbed in a region near the electrode, which may be referred to as the *reaction layer*, analogous to the diffusion layer; this represents the region in which the concentration of reactant differs from the bulk concentration because of the finite rate of reaction. The effective thickness of the reaction layer, x_R, is much less than the thickness of the diffusion layer, x_D.

Under the conditions specified, the protons that are reduced come mostly from HA dissociating in the reaction layer. Thus one is interested in the flux of HA through the diffusion layer. At steady state, this flux is

$$j_L = \frac{D_{HA}[c_{HA}(x_D) - c_{HA}(x_R)]}{x_D - x_R}$$

where D_{HA} is the diffusion coefficient of HA and c_{HA} in the diffusion-layer model is supposed to be constant for $x > x_D$. For $x < x_R$, in the reaction layer, $c_{HA}(x)$ changes because of (a) diffusion, (b) the dissociation of HA with rate constant k_1, and (c) the recombination of $H^+ + A^-$ with rate constant k_{-1}, where $k_1/k_{-1} = K_A$;

$$\frac{\partial c_{HA}}{\partial t} = D_{HA} \frac{\partial^2 c_{HA}}{\partial x^2} - k_1 c_{HA} + k_{-1} c_{H^+} c_{A^-} = 0 \quad [71]$$

in the steady state. At the electrode, $\partial c_{HA}/\partial x = 0$ because HA does not react at the electrode. For the protons, the steady-state condition is

$$D_{H^+} \frac{\partial^2 c_{H^+}}{\partial x^2} + k_1 c_{HA} - k_{-1} c_{H^+} c_{A^-} = 0 \quad [72]$$

with $c_{H^+}(0) = 0$ since H^+ is consumed immediately at the electrode.

Except in the reaction layer, the equilibrium [70] means that $-k_1 c_{HA} + k_{-1} c_{H^+} c_{A^-} = 0$. It is convenient to define

$$c' = (K_A c_{HA}/c_{A^-}) - c_{H^+}$$

so that $c' = 0$ outside the reaction layer and $c'(0) = K_A c_{HA}(0)/c_{A^-}$, where c_{A^-} is essentially independent of x. Now $k_1 c_{HA} - k_{-1} c_{H^+} c_{A^-}$ is $k_{-1} c_{A^-} c'$. Thus, on multiplying equation [71] by $K_A/(c_{A^-} D_{HA})$, dividing equation [72] by D_{H^+}, and subtracting, one has

$$\frac{K_A}{c_{A^-}} \frac{\partial^2 c_{HA}}{\partial x^2} - \frac{\partial^2 c_{H^+}}{\partial x^2} - \frac{K_A}{D_{HA}} k_{-1} c' - \frac{1}{D_{H^+}} k_{-1} c_{A^-} c' = 0$$

or, since c_{A^-} is constant,

$$\frac{\partial^2 c'}{\partial x^2} = \left(\frac{K_A}{c_{A^-} D_{HA}} + \frac{1}{D_{H^+}} \right) k_{-1} c_{A^-} c'$$

Since K_A/c_{A^-} is small because of the presence of the A^- salt, this equation is easily solved to give

$$c' = c'(0) \exp\left(-\frac{x}{x_R} \right) \qquad [73]$$

where $x_R = (D_{H^+}/k_{-1} c_{A^-})^{1/2}$ is the thickness of the reaction layer. For typical conditions, it is shown (Albery 1975, p. 129) that x_R is of the order of 1 Å.

The flux of H^+ at the electrode surface is

$$j = D_{H^+} \left(\frac{\partial c_{H^+}}{\partial x} \right)_0 = -D_{H^+} \left(\frac{\partial c'}{\partial x} \right)_0$$

$$= \frac{D_{H^+} c'(0)}{x_R} = \left(\frac{D_{H^+} k_{-1}}{c_{A^-}} \right)^{1/2} K_A c_{HA}(0) \qquad [74]$$

using equation [73], or $j = k'_R c_{HA}(0)$. Here k'_R is an effective rate constant, giving the proportionality of the flux of H^+ toward the electrode (the reduction current) to the concentration of HA in the reaction layer. One can also say that $j = j_L$, the flux of HA through the diffusion layer, since the H^+ is brought in as HA, which dissociates near the electrode, and, because x_R is much smaller than x_D, we have

$$j_L = \frac{D_{HA} [c_{HA}(\infty) - c_{HA}(0)]}{x_D}$$

Combining these relations, we get

$$j = k'_R c_{HA}(0) = j_L = \frac{D_{HA} [c_{HA}(\infty) - c_{HA}(0)]}{x_D}$$

which gives

$$c_{HA}(0) = \frac{D_{HA}c_{HA}(\infty)}{k'_R x_D + D_{HA}} \quad [75]$$

and finally gives the flux of H^+ in terms of $c_{HA}(\infty)$:

$$j^{-1} = \frac{k'_R x_D + D_{HA}}{D_{HA}c_{HA}(\infty) k'_R}$$

$$= [k'_D c_{HA}(\infty)]^{-1} + [k'_R c_{HA}(\infty)]^{-1} \quad [76]$$

on combining equations [74] and [75], where $k'_D = D_{HA}/x_D$ is the effective rate constant for diffusion.

Depending on the relative sizes of k'_D and k'_R, the rate can be diffusion-controlled or reaction-controlled. The expression of j^{-1} as a sum of inverses of rates is a consequence of the overall process involving two steps in series. Equation [76] is the basis for the electrochemical measurement of the rates of fast homogeneous reactions (Albery 1975, pp. 131–132). Since $k'_R = K_A(D_{H^+}/c_{A^-})^{1/2}(k_{-1})^{1/2}$, measurements of k'_R and $K = k_1/k_{-1}$ allow the determination of the forward and reverse rate constants, k_1 and k_{-1}.

P. DISCHARGE POTENTIALS

If diffusion and reaction overpotentials are small, the current flowing through a cell will be proportional to the overpotential. Suppose an electrochemical cell is at equilibrium with a potential difference in the external circuit exactly equal to the cell emf $\mathcal{E}$. No current is flowing and the electrode–solution potential difference at each electrode is equal to the reversible potential. Increase of the external potential to $\mathcal{E}'$ will lead to electrolysis and a current

$$I = \frac{\mathcal{E}' - \mathcal{E}}{R}$$

where R is the resistance of the circuit. However, the electrodes will operate reversibly and remain unpolarized only for low currents. As $\mathcal{E}'$ and I increase, polarization will become important and the cell emf will exceed the difference of the reversible potentials of the two electrodes by the sum of their overpotentials. Then a plot of I vs. the imposed potential $\mathcal{E}'$ will no longer be linear, and the current will fall below the straight line with slope $1/R$.

If the cell is not initially at equilibrium, however, I will not increase linearly with $\mathcal{E}'$, even for small changes in $\mathcal{E}'$. For a Pt electrode in a chloride-containing solution, for example, the absence of Cl_2 means that the Cl_2/Cl^- electrode will not be in equilibrium. A small applied emf will cause migration of Cl^- to the anode and oxidation of Cl^- to Cl_2. The appearance of Cl_2 establishes the reversible potential according to

$$\mathcal{E}_{1/2}(Cl_2/Cl^-) = \mathcal{E}^0_{1/2}(Cl_2/Cl^-) - \frac{RT}{\mathcal{F}} \ln \frac{a_{Cl^-}}{p_{Cl_2}^{1/2}}$$

Thus the initial change in $\mathcal{E}'$ should lead to a transient current until equilibrium is established and the cell emf is equal to $\mathcal{E}'$. In addition to a *faradaic current* associated with the electrochemical reaction which establishes equilibrium, the transient current will include a *nonfaradaic current* associated with the formation of the electrical double layer at the electrode–solution interface.

As the Cl_2 diffuses away from the electrode, the cell emf $\mathcal{E}$ will start to differ from $\mathcal{E}'$, leading to further current flow. The higher the value of $\mathcal{E}'$, the higher the pressure of Cl_2 required to make $\mathcal{E}' = \mathcal{E}$ and the larger this *residual current* must be. However, if p_{Cl_2} exceeds 1 atm, Cl_2 will be continuously evolved at the electrode. When $\mathcal{E}'$ becomes large enough so the electrode–solution potential exceeds that of a Cl_2 half-cell with Cl_2 at 1 atm, an electrolysis current flows, with the applied emf being given by the *IR* drop plus the sum of the operating potentials of the electrodes, which include the overpotentials associated with the electrode processes. At this point we have reached the *decomposition voltage*.

In this region of potential, the cell emf varies linearly with current, and an extrapolation to $I = 0$ gives a value for the decomposition voltage. The potentials of the electrodes are called *discharge potentials* or (in the case of metal deposition) *deposition potentials*. Each discharge potential equals the reversible potential plus an overpotential.

Q. POLAROGRAPHY

Polarographic techniques for analysis of dissolved species are based on the proportionality of the limiting diffusion current I_l to the bulk concentration of the electroactive species in solution. To analyze for cations, the solution is electrolyzed between mercury electrodes, in the presence of an excess of electrolyte with non-electroactive cations to make the transport number of the electroactive species close to zero. The cathode is often a drop of mercury hanging from the tip of a capillary tube connected to a mercury reservoir; the size of the drop is made to increase slowly until it falls off and a new drop starts to form. Renewal of the cathode surface keeps it clean and free from electrolysis products. The anode is a pool of mercury with large surface area, often below the cathode to receive the falling drops. The large surface area keeps the current density low and minimizes polarization of the anode. Anodic polarography, in which substances are oxidized at a rotating platinum anode, is also performed.

In mercury-drop polarography, the cathode is brought to increasingly negative potentials, and metal ions of one species after another are discharged to form amalgams with the mercury. The current resulting from each such process increases until the diffusion current of the ion reaches its limiting value. A polarogram is a plot of cell current vs. cell potential and consists of a series of *polarographic waves*, each corresponding to the reduction of one of the cations present. (There is also a small current, increasing with increasingly cathodic potentials, which is due partly to impurities and partly to the residual current discussed in Section P, and which must be corrected for.) The height of the wave on a plot of current vs. cell potential should be proportional to the concentration of the cation. We now obtain an equation for the polarographic wave.

The current for the reduction reaction, $O + ze^- \rightarrow R$, if the rate-determining step is the diffusion process, is given by [67] with $t_+ = 0$, which we now write as

$$I = k_1([O] - [O]_e)$$

where k_1 includes the diffusion layer thickness and the diffusion constant for O, $[O]$ is the bulk solution concentration of O, and $[O]_e$ is the concentration at the electrode surface. The limiting current I_l for this bulk concentration is obtained for $[O]_e = 0$, so

$$I - I_l = -k_1[O]_e \qquad [77]$$

At the electrode surface, O is reduced to R which, if it is soluble in mercury or water, will diffuse away from the electrode surface. Assuming fast electrode reactions, the rate of production of R is determined by the current and, in steady state, equals the rate at which it diffuses away. Thus

$$I = k_2[R] \qquad [78]$$

where k_2 characterizes the diffusion of R. The potential of the electrode is

$$E = E^0 + \frac{RT}{z\mathcal{F}} \ln \frac{a_O}{a_R}$$

(no activation polarization), where a_O and a_R refer to bulk solution. We write $a_O = \gamma_O[O]$ and $a_R = \gamma_R[R]$, introducing activity coefficients, and substitute equations [77] and [78] to get

$$E = E^0 + \frac{RT}{z\mathcal{F}} \ln \left\{ \frac{\gamma_O k_2}{\gamma_R k_1} \frac{I_l - I}{I} \right\}$$

It can be shown that k_2/k_1 is equal to the ratio of the square roots of the diffusion coefficients for R and O. The activity coefficients are independent of the concentrations of R and O because the former are determined by the excess of indifferent electrolyte present. Then

$$E = E^0 + C + \frac{RT}{z\mathcal{F}} \ln \frac{I_l - I}{I} \qquad [79]$$

where C is a constant. This is the equation for a *reversible wave*, since we have assumed that mass transport is the rate-determining step, that is, that equilibrium with respect to the electrode reaction is always maintained.

Let $E_{1/2}$ be the potential when $I = I_l/2$, so the logarithm in [79] vanishes. $E_{1/2} = E^0 + C$ is called the *half-wave potential*. The equation

$$E = E_{1/2} + \frac{RT}{z\mathcal{F}} \ln \frac{I_l - I}{I} \qquad [80]$$

is plotted in Figure 5; E has a point of inflection at $I_l/2$. The half-wave potential is independent of ion concentration (but not independent of the nature and concentration of the supporting electrolyte used) and hence characterizes the species involved in the electrode reaction, allowing identification of species to be made from a polarogram. A plot of E vs. $\ln [(I_l - I)/I]$ should be linear with slope $RT/z\mathcal{F}$, giving the number of electrons involved in the reaction, and intercept $E_{1/2}$.

If several ions are present, there will be a succession of polarographic waves.

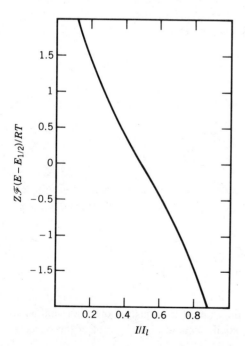

FIGURE 5. Reversible polarographic wave. Cell potential is plotted against current.

At a cathode potential more positive than the half-wave potential of any ion, one will have only the residual current. As the potential is made more negative, one will see a rapid increase in current when discharge of the ion with the highest half-wave potential begins (the current describes a wave such as in Figure 5), with the increase in I from one end to the other giving I_l. I_l is proportional to the concentration of this ion, and concentrations down to 10^{-6} mol/dm^3 can be measured by this technique. Further decrease in cathode potential will result in only a small increase in current until the discharge potential of the next dischargeable ion is approached. A second wave will then be added to the plot of I vs. potential, after which the current will be the residual current plus the sum of the limiting currents for the two ions. With more ions present, this behavior will be repeated. The range of potential for a mercury electrode is limited to between about $+0.25$ V, at which mercury will dissolve, and about -1.60 V (or -2.0 V in alkaline solution), at which hydrogen will be evolved.

The processes discussed may also be used in titration. Suppose a mercury drop electrode forms a cell with a nonpolarizable (reversible) electrode. Then in the presence of a reducible substance a plot of I vs. potential E' will be a polarographic wave. Now suppose the mercury electrode dips into a solution containing two reducible substances 1 and 2 with half-wave potentials E_1 and E_2, where $E_1 > E_2$. If the potential of the mercury cathode is fixed at a value between E_1 and E_2, the current at steady state will correspond to reduction of the first substance only. If substance 1 is present and substance 2 reacts with it to form a nonionizing species, addition of 2 will decrease I until all of 1 is consumed; further addition of 2 will give no additional current. If the potential of the cathode is fixed at a value below E_2, addition of 2 to the solution of 1 will result in a linear decrease of I with the amount of 2 added until all of 1 is consumed (the end point), but subsequent addition of 2 will give a linear increase in I. Again, the break in a plot of I vs. amount of 2 added indicates the end point of the titration.

2

THERMODYNAMICS AND ELECTROSTATICS

A. SURFACE QUANTITIES

Electrochemical processes occur mostly at the interface between two phases. On a microscopic level, the interface is a three-dimensional region in which intensive properties, such as concentrations of different chemical species, vary rapidly over small distances. On a macroscopic level, it may be considered as a mathematical surface of two dimensions. Indeed, the commonly used Gibbs thermodynamic description, which we now present, ignores the structure of the interfacial region and assigns it zero volume. It defines its properties in terms of the defaults between the properties of the actual system, containing an interface, and the properties of a hypothetical reference system of two phases, each of which remains homogeneous up to a mathematical surface of discontinuity or boundary. The location of the boundary of course cannot be specified by the theory, which deals with macroscopic quantities only and not with those on a molecular scale. Correspondingly, the values of certain properties of the interface are not fixed, but instead depend on the location of the boundary. They can thus be calculated, but not measured. A fully molecular description defines them as integrals over the direction perpendicular to the phase boundary.

Consider a system of cross-sectional area $\mathcal{A}$ and height, in the z direction, equal to L, containing an interface near $z = 0$. Let the system be bounded above $z = 0$ by the plane $z = a$ and below by the plane $z = -b$, so that the volume is $\mathcal{A}L$, where $L = a + b$. Now imagine two homogeneous phases meeting at $z = 0$, giving phase P the volume $V_P = \mathcal{A}a$ and phase Q the volume $V_Q = \mathcal{A}b$. Since $V = V_P + V_Q$, no volume is ascribed to the interface. This is the reference system (see Figure 6).

Now, considering numbers of particles, let n_i^P represent the number density of chemical species i in phase P and let n_i^Q represent the corresponding quantity in

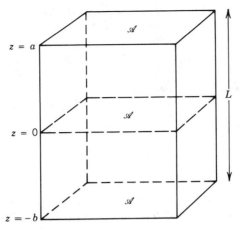

FIGURE 6. Volume of thermodynamic system, showing Gibbs dividing surface at $z = 0$.

phase Q. The presence of an interfacial region reveals itself in the fact that the total number of particles of species i in the system, N_i, is not the sum of $n_i^P V_P$ and $n_i^Q V_Q$, the numbers of such particles in the two homogeneous phases. The difference is assigned to the interface or surface phase, denoted by a superscript σ.

$$N_i^\sigma = N_i - n_i^P V_P - n_i^Q V_Q \qquad [1]$$

By putting $V = V_P + V_Q$, we have given zero thickness and zero volume to the interface, so that one cannot sensibly assign to it volume number densities for particles of various species. One can only assign a surface density

$$N_A \Gamma_i = \frac{N_i^\sigma}{\mathcal{A}} \qquad [2]$$

where the introduction of Avogadro's number N_A makes Γ_i a molar quantity. In reality, n_i varies continuously as a function of z. The surface density is the integral, over z, of the difference between $n_i(z)$ and the assumed discontinuous variation of the density:

$$N_A \Gamma_i = \int dz \left[n_i(z) - n_i^P \beta(z) - n_i^Q \beta(-z) \right] \qquad [3]$$

Here, $\beta(x)$ is the unit step function: $\beta(x) = 0$ for $x < 0$ and $\beta(x) = 1$ for $x > 0$.

Since particles in the interface are exposed to a different environment from particles in a homogeneous phase, the energy of the system would differ from the sum of energies of two homogeneous phases, even if the n_i^σ were zero. One can expect the difference to be proportional to the interfacial area, so

$$\mathcal{A}u^\sigma = U - u^P V_P - u^Q V_Q \qquad [4]$$

where u^P and u^Q represent the internal energy densities for the two homogeneous phases and U represents the total internal energy; u^σ is the surface energy density. Similarly, we define a surface free-energy density by

$$\mathcal{A}a^\sigma = A - a^P V_P - a^Q V_Q \qquad [5]$$

where A is the total Helmholtz free energy and a^P is the Helmholtz free energy per unit volume in homogeneous phase P. This suggests that work must be done to create surface area, the origin of the work being the need to bring molecules from the interior of bulk phase to the surface, where they experience different forces.

The asymmetry of forces acting on particles in the surface region (in addition to the choice of phase boundary in the reference system) determines the values of quantities like u^σ and a^σ. Their values should be calculable in terms of the interparticle forces and the arrangements of particles in the region of the interface, which is itself a consequence of these forces. (Of course such a calculation is outside the scope of macroscopic thermodynamics.) Because the forces between particles are also responsible for the properties of the homogeneous phases, the interfacial structure and hence all interfacial properties are assumed to be fixed when the homogeneous phases on either side of the interface are fixed.

The laws of thermodynamics can be applied to the actual two-phase system and to the reference system of two phases, each assumed to be homogeneous up to its boundaries. Then, by subtraction, one obtains relations between the surface quantities. Thus the surface or interfacial entropy s^σ may be defined by

$$\mathcal{A}s^\sigma = S - s^P V_P - s^Q V_Q \qquad [6]$$

and may be related to u^σ and a^σ by invoking

$$A = U - TS$$
$$A_P = a^P V_P = U_P - TS_P = u^P V_P - Ts^P V_P$$

and

$$A_Q = a^Q V_Q = u^Q V_Q - Ts^Q V_Q$$

Subtracting A_P and A_Q from A, we have

$$\mathcal{A}a^\sigma = \mathcal{A}(u^\sigma - Ts^\sigma)$$

since T is the same throughout a system at equilibrium.

Similarly, if any substance is present in both phases P and Q, its chemical potential μ_i must be the same in both at equilibrium. The chemical potential of any substance assigned to the surface phase likewise has the same value in the surface phase as in either homogeneous bulk phase. The value relates to the structure of the surface phase and the bulk phases, but the former depends on the latter.

B. SURFACE TENSION

The change in the Helmholtz free energy of a system not containing an interface depends in the usual way on T, V, and mole numbers:

$$dA = -S\, dT - p\, dV + \sum \mu_i \frac{dN_i}{N_A} \qquad [7]$$

where N_A is Avogadro's number and μ_i is the chemical potential. Thus the equation for the homogeneous phase P is

$$d(a^P V_P) = -s^P V_P\, dT - p\, dV_P + \sum \mu_i \frac{d(n_i^P V_P)}{N_A} \qquad [8]$$

and similarly for phase Q. For a system containing an interface, a term representing work of creating surface must be added to equation [7]:

$$dA = -S\, dT - p\, dV + \sum \mu_i \frac{dN_i}{N_A} + \gamma\, d\mathcal{A} \qquad [9]$$

where γ is the surface or interfacial tension, which has now been defined as

$$\gamma = \left(\frac{\partial A}{\partial \mathcal{A}}\right)_{T,V,N_i}$$

This quantity is positive, since increasing the area under these conditions requires bringing molecules from bulk-phase regions to the surface, which involves positive work because their potential energy is higher at the surface; if the potential energy were not higher there, the two phases would mix rather than coexist. For a solid surface, some modification of equation [9] is necessary.

Subtracting equation [8] and the corresponding equation for phase Q from equation [9], we have

$$d(\mathcal{A} a^\sigma) = -\mathcal{A} s^\sigma\, dT + \sum \mu_i d(\mathcal{A} \Gamma_i) + \gamma\, d\mathcal{A} \qquad [10]$$

We may integrate this by starting from a very small surface and increasing the area $\mathcal{A}$ keeping all intensive variables constant (a^σ, μ_i, s^σ, T, p, γ, Γ_i). Then

$$\mathcal{Q}a^\sigma = \sum \mu_i \mathcal{Q}\Gamma_i + \gamma\mathcal{Q}$$

Taking the differential of this and subtracting equation [10], we find

$$0 = \mathcal{Q}s^\sigma\, dT + \sum \mathcal{Q}\Gamma_i\, d\mu_i + \mathcal{Q}\, d\gamma \qquad [11]$$

The Gibbs free energy for a homogeneous bulk phase is defined as $G = A + pV$; for a phase including a surface the presence of $\gamma\, d\mathcal{Q}$ in dA suggests defining the Gibbs free energy by

$$G = A + pV - \gamma\mathcal{Q} \qquad [12]$$

As for the other properties, we write

$$G - g^P V_P - g^Q V_Q = \mathcal{Q} g^\sigma \qquad [13]$$

for the surface Gibbs free energy density, where

$$g^P = a^P + p$$

and p must have the same value in both homogeneous phases at equilibrium. Therefore we have

$$\begin{aligned}\mathcal{Q}g^\sigma &= A + pV - \gamma\mathcal{Q} - (a^P + p)V_P - (a^Q + p)V_Q \\ &= \mathcal{Q}a^\sigma - \gamma\mathcal{Q}\end{aligned} \qquad [14]$$

so that $g^\sigma = a^\sigma - \gamma$.

Differential relations between surface thermodynamic properties can now be derived. Using equation [10] in equation [14], we get

$$d(\mathcal{Q}g^\sigma) = -\mathcal{Q}s^\sigma\, dT + \sum \mu_i\, d(\mathcal{Q}\Gamma_i) - \mathcal{Q}d\gamma \qquad [15]$$

The integration of this, keeping all intensive variables constant, yields

$$\mathcal{Q}g^\sigma = \sum \mu_i \mathcal{Q}\Gamma_i$$

Subtracting equation [15] from the differential of this leads to a Gibbs–Duhem equation for the surface phase:

$$0 = \mathcal{Q}\left(s^\sigma\, dT + \sum \Gamma_i\, d\mu_i + d\gamma\right) \qquad [16]$$

It follows from equation [10] that

$$\left(\frac{\partial a^\sigma}{\partial T}\right)_{\mathcal{A},\Gamma_i} = -s^\sigma \qquad [17]$$

which is analogous to $(\partial A/\partial T)_{V,N_i} = -S$ for a bulk phase. Similarly, from equation [10] the interfacial tension γ obeys

$$\mathcal{A}\left(\frac{\partial a^\sigma}{\partial \mathcal{A}}\right)_{T,\Gamma_i} + a^\sigma = \sum \mu_i \Gamma_i + \gamma$$

and the first term vanishes since at constant n_i^σ and T the surface free-energy density a^σ is independent of surface area. Thus

$$\gamma = a^\sigma - \sum \mu_i \Gamma_i \qquad [18]$$

Equation [18] thus relates γ and the surface free-energy density.

According to equation [9], the work that must be done by the surroundings to reversibly deform a system containing a surface, keeping the temperature and the numbers of particles of each species constant, is $-p\,dV + \gamma\,d\mathcal{A}$. Thus γ is the work required to create unit surface area, keeping the temperature, total volume, and mole numbers of a multiphase system constant. It differs from the surface free-energy density a^σ, according to equation [18], by $\sum \mu_i n_i^\sigma / N_A$. Identifying the latter with the work required to bring molecules from the bulk phases to the surface, we take γ to be a measure of the mechanical work of deformation.

Here it must be emphasized that the values of surface properties, such as Γ_i or u^σ, depend on where the Gibbs geometrical surface, which separates the homogeneous phases, is assumed to be placed. This is not the case for γ: Let the Gibbs surface (Figure 6) be moved from $z = 0$ to $z = m$, thus increasing the Helmholtz free energy of the reference system by $m\mathcal{A}(a^Q - a^P)$ and the number of particles of species i in the reference system by $m\mathcal{A}(n_i^Q - n_i^P)$. The quantities $\mathcal{A}a^\sigma$ and $\mathcal{A}n_i^\sigma$ are decreased by the same amount. According to equation [18], γ is changed by

$$-m\left(a^Q - a^P - \sum \frac{\mu_i n_i^Q}{N_A} + \sum \frac{\mu_i n_i^P}{N_A}\right)$$

which vanishes since, for a homogeneous phase, $aV = \sum \mu_i n_i V/N_A$. Similarly, one can show that thermodynamic relations between surface properties are invariant to the placement of the Gibbs surface. We may also note that the surface properties n_i^σ, a^σ, and so on, are sometimes referred to as surface excess, excess surface free energy, and so forth, although their values may very well be negative.

Often, the location of the Gibbs surface is defined by the condition

$$\Gamma_j = 0$$

for some component j, usually the major component. If this is done for a one-component system, the surface tension γ becomes equal to the surface Helmholtz free energy. Equation [16], in this case, becomes

$$d\gamma = -s^\sigma \, dT$$

so the surface entropy density is equal to the negative of the variation of surface tension with temperature.

In general, equation [16] gives, at constant temperature,

$$d\gamma = -\sum \Gamma_i \, d\mu_i \qquad [19]$$

which seems to permit the determination of the surface concentration by variation (at constant T) of the chemical potential μ_i, which would be done by changing the concentration of the component i in the bulk phases that are in equilibrium with the surface phase. We have emphasized the arbitrariness of the actual values of the surface concentrations, and it would be remarkable if their values could actually be determined. However, they cannot be determined; what prevents it is the impossibility of varying one chemical potential without changing the others (or changing the concentration of only one component of the mixture). The chemical potentials of a bulk phase are related by the Gibbs–Duhem equation

$$\sum x_i \, d\mu_i = 0$$

where the x_i are mole fractions. For an n-component system, there are only $n - 1$ independently variable μ_i, so only certain combinations of the Γ_i can be obtained.

Thus *relative* surface concentrations, sometimes called *surface excesses*, can be determined. Using the Gibbs–Duhem relation to express dx_1 in terms of the other dx_i, we have from equation [19]

$$d\gamma = -\sum_i \Gamma_i \, d\mu_i = -\left[\sum_i^{(i \neq 1)} \Gamma_i \, d\mu_i + \Gamma_1 \, d\mu_1 \right]$$

$$= -\sum_i d\mu_i \left[\Gamma_i - \frac{n_i}{n_1} \Gamma_1 \right]$$

where the ratio of number densities n_i/n_1 has been substituted for x_i/x_1. This is the Gibbs adsorption equation. The quantity in square brackets is the surface excess of component i relative to component 1. Its definition depends on the ratio of concentrations of components i and 1 in the bulk phase. Relative surface excesses are easily seen from their definition to be invariant to a shift in the phase boundary of the reference system. If we choose the boundary's position so one n_i^σ vanishes, we can determine values for the others, relative to this choice.

The chemical potential μ_i refers to the work accompanying the introduction of substance i into the system. Although it has been tacitly assumed up to now that each substance is present in both homogeneous phases, it is not necessary. If substance j is present in phase P only, then $n_j^Q = 0$, and the surface concentration Γ_j is still given by equation [3].

C. ADSORPTION

The Gibbs adsorption isotherm, which relates the surface tension to the surface excesses of the various components, may be written either as equation [19] or as

$$d\gamma = -\sum_i d\mu_i \Gamma_{i(1)} \quad (T \text{ constant}) \qquad [20]$$

where we have introduced a conventional abbreviation for the relative surface excess. Since $\Gamma_{(1)1}$ vanishes by its definition, component 1 does not appear in the summation. Note that $\Gamma_{i(1)}$ vanishes if the ratio of the surface concentrations n_i^σ/n_1^σ is the same as the ratio of the bulk concentrations n_i/n_1. Also, if n_i/n_1 is very small (dilute solution), $\Gamma_{i(1)}$ becomes the same as Γ_i. A positive value of $\Gamma_{i(1)}$ reflects a tendency of component i to concentrate in the surface region. According to equation [20], increasing μ_i (or the concentration of substance i in a bulk phase) will then decrease the surface tension.

For a two-component system we have

$$d\gamma = -s^\sigma dT - \Gamma_{2(1)} d\mu_2 \qquad [21]$$

according to equation [16]. The chemical potential of substance 2 in homogeneous phase P depends upon the three independent intensive variables T, p, and x_2^P with

$$\frac{\partial \mu_2^P}{\partial T} = -\frac{\partial S_P}{\partial n_2} \equiv -\bar{s}_2^P$$

$$\frac{\partial \mu_2^P}{\partial P} = \frac{\partial V_P}{\partial n_2} \equiv \bar{v}_2^P$$

Therefore

$$d\mu_2 = \left[-\bar{s}_2^P + \bar{v}_2^P \left(\frac{\partial P}{\partial T} \right)' \right] dT + \frac{\partial \mu_2}{\partial x_2^P} dx_2$$

where $(\partial P/\partial T)'$ is the change of pressure with temperature along the coexistence curve of phases P and Q (vapor–pressure curve if one phase is liquid and the other vapor). Inserting $d\mu_2$ in equation [21] yields an expression for $-(\partial \gamma/\partial T)_{x_2}$, which may be shown (Lewis and Randall 1961, p. 475) to be the surface entropy, and the equation

$$\left(\frac{\partial \gamma}{\partial x_2^P}\right)_T = -\Gamma_{2(1)} \left(\frac{\partial \mu_2}{\partial x_2^P}\right)_T' \quad [22]$$

The primed derivative is the variation of chemical potential μ_2 with the composition of phase P, with the temperature fixed and the pressure such as to maintain equilibrium between P and Q. Equation [22] allows the experimental determination of relative surface excess. If P were an ideal solution, the primed derivative would be RT/x_2 so that

$$\Gamma_{2(1)} = -(RT)^{-1} x_2 \left(\frac{\partial \gamma}{\partial x_2}\right)_T$$

For an n-component system, similar manipulations may be performed, leading again to equation [20] (Lewis and Randall 1961, p. 478). To obtain the $n - 1$ surface excesses one requires the variation of interfacial tension and chemical potentials μ_i with $n - 1$ composition variables, with temperature constant and pressure such as to maintain phase coexistence.

The adsorption equations hold for charged particles, even though it is impossible to add a single charged species to a macroscopic system. Although one may discuss, and even calculate, chemical potentials for charged species, they are not experimentally measurable. Only electrically neutral combinations of such species can be dealt with experimentally, so that one eventually combines properties calculated for individual ions into properties of neutral "salts." As we discussed in Chapter 1, for charged particles one needs to consider electrochemical potentials $\tilde{\mu}_i$ instead of chemical potentials μ_i. The Gibbs adsorption equation (equation [20]) is still correct, with electrochemical potentials instead of chemical potentials. The value of the electrochemical potential $\tilde{\mu}_i$ depends on the (inner) electrical potential of the phase in which i is located.

At equilibrium, the surface (or surface region) is adjoined by homogeneous phases free of electric fields, so it must be electrically neutral. However, the existence of charged species raises the possibility that, within the surface region, there are regions of nonzero electrical charge density, that is, there may exist a nonzero charge density function $\rho(z)$ in a macroscopically very thin (in the z direction) region. The overall electrical neutrality requires that

$$\int \rho(z)\, dz = 0$$

but the existence of $\rho(z)$ affects the homogeneous phases because it leads to a difference of electrical potential in the two phases. Assuming for simplicity a dielectric constant ϵ pervading the system, the electrostatic potential ϕ obeys (in electrostatic units)

$$\frac{d^2\phi}{dz^2} = -\frac{\rho(z)}{\epsilon} \qquad [23]$$

The electric field then obeys

$$E(z) = -\frac{d\phi}{dz} = \int^z dz' \frac{\rho(z')}{\epsilon} + E(-\infty)$$

so that, if the field vanishes in the region at $z \to -\infty$, it also vanishes at $z \to +\infty$. Assuming $E(-\infty) = 0$ and integrating once more, we have

$$\phi(z) - \phi(\infty) = -\int^z dz'' \int^{z''} dz' \frac{\rho}{\epsilon}$$

so the total potential drop across the interface is

$$\Delta\phi = \phi(\infty) - \phi(-\infty) = -\int^\infty dz'' \int^{z''} dz' \frac{\rho(z')}{\epsilon}$$

Integrating by parts, we find

$$\Delta\phi = \int^\infty dz'' \, z'' \frac{\rho(z'')}{\epsilon} \qquad [24]$$

where the integral of $z\rho(z)$ is the surface dipole moment of the interfacial charge distribution. The existence of such a surface dipole term together with the overall electroneutrality condition suggests the common term *double layer* to describe the situation, that is, it is as if the interface contained equal and opposite electrically charged layers.

The range of z over which $\rho(z)$ is nonzero is the thickness of the interfacial region, which is supposed to be very small. The integral in equation [24] extends only over this region, so that $\Delta\phi$ is the difference of inner potentials between two phases in contact. The value of $\Delta\phi$ depends on the nature of the two phases. If charge transfer occurs across the interface, the value of $\Delta\phi$ is fixed by the equilibrium (Chapter 1, Section C). However, suppose that each charged component exists only in one of the two homogeneous phases and that no chemical reaction between charged components is possible. This corresponds to the ideally polarizable electrode of Chapter 1, Section E.

When we introduce electrochemical potentials into equation [19] and write the electrostatic parts explicitly we get

$$d\gamma = -\sum^{P} \Gamma_i \, d(\mu_i + z_i \mathcal{F} \phi^P)$$
$$- \sum^{Q} \Gamma_j \, d(\mu_j + z_j \mathcal{F} \phi^Q) - \sum \Gamma_k \, d\mu_k \quad [25]$$

where the first summation is over charged species found only in P, the second is over charged species found only in Q, and the third is over electrically neutral species. Since the system containing the interface, as well as the reference system, is electrically neutral overall, we have

$$\sum^{P} \Gamma_i z_i \mathcal{F} + \sum^{Q} \Gamma_j z_j \mathcal{F} = 0$$

and we may identify the first summation with q^P, the surface charge density on the P-side of the interface, and identify the second with $q^Q = -q^P$, thus recovering a double layer. Now equation [25] becomes

$$d\gamma = -q^P \, d(\phi^P - \phi^Q) - \sum^{P} \Gamma_i \, d\mu_i - \sum^{Q} \Gamma_j \, d\mu_j - \sum \Gamma_k \, d\mu_k \quad [26]$$

The potential difference across the interface is $\Delta\phi = \phi^P - \phi^Q$. We see that the variation of the surface tension with this potential difference, keeping the composition of phases Q and P fixed ($d\mu_i = d\mu_j = d\mu_k = 0$), yields the surface charge. The value of $\Delta\phi$ (a difference of inner potentials between chemically different phases) is not experimentally measurable, but the change in $\Delta\phi$ is accessible under certain circumstances.

D. LIPPMANN EQUATION

In reality, the polarizable interface must be included in an electrochemical cell. For example, let the left-hand electrode be a reference electrode reversible with respect to the anion A^-, such as a metal–metal salt electrode with half-cell reaction $MA + e^- \rightarrow M + A^-$, and let the right-hand electrode be the polarizable electrode, with metal phase N, dipping into the same solution S, which contains A^-. The cell is represented by

$$L \mid M \mid MA \mid S \mid N \mid R$$

The potential across the leads R and L is

$$\mathcal{E} = \phi^{(R)} - \phi^{(L)} = \frac{-\tilde{\mu}_{e^-}^{(R)} + \tilde{\mu}_{e^-}^{(L)}}{\mathcal{F}}$$

D. LIPPMANN EQUATION

since the leads are made of the same material. We write it as

$$\mathcal{E} = (\phi^{(R)} - \phi^{(N)}) + (\phi^{(N)} - \phi^{(S)}) + (\phi^{(S)} - \phi^{(M)}) + (\phi^{(M)} - \phi^{(L)})$$

The values of the first and fourth terms are fixed, at any temperature, by the nature of the metals, corresponding to the equilibrium for electron transfer between metals in contact; the value of the third term is fixed by the equilibrium of the half-cell reaction:

$$\mu_{A^-}^{(S)} - \mathcal{F}\phi^{(S)} + \mu_M^{(M)} = \mu_{e^-}^{(M)} - \mathcal{F}\phi^{(M)} + \mu_{MA}^{(MA)}$$

or

$$-\mathcal{F}(\phi^{(S)} - \phi^{(M)}) = \mu_{e^-}^{(M)} + \mu_{MA}^{(MA)} - \mu_M^{(M)} - \mu_{A^-}^{(S)}$$

Note that we have used the equilibrium condition for an interphase charge-transfer reaction to express the difference in electrical potentials between phases in terms of a combination of chemical potentials. We now have

$$\mathcal{E} = E + \frac{\mu_{A^-}^{(S)}}{\mathcal{F}} + (\phi^{(N)} - \phi^{(S)}) \qquad [27]$$

where E combines all the chemical potentials and electrical potential differences whose values must remain constant. Then

$$d\mathcal{E} = d\frac{\mu_{A^-}^{(S)}}{\mathcal{F}} + d(\phi^{(N)} - \phi^{(S)}) \qquad [28]$$

at constant temperature.

Assuming that the charge on the metal side of the polarizable interface (N) consists of electrons and metal ions and that the charge on the solution side consists of charged species i (including A^-), equation [26] is

$$d\gamma = -q^N d(\phi^{(N)} - \phi^{(S)}) - \Gamma_{e^-}^{(N)} d\mu_{e^-}^{(N)} - \Gamma_{N^+}^{(N)} d\mu_{N^+}^{(N)} - \sum \Gamma_i d\mu_i^{(S)} - \sum \Gamma_k d\mu_k^{(S)}$$

We have dropped neutral species of the metal from the last summation, since their chemical potentials are invariable at constant temperature. The same goes for the chemical potentials of electrons and metal ions in metal N, so their differentials may also be dropped. Then inserting equation [28] we have

$$d\gamma = -q^N d\mathcal{E} + q^N d\frac{\mu_{A^-}^{(S)}}{\mathcal{F}} - \sum \Gamma_i d\mu_i^{(S)} - \sum \Gamma_k d\mu_k^{(S)} \qquad [29]$$

This contains the Lippmann or Gibbs–Lippman equation.

According to this equation, varying the potential across the cell with the concentration of the solution phase held constant (constant chemical potentials of solution species) yields the surface charge on the metal side of the polarizable interface:

$$\left(\frac{\partial \gamma}{\partial \mathcal{E}}\right)_{p,T,\text{comp}} = -q^N \qquad [30]$$

The variation of interfacial tension with electrode potential is called an *electrocapillary curve*. Electrocapillary curves typically resemble inverted parabolas. When the surface tension is a maximum (*electrocapillary maximum*), the charge density for either side of the interface is zero. The corresponding potential is called the *potential of zero charge* (pzc). Of course, its value depends on the reference electrode in the cell.

The second derivative of the electrocapillary curve is the derivative of surface charge with potential, that is, the differential capacitance C, which also depends on the electrode potential. The integral capacitance K is an average of C, which is the integral of $C\,d\mathcal{E}$ from the pzc to E, divided by $E - \mathcal{E}_{\text{pzc}}$. The integral of $C\,d\mathcal{E}$ is easily shown to be equal to the surface charge at potential E.

The derivative of surface tension with activities of ions in solution can be used to obtain surface excesses Γ_i. Suppose only the electrolyte C^+A^- is present in solution. The only neutral species is water and $n_{H_2O}d\mu_{H_2O} = -n_A\cdot d\mu_{A^-} - n_{C^+}d\mu_{C^+}$, so equation [29] reduces to

$$d\gamma = -q^N\,d\mathcal{E} + q^N \frac{d\mu_{A^-}^{(S)}}{\mathcal{F}}$$
$$- \Gamma_{A^-(1)}\,d\mu_{A^-}^{(S)} - \Gamma_{C^+(1)}\,d\mu_{C^+}^{(S)} \qquad [31]$$

where the surface excesses of ions relative to water appear. The surface charge density on the solution side is

$$q^S = \mathcal{F}(\Gamma_{C^+} - \Gamma_{A^-}) = \mathcal{F}(\Gamma_{C^+(1)} - \Gamma_{A^-(1)}) = -q^{(N)}$$

Using this to substitute for q^N in the second term on the right-hand side of equation [31], we get

$$d\gamma = -q^N\,d\mathcal{E} - \Gamma_{C^+}(d\mu_{A^-}^{(S)} + d\mu_{C^+}^{(S)}) \qquad [32]$$

From equation [32] through the end of this section, Γ_{A^-} and Γ_{C^+} stand for the relative surface excesses $\Gamma_{A^-(1)}$ and $\Gamma_{C^+(1)}$.

Since

$$d\mu_{A^-}^{(S)} + d\mu_{C^+}^{(S)} = d(RT \ln a_{A^-}^{(S)} + RT \ln a_{C^+}^{(S)})$$
$$= RT\,d \ln (a_\pm)^2 \qquad [33]$$

equation [32] may be written

$$d\gamma = -q^N \, d\mathcal{E} - RT \, \Gamma_{C^+} \, d \ln (a_\pm)^2 \qquad [34]$$

The surface excess of C^+ is thus

$$\Gamma_{C^+} = -(RT)^{-1} \left(\frac{\partial \gamma}{\partial \ln a_\pm^{(2)}} \right)_\mathcal{E} \qquad [35]$$

and is measurable by varying the activity of C^+A^-, keeping constant the potential of the polarizable electrode relative to the reference electrode which is reversible to anion in this example.

If a reversible C^+/C electrode replaces the M/MA electrode, creating the cell L|C|S|N|R, the potential $\mathcal{E}_C$ of the polarizable electrode relative to it is

$$\mathcal{E}_C = (\phi^{(N)} - \phi^{(S)}) - (\phi^{(C)} - \phi^{(S)}) + (\phi^{(R)} - \phi^{(N)}) + (\phi^{(C)} - \phi^{(L)})$$

the last terms being constants. Writing $\tilde{\mu}_{C^+}^{(S)} + \tilde{\mu}_{e^-}^{(C)} = \mu_C^{(C)}$ in terms of electrical potentials, we have that $d(\phi^{(C)} - \phi^{(S)}) = d\mu_{C^+}/\mathcal{F}$ so that

$$d\mathcal{E}_C = d(\phi^{(N)} - \phi^{(S)}) - \frac{d\mu_{C^+}}{\mathcal{F}}$$

$$= d\mathcal{E} - \frac{d\mu_{A^-}^{(S)}}{\mathcal{F}} - \frac{d\mu_{C^+}^{(S)}}{\mathcal{F}}$$

where equation [28] has been used. Using this, equation [31] becomes

$$d\gamma = -q^N \, d\mathcal{E}_C - q^N \frac{d\mu_{C^+}^{(S)}}{\mathcal{F}} - \Gamma_{A^-} \, d\mu_{A^-}^{(S)} - \Gamma_{C^+} \, d\mu_{C^+}^{(S)}$$

$$= -q^N \, d\mathcal{E}_C - \Gamma_{A^-} (d\mu_{C^+}^{(S)} + d\mu_{A^-}^{(S)})$$

Again using equation [33], we have for the surface excess of A^-

$$\Gamma_{A^-} = -(RT)^{-1} \left(\frac{\partial \gamma}{\partial \ln a_\pm^{(2)}} \right)_{\mathcal{E}_C} \qquad [36]$$

so this quantity is measurable by varying the activity of C^+A^-, keeping constant the potential of the polarizable electrode relative to the cationic reference electrode C^+/C.

Thus the surface excess of anion or cation may be determined, depending on whether the electrode coupled in the cell to the polarizable electrode is reversible to cation or anion, respectively. It may also be noted that if $\mathcal{E}_C$ is the potential of the electrocapillary maximum (ecm), $q^S = -q^N = 0$ or $\Gamma_{A^-} = \Gamma_{C^+}$, and from equation [34] we get

$$\left(\frac{\partial \gamma}{\partial \ln a_\pm}\right)_{ecm} = -2RT\Gamma_{A^-} = -RT(\Gamma_{A^-} + \Gamma_{C^+}) = -2RT\Gamma_{AC} \quad [37]$$

In equation [37], the potential is kept at the ecm when the activity of C^+A^- is varied.

The mixed second derivative of γ with respect to $\mathcal{E}_C$ and $\ln(a_\pm)^2$ is, according to equations [30] and [36],

$$-\frac{\partial q^N}{\partial \ln(a_\pm)^2} = -RT\frac{\partial \Gamma_{A^-}}{\partial \mathcal{E}_C}$$

Differentiating with respect to $\mathcal{E}_C$ we have

$$\frac{\partial C}{\partial \ln(a_\pm)^2} = RT\frac{\partial^2 \Gamma_{A^-}}{\partial \mathcal{E}_C^{(2)}}$$

where $C = (\partial q^N / \partial \mathcal{E}_C)_{a_\pm}$ is the differential capacity; one can thus obtain Γ_{A^-} from capacitance measurements if two constants of integration are available. To get these, one sometimes assumes that for potentials sufficiently negative with respect to the ecm there is no specific adsorption of alkali or alkaline-earth metal ions, so that Γ_{A^-} can be calculated from diffuse-layer theory (see Chapter 4, Section B).

These thermodynamic relations are used to obtain information about Γ_+ and Γ_- and how they vary with chemical potential and with electrode potential. By using the cross-derivative relationships, we have shown how one can use measurements of surface tension, surface charge, and capacity for this purpose. Direct measurements of $\Gamma_\pm$, such as by radiotracers, are also possible. Reeves (1980, Section 2) discusses and compares the direct and thermodynamic measurements, giving the advantages and problems of each.

We return to a cell $L|M|S|N|R$, where L and R are leads of the same metal, and we again consider the surface tension of the N/S interface. Now, let the cation have charge $+z$, so that the electrolyte is $C^{z+}(A^-)_z$. Then, when we drop differentials of invariant chemical potentials, equation [26] becomes

$$d\gamma = -q^N d(\phi^{(N)} - \phi^{(S)}) - \Gamma_{C^{z+}} d\mu^{(S)}_{C^{z+}} - \Gamma_{A^-} d\mu^{(S)}_{A^-} \quad [38]$$

The surface excesses are actually relative surface excesses with respect to water. The change in the cell potential $\mathcal{E}$ differs from $d(\phi^{(N)} - \phi^{(S)})$ by $d(\phi^{(M)} - \phi^{(S)})$. The potential difference $\phi^{(M)} - \phi^{(S)}$, according to the equilibrium at the reversible electrode, will change at constant temperature only by virtue of the change of the chemical potential of A^- or of C^{z+}. Therefore

D. LIPPMANN EQUATION

$$d(\phi^{(M)} - \phi^{(S)}) = \frac{d\mu_r}{z_r \mathcal{F}}$$

where r is C^{z+} or A^- and z_r is z_+ or -1, and $d(\phi^{(N)} - \phi^{(S)}) = d\mathcal{E} + d\mu_r/(z_r\mathcal{F})$. Equation [38] may now be written

$$d\gamma = -q^N d\mathcal{E} - q^N (z_r\mathcal{F})^{-1} d\mu_r - \Gamma_{C^{z+}} d\mu_{C^{z+}} - \Gamma_{A^-} d\mu_{A^-} \qquad [39]$$

If the charge on the solution side of the N/S interface is formed by C^{z+} and A^- (as implied by the fact that these are the only charged species in equation [38]),

$$q^N = -q^S = -(z\Gamma_{C^{z+}} - \Gamma_{A^-})\mathcal{F} \qquad [40]$$

On substituting equation [40] into the second term of equation [39], we get

$$d\gamma = -q^N d\mathcal{E} + \frac{\Gamma_{C^{z+}}(z d\mu_r - z_r d\mu_{C^{z+}})}{z_r} - \frac{\Gamma_{A^-}(d\mu_r + z_r d\mu_{A^-})}{z_r}$$

If M/S is reversible to A^- ($z_r = -1$),

$$d\gamma = -q^N d\mathcal{E} - \Gamma_{C^{z+}}(z d\mu_{A^-} + d\mu_{C^{z+}})$$

while, if M/S is reversible to C^{z+} ($z_r = z$),

$$d\gamma = -q^N d\mathcal{E} - \frac{\Gamma_{A^-}(d\mu_{C^{z+}} + z d\mu_{A^-})}{z}$$

The quantity in parentheses in each of these two equations is $d\mu_{CA_z}$.

From the variation of surface tension with the chemical potential of the electrolyte CA_z at constant cell potential, one thus obtains one surface excess concentration: that of the cation if the reversible electrode is reversible to anions, and vice versa. In terms of the mean ionic activity $a_\pm$, we have $d\mu_{CA_z} = (z + 1)RT \ln a_\pm$. Therefore

$$\left(\frac{\partial \gamma}{\partial \ln a_\pm}\right)_{T,p,\mathcal{E}_\pm} = -\frac{RT(z+1)}{|z_r|} \Gamma_\mp$$

where $\mathcal{E}_+$ means the cell potential is held constant with M/S reversible to cations and $\mathcal{E}_-$ means $\mathcal{E}$ is held constant with M/S reversible to anions. Since q^N is obtainable as $-(\partial \gamma / \partial \mathcal{E})_\mu$, one can then obtain the surface excess concentration of the second ion when that of the first is known.

As indicated after equation [37], one can use cross-derivative relationships of the type $(\partial \Gamma / \partial \mathcal{E})_{T,\mu} = (\partial q^N / \partial \mu_{CA_z})$ to derive information about surface excesses from measured charges. Similarly one can use measured values of the capacitance $C = (\partial q^N / \partial \mathcal{E})_{T,\mu}$. The formulas needed for this and related calculations are given by Parsons (1980, Section 1.5), who also discusses more complicated situations, for example, two salts in solution and alloy electrodes.

E. ADSORPTION OF NEUTRALS

In the Gibbs–Lippman equation [29], the sum over k includes electrically neutral species in the solution as well as solvent. The adsorption of such species affects electrocapillary quantities. For instance, some neutrals are surface-active in electrolyte solution because they are less polar than solvent molecules; an increase in the magnitude of the electric field at the electrode, by attracting solvent, leads to desorption of the surface-active species by displacement.

Subtracting equation [29] in the absence of the neutral substance k from the corresponding equation in which k is present, we have

$$d(\gamma' - \gamma) = -(q^{M'} - q^M) \frac{d\mathcal{E} - d\mu_{A^-}^{(S)}}{\mathcal{F}}$$

$$- \sum (\Gamma_i' - \Gamma_i) d\mu_i^{(S)} + \Gamma_k d\mu_k$$

where the primes identify quantities in the absence of k and the Γ_k are surface excesses relative to solvent. The *surface pressure* of adsorbate π is defined by

$$\pi = \gamma' - \gamma$$

so that

$$\left(\frac{\partial \pi}{\partial \mu_k} \right)_{T,p,\mathcal{E},\mu_i} = \Gamma_k \qquad [41]$$

If the dilute-solution law, $\mu_k = \mu_k^0 + RT \ln c_k$, holds, equation [41] gives

$$\Gamma_k = (RT)^{-1} c_k \frac{\partial \pi}{\partial c_k}$$

In general, the adsorption isotherm for an ion or molecule in solution follows from the equilibrium condition

$$\tilde{\mu}_i^{0(s)} + RT \ln a_i = \tilde{\mu}_i^{0(a)} + RT \ln (\Gamma_i g_i)$$

where a_i is activity, Γ_i is surface concentration of adsorbed species, and g_i is a kind of activity coefficient. Then

E. ADSORPTION OF NEUTRALS

$$\frac{\Gamma_i}{a_i} = g_i^{-1} \exp\left(\frac{\tilde{\mu}_i^{0(a)} - \tilde{\mu}_i^{0(s)}}{RT}\right) = g_i^{-1} \exp\left(\frac{-\Delta G_{\text{ads}}^0}{RT}\right)$$

If species i is charged, the electrical state of the surface must affect its adsorption through ΔG_{ads}^0. Let $\Delta G_{\text{ads}}^0 = A + B\Omega$, where the electrical variable Ω may be surface charge or potential, and let g_i be independent of Ω. Then

$$\frac{\Gamma_i}{a_i} = g_i^{-1} \exp\left(-\frac{A}{RT}\right) \exp\left(-\frac{B\Omega}{RT}\right)$$

This means that all the adsorption isotherms (Γ_i as a function of a_i at constant T) for different values of Ω have the same shape, differing only by a factor depending on Ω. This has been referred to as *congruence* (Hurwitz 1978b, Section 2). The electrical variable that makes isotherms congruent for the adsorption of ions seems to be the surface charge: for the adsorption of organic molecules, the electrode potential often works. There is no theoretical proof of either, so the congruence may be only approximate, disappearing when the precision of the experiments improves.

If only a monolayer of neutral adsorbate can form at the electrode, it is convenient to consider the fractional coverage

$$\theta_k = \frac{\Gamma_k}{\Gamma_k^{\max}} \qquad [42]$$

Perhaps the simplest relation between θ_k and the bulk-phase concentration c_k is the Langmuir adsorption isotherm

$$Kc_k = \frac{\theta_k}{1 - \theta_k} \qquad [43]$$

where K is a constant. The Langmuir theory assumes a limited number of adsorption sites per unit area, n, and no interaction between adsorbed molecules, so that, with only species k adsorbing, the adsorption rate per unit area is $k_a c_k (1 - \theta_k) n$ and the desorption rate per unit area is $k_d \theta_k n$. At equilibrium, the rates are equal, which gives equation [43] with $K = k_a/k_d$.

A statistical mechanical derivation calculates the chemical potential of species k in solution,

$$\mu_k^{(S)} = \mu_k^0 + RT \ln c_k$$

and equates it to the chemical potential of adsorbed species k. For m molecules adsorbed on unit area, the partition function is

2. THERMODYNAMICS AND ELECTROSTATICS

$$Q = \frac{(An)!\, q^{Am}}{(Am)!\, (A(n-m))!} \quad [44]$$

where A is the surface area and q is an internal molecular partition function, so the chemical potential of adsorbed molecules is

$$\mu_k = -N_A kT \left(\frac{\partial \ln Q}{\partial (Am)}\right)_{A,n,T} = -RT\left[\ln q - \ln \frac{m}{n-m}\right]$$

Equating to $\mu_k^{(S)}$ gives

$$\ln \frac{c_k(n-m)}{m} = -\frac{\mu_k^0}{RT} - \ln q = -\ln K \quad [45]$$

Since $(n-m)/m = \theta_k^{-1} - 1$, K is identical to that of equation [43].

The constant K is related to the free energy of adsorption for a single molecule:

$$-RT \ln K = -\mu_k^0 + \mu_k^{(a)} = \Delta G_{\text{ads}} \quad [46]$$

When we replace the concentration c_k in equation [43] by the activity, we have for equation [46]

$$\frac{\theta_k}{1-\theta_k} = a_k \exp\left(-\frac{\Delta G_{\text{ads}}}{RT}\right) \quad [47]$$

The factor of $1 - \theta_k$ in the denominator of equation [47] results from the limitation to monolayer adsorption. Without this restriction, it makes more sense to write $A_k \Gamma_k$ for θ_k, where A_k is the area taken by an adsorbed molecule (in cm²/mole), so

$$A_k \Gamma_k = a_k \exp\left(\frac{-\Delta G_{\text{ads}}}{RT}\right)$$

Improvements on the Langmuir isotherm involve considering interactions between adsorbed molecules.

The Frumkin isotherm takes this energy of interaction into account by multiplying the partition function (equation [44]) by $\exp[\tfrac{1}{2}aA(\theta_k)^2]$, where a is positive for intermolecular attraction, which favors adsorption, and negative for intermolecular repulsion. This adds to $\mu_k^{(a)}$ a term $-aRT\theta$ and multiplies the right-hand side of equation [43] by $e^{-a\theta}$, thus giving the Frumkin isotherm:

$$K'c = \frac{\theta e^{-a\theta}}{1-\theta} \quad [48]$$

Other ways of introducing the effect of interparticle interactions on the surface have been proposed. These include making $\Delta G_{ads}/RT$ in equation [47] a function of the coverage Γ_k. Of course the limitation on the number of adsorption sites is a way of representing the repulsive interaction between adsorbed particles, which prevents two particles from occupying the same space.

The applicability of the notion of adsorption sites, which is inherent in the notion of the Langmuir equation and its modifications, is somewhat questionable for the surface of a liquid (e.g., the mercury electrode). Models based on mobile adsorption rather than localized adsorption are perhaps more appropriate. The simplest model of this kind represents the adsorbed species as a two-dimensional gas. The surface concentration Γ (number of adsorbed molecules per unit area) would then be p'/RT, where p' is the two-dimensional pressure. The equilibrium between adsorbed molecules, with chemical potential $\mu^{(a)} + RT \ln p'$, and molecules in the solution, with chemical potential $\mu^{(s)} + RT \ln a$, makes p' proportional to a. The adsorption isotherm for this model is then

$$\Gamma = \frac{a}{RT} \exp\left(\frac{\mu^{(s)} - \mu^{(a)}}{RT}\right)$$

and one can identify $(\mu^{(a)} - \mu^{(s)})$ with the free energy of adsorption ΔG_{ads}.

Interactions between adsorbed molecules are ignored in the simple two-dimensional gas model, so there is no limit on the number of molecules that can be adsorbed per unit area, with Γ being proportional to a at all adsorbate activities. To correct the situation, one can introduce a virial expansion as for ordinary gases. With one virial coefficient B,

$$\Gamma + B\Gamma^2 = \frac{p'}{RT} = \frac{a}{RT} \exp\left(\frac{\Delta G^0_{ads}}{RT}\right) \qquad [49]$$

where $\mu^{(a)} - \mu^{(s)}$ has now been taken as a standard free energy of adsorption. If $B > 0$ (intermolecular repulsion), Γ is proportional to a at low surface coverage but deviates negatively from the linear relation as Γ increases. More terms in the virial expansion can be added.

The effect of electric field on desorption of nonpolar substance, due to the increased adsorption of polar solvent, can be expressed by writing ΔG_{ads} as a function of electrode potential or electric field with a minimum at some value E_m. The simplest such representation is

$$-\Delta G_{ads} = b - a(E - E_m)^2$$

where b and a are constants. Then equation [46] becomes

$$K = K_0 \exp\left[-\frac{a(E - E_m)^2}{RT}\right] \qquad [50]$$

where K_0 is the value of K at $E = E_m$. According to equation [29] with one neutral surfactant k, we have

$$\left(\frac{\partial q^M}{\partial \mu_k}\right)_{E,\mu_i} = (\partial \Gamma_k / \partial E)_{\mu_i, \mu_k}$$

and a second differentiation gives

$$\frac{\partial C}{\partial \mu_k} = \frac{\partial^2 \Gamma_k}{\partial E^2} \qquad [51]$$

where C is the capacitance. For example, if Γ_k is proportional to concentration in solution as in equation [43] or equation [48] for low θ, and if the ideal solution law holds, equation [51] becomes

$$\frac{c_k}{RT}\frac{\partial C}{\partial c_k} = c_k \frac{\partial^2 K}{\partial E^2}$$

Dividing by c_k and integrating from $c_k = 0$ at constant E gives

$$\frac{C - C'}{RT} = c_k \frac{\partial^2 K}{\partial E^2} \qquad [52]$$

where C' is capacitance in the absence of surfactant. The simple dependence of K on E of equation [50] makes

$$\frac{\partial^2 K}{\partial E^2} = K\left[\frac{2a(E - E_m)}{RT}\right]^2 - \frac{2aK}{RT} \qquad [53]$$

Thus the capacity should decrease with increased surfactant concentrations for a range of electrode potentials symmetrical about E_m. Furthermore, differentiating equation [52] with respect to E gives

$$\left(\frac{\partial C}{\partial E}\right)_c = RTc_k K(8a^2)(E - E_m)\left[-a(E - E_m)^2(RT)^{-3} + \tfrac{3}{2}(RT)^{-2}\right]$$

which means there are two maxima in a C vs. E curve, at each of which capacitance is proportional to concentration. This is qualitatively true for certain alcohols in NaF solution, although the positions of the maxima do vary somewhat with concentration.

Since the adsorption of organic molecules onto an electrode is the replacement of water by organic substance, it means a lowering of the dielectric constant from ϵ_w to ϵ_{org}. A lower capacity corresponds to the lower dielectric constant. If the replacement occurs at constant potential, there must be a charging of the double layer. Letting p be the difference between $\Delta_S^M \phi$ and $(\Delta_S^M \phi)_{pzc}$, at the potential of zero charge (pzc), the change in surface charge is $(C_{org} - C_w)p$, where C_{org} and C_w are the capacitances corresponding to ϵ_w and ϵ_{org}. Removal of a charge density $(C_w - C_{org})p$ requires an energy $\frac{1}{2}(C_w - C_{org})p^2$, to which one adds the molar energy of adsorption U_0, associated with the movement of organic substance from bulk solution to the electrode, and an energy $-C_{org}E_d p$, with E_d the shift of the electrocapillary maximum or point of zero charge caused by replacement of water by organic in the compact layer (Antropov 1972, pp. 259–262). E_d is the potential difference due to the dipole moment of the organic species.

In this model, the double layer is considered as two capacitors in parallel so that, if θ is the fractional surface coverage by organic substance, the electrode charge density is $C_w p(1 - \theta) + C_{org} p \theta$. To discuss the dependence of surface tension, adsorption, and so on, on bulk concentration and electrode potential, the parallel-capacitor model is combined with the Frumkin isotherm (equation [48]). Now, the equilibrium constant K' is dependent on p according to $K' = k e^{U/RT}$, where k is a constant and

$$U = U_0 - [\tfrac{1}{2}(C_w - C_{org})p^2 + C_{org}E_d p]\Omega_{org}$$

where Ω_{org} is the surface area per mole of adsorbed organic. The results of comparison of experimental electrocapillary curves with those calculated from this and other models are discussed by Antropov (1972, pp. 262–263).

F. SOLID SURFACES

The interfacial tension γ was defined by equation [9]. In what followed, a liquid surface was tacitly assumed, and electrocapillary measurements usually deal with liquid metal surfaces. For a solid surface, the meaning of γ is not evident and is indeed somewhat controversial. One can consider γ as the reversible work that must be done on a system to create unit interfacial area, keeping T, V, and mole numbers constant. For a crystal, the increase in interfacial area is done by cleavage. One can also consider the work necessary to produce unit area by stretching, keeping other variables constant. Increasing the area of a solid surface in this way entails doing work against intermolecular forces, but not bringing molecules from bulk regions to the surface. The structure of the surface is changed, so that γ for a solid depends on area, which it does not for a liquid. The relation between the work of stretching and the work of cleavage is considered at the end of this section.

The thermodynamics of the surface tension can still be applied to solid substances (Hurwitz 1978a, Section 4.3). For instance, consider a cell formed by two platinum electrodes, at one of which there is equilibrium with respect to the reaction $H^+ + e^- \rightarrow H$, where H is adsorbed hydrogen. The equilibrium between H and

$H_2(g)$ at a given pressure fixes the chemical potential of H. With the electrolytes HX and CX in solution, the cell is

$$L|H^+, C^+, X^-|H|Pt$$

where L is also made of platinum. With γ referring to the H/Pt electrode, the Gibbs equation is

$$d\gamma = -\Gamma_{Pt}^{(Pt)} d\mu_{Pt} - \Gamma_{Pt^{n+}}^{(Pt)} d\tilde{\mu}_{Pt^{n+}} - \Gamma_{e^-}^{(Pt)} d\tilde{\mu}_{e^-} - \Gamma_H^{(S)} d\mu_H$$
$$- \Gamma_{H^+}^{(S)} d\tilde{\mu}_{H^+} - \Gamma_{C^+}^{(S)} d\tilde{\mu}_{C^+} - \Gamma_{X^-}^{(S)} d\tilde{\mu}_{X^-} \quad [54]$$

at constant temperature. Superscript "s" refers to surface (adsorbed) species.

The electrochemical potential $\tilde{\mu}_{e^-}$ can be replaced by $\mu_H - \tilde{\mu}_{H^+}$ because of the equilibrium condition on $H^+ + e^- \to H$,

$$\tilde{\mu}_{H^+}^{(S)} + \tilde{\mu}_{e^-}^{(Pt)} = \mu_H^{(s)} \quad [55]$$

and the electrochemical potential $\tilde{\mu}_{Pt^{n+}}$ can be replaced by $\mu_{Pt} - n\tilde{\mu}_{e^-}$ because of the equilibrium condition on $Pt^{n+} + ne^- \to Pt$. Then, using the invariance of the chemical potential μ_{Pt} in Pt, we reduce equation [54] to

$$d\gamma = \frac{q^M}{\mathcal{F}} d(\mu_H - \tilde{\mu}_{H^+}) - \Gamma_H^{(S)} d\mu_H - \Gamma_{H^+}^{(S)} d\tilde{\mu}_{H^+}^{(S)}$$
$$- \Gamma_{C^+}^{(S)} d\tilde{\mu}_{C^+}^{(S)} - \Gamma_{X^-}^{(S)} d\tilde{\mu}_{X^-}^{(S)}$$

where the surface charge density on the metal side of the interface, q^M, is $\mathcal{F}(n\Gamma_{Pt^{n+}}^{(Pt)} - \Gamma_{e^-}^{(Pt)})$. The surface charge density on the electrolyte side is

$$q^S = \mathcal{F}(\Gamma_{H^+}^{(S)} + \Gamma_{C^+}^{(S)} - \Gamma_{X^-}^{(S)}) = -q^M$$

so that, substituting the above for q^M, we obtain

$$d\gamma = (-\Gamma_{H^+}^{(S)} - \Gamma_{C^+}^{(S)} + \Gamma_{X^-}^{(S)} - \Gamma_H^{(S)}) d\mu_H - \Gamma_{C^+}^{(S)} d\tilde{\mu}_{C^+}^{(S)}$$
$$- \Gamma_{X^-} d\tilde{\mu}_{X^-}^{(S)} + (\Gamma^{(S)[rbc^+} - \Gamma_{X^-}^{(S)}) d\tilde{\mu}_{H^+}^{(S)}$$
$$= -\left(\frac{q^S}{\mathcal{F}} + \Gamma_H\right) d\mu_H - \Gamma_{X^-}(d\tilde{\mu}_{X^-} + d\tilde{\mu}_{H^+}) + \Gamma_{C^+}(d\tilde{\mu}_{H^+} - d\tilde{\mu}_{C^+})$$

The combinations of electrochemical potentials in the last member of the equation are independent of the inner potential of the solution phase. In fact, they can be replaced by chemical potentials of neutral species: $\mu_{HX} = \tilde{\mu}_{H^+} + \tilde{\mu}_{X^-}$ and $\mu_{CX} = \tilde{\mu}_{C^+} + \tilde{\mu}_{X^-}$.

Thus one can write

$$d\gamma = -\left(\frac{q^S}{\mathcal{F}} + \Gamma_H\right) d\mu_H - \Gamma_{X^-} d\mu_{HX} + \Gamma_{C^+}(d\mu_{HX} - d\mu_{CX}) \quad [56]$$

so that

$$\left(\frac{\partial \gamma}{\partial \mu_{HX}}\right)_{a_H, a_{CX}} = -\Gamma_{X^-}^{(S)} + \Gamma_{C^+}^{(S)} = -\Gamma_{HX}^{(S)} \quad [57]$$

$$\left(\frac{\partial \gamma}{\partial \mu_{CX}}\right)_{a_H, a_{HX}} = -\Gamma_{C^+}^{(S)} = -\Gamma_{CX} \quad [58]$$

The last members of these equations are appropriate if one wants to interpret surface excesses in terms of the neutral components and H^+. Both quantities are experimentally measurable if γ can be measured for the solid platinum surface. Using Γ_{HX} as defined in equation [57], we get

$$\left(\frac{\partial \gamma}{\partial \mu_H}\right)_{a_{HX}, a_{CX}} = -(\Gamma_{H^+} - \Gamma_{HX} + \Gamma_H) \quad . \quad [59]$$

One obtains a surface excess of H which is composed of hydrogen atoms and hydrogen ions not included in the neutral HX. One could separate the ionic and neutral parts of this surface excess if the surface charge could be measured from an electrocapillary curve, since $q^S = \mathcal{F}(\Gamma_{H^+} - \Gamma_{Hx})$, but the equilibrium [55] means that the electrode potential cannot be varied.

A cell containing the H_2/Pt electrode and a reversible electrode would have its potential given by a Nernst equation. If there is no equilibrium between the gas and the solution, the electrode potential can be varied and the surface charge can be determined via the Lippmann equation. For oxygen on platinum, it seems that the adsorbed oxygen does not equilibrate with molecular oxygen in the gas phase, although there is equilibrium between adsorbed oxygen and solution species.

Now we consider the relation between the surface tension and the work of stretching (Clavier and Van Laethem-Meuree 1978, Chapter VIII). If x and y are in the plane of the surface, consider a line segment of length dx in the x-direction. A tensor of components g_{ij} is defined such that, to move the line segment a distance dy, and hence increase the surface area by $d\Omega = dx\, dy$, the work is $dy(g_{xy} dx)$, that is, g_{xy} is the force in the y-direction per unit length of line in the x-direction. If γ is the surface free-energy density (surface tension for a liquid), we have

$$g_{ij} = \gamma \delta_{ij} + \frac{\partial \gamma}{\partial \epsilon_{ij}} \quad [60]$$

where ϵ_{ij} is a fractional (dimensionless) deformation: the free energy is changed because of the increase in area ($\gamma \delta_{ij}$) and because γ is changed. For an isotropic solid, $g = \gamma + \Omega(d\gamma/d\Omega)$; for a liquid, $d\gamma/d\Omega = 0$. A suggested nomenclature, *surface stress* for g and *specific surface work* for γ (Parsons 1980, Section 1.4), emphasizes the distinction. It has been suggested that $d\gamma$ in equation [41] and thereafter should be replaced by $d\gamma + (\gamma - g_{ij}) d\epsilon$, but that the second term is usually negligible.

Problems with the experimental measurement of g and γ, as well as different experimental techniques for related electrochemical properties, are discussed by Clavilier and Van Laethem-Meuree (1980). The surface tension of a metal surface immersed in electrolyte has been measured via contact angle measurements for a gas bubble on the surface, as well as by mechanically stressing the metal (Habib and Bockris 1980, Section 4.5.4).

Models for the solid–liquid interface are similar to those for the liquid–liquid interface. Because of the problems with the measurement of surface tension, one generally relies on capacitance measurements. Thus, if only anions are specifically adsorbed, $-q^M = q^S = q^i_- + q^d_- + q^d_+$, where i and d refer to inner-layer and diffuse-layer contributions (see Chapter 4). If one makes the metal sufficiently negatively charged to remove anions from the inner layer, leaving charge in the diffuse layer only, one can calculate q^S by using diffuse-layer theory. Then integration of a capacity–potential curve gives q^S as a function of potential. The same is done for a liquid metal electrode. Subtracting the diffuse-layer contribution, as calculated from Gouy–Chapman theory (Chapter 4), gives the inner-layer contribution q^i_-.

G. ELECTROSTATICS

Here we review the laws of electrostatics relevant to a description of an electrochemical interface, starting from the Coulomb law of interaction between two stationary point charges in vacuum:

$$U = \frac{q_1 q_2}{r_{12}} \qquad [61]$$

where U is the energy of interaction, q_1 and q_2 are the charges, and r_{12} is the distance between them. The system of electrostatic (or Gaussian) units is used: The charges are in esu or statcoulombs, the distance is in centimeters, and the interaction energy is in ergs. In the mks system, one writes

$$U = \frac{q_1 q_2}{4\pi\epsilon_0 r_{12}} \qquad [62]$$

where charges are in coulombs, distance is in meters, and ϵ_0, the permittivity of free space, equals 8.854×10^{-12} C/(V · m), so that U is in joules: 1 J = 1

G. ELECTROSTATICS

coulomb (C) × 1 volt (V). We will usually write formulas in electrostatic units; the conversion to the mks system generally is made by replacing 4π by ϵ_0^{-1}. The mks equivalents of electrostatic units are as follows: 1 esu = 3.336×10^{-10} C, 1 statvolt = 1 esu/cm = 299.8 V. The electronic charge is 4.774×10^{-10} esu or 1.592×10^{-19} C.

According to equation [61] the force on a charge q_1 at $\mathbf{r}_1$ due to a charge q_2 at $\mathbf{r}_2$ is

$$\mathbf{F}_{21} = -q_1 q_2 \nabla_1 (r_{12})^{-1} = \frac{q_1 q_2 \mathbf{r}_{21}}{(r_{12})^3}$$

where $\mathbf{r}_{21}$ is a vector from $\mathbf{r}_2$ to $\mathbf{r}_1$. The force due to a collection of charges, with charge q_i at $\mathbf{r}_i$, is

$$\mathbf{F}_1 = \frac{q_1 \sum_i q_i \mathbf{r}_{i1}}{(r_{i1})^3} = q_1 \mathbf{E}(\mathbf{r}_1) \qquad [63]$$

The second equality defines the electric field at $\mathbf{r}_1$, $\mathbf{E}(\mathbf{r}_1)$. The field of a continuous distribution of charge described by a charge density $\rho(\mathbf{r}_2)$ (in esu/cm^3) is given by

$$\mathbf{E}(\mathbf{r}_1) = \int \frac{d\mathbf{r}_2 \rho(\mathbf{r}_2) \mathbf{r}_{21}}{(r_{12})^3} \qquad [64]$$

and $\mathbf{F}_1 = q_1 \mathbf{E}(\mathbf{r}_1)$. The electrical potential $\phi(\mathbf{r}_1)$ has the property that $\mathbf{E}(\mathbf{r}_j) = -\nabla_j \phi(\mathbf{r}_j)$ so that

$$\phi(\mathbf{r}_j) = \sum \frac{q_i}{r_{ij}}$$

for a collection of discrete charges or

$$\phi(\mathbf{r}_j) = \int \frac{d\mathbf{r}_i \rho(\mathbf{r}_i)}{r_{ij}} \qquad [65]$$

for a continuous distribution of charge. Equation [65] is mathematically equivalent to the Poisson equation for charges in a vacuum:

$$\nabla^2 \phi(\mathbf{r}) = -4\pi \rho(\mathbf{r}) \qquad [66]$$

In mks units, this equation is

$$\nabla^2 \phi(\mathbf{r}) = -\frac{\rho(\mathbf{r})}{\epsilon_0} \qquad [67]$$

where ϕ is in volts and ρ is in coulombs per cubic meter. In terms of the electrostatic field, the left-hand side of equation [66] or equation [67] is $-\nabla_j \cdot \mathbf{E}(\mathbf{r}_j)$.

In addition to volume distributions of charge described by $\rho(\mathbf{r})$ and point charges, we often deal with surface charge distributions. If $w(\mathbf{s})$ represents the charge per unit area at point $\mathbf{s}$ on a surface S, the potential at point $\mathbf{r}$ due to these charges is

$$\phi(\mathbf{r}) = \int w(\mathbf{r}') |\mathbf{r} - \mathbf{r}'|^{-1} \, dS'$$

where dS' is an infinitesimal surface area located at $\mathbf{r}'$, on the surface. For a planar surface with a uniform charge density w, the potential at a point a distance t away is

$$\phi(t) = w \int dx \, dy \, (x^2 + y^2 + t^2)^{-1/2}$$

where the origin of coordinates is chosen so that the observation point is $(0, 0, t)$ and the integral extends over the charged surface. If the edges of the surface are far enough away compared to t, the limits of the integral may be taken as infinite. The magnitude of the electric field, which will be uniform and perpendicular to the plane, is

$$E(t) = -\frac{d\phi}{dt} = w \int dx \, dy \, t(x^2 + y^2 + t^2)^{-3/2}$$
$$= w \int d\theta \int_0^\infty r \, dr \, t(r^2 + t^2)^{-3/2} = 2\pi w$$

where cylindrical coordinates r and θ have been used.

If we have two parallel planes with surface charge densities w and $-w$ such that the distance d between them is small enough compared to their lateral dimensions to ignore edge effects (parallel-plate capacitor), the electric field will be 0 outside and $4\pi w$ (adding the fields of the two plates) inside. The difference of electrical potential between the plates, obtained by integrating $-E$ from 0 to d, is

$$\phi_2 - \phi_1 = 4\pi w d \qquad [68]$$

The charges on the plates are $\pm wS$, where S is the lateral area. The capacitance, defined as the ratio of the charge on one plate to the potential difference, is $S/4\pi d$, and the capacitance per unit area is $(4\pi d)^{-1}$. The units of capacitance are esu/statvolt or cm (sometimes called statfarad). In the mks system, the capacitance per unit area is ϵ_0/d, the units being farad/m^2; 1 farad = 8.99×10^{11} cm. In a less ideal situation, the potential drop V across a circuit element may change with the charge Q on either side, without being directly proportional. The differential capacitance is then defined as dV/dQ and may not be a constant independent of Q or V.

If two capacitors are arranged so that one plate of each is joined by a conducting path, and the other two plates are also joined by a conducting path, they are said to be connected in parallel. The connected plates must come to the same potential, so that the potential difference across each capacitor is the same, say V. If the charges on the two capacitors are Q_1 and Q_2, the charge on each side of the combined system is $\pm(Q_1 + Q_2)$, and the capacitance is $d(Q_1 + Q_2)/dV = C_1 + C_2$, where $C_1 = dQ_1/dV$ is the capacitance of capacitor 1 and $C_2 = dQ_2/dV$ is that of capacitor 2. Thus two capacitors in parallel have a capacitance that is the sum of the individual capacitances. If one plate of one capacitor is connected to one of the other, but the other two plates are not connected, we have a series connection. The potential drop across this system is the sum of the potential drops across the individual capacitors. Accompanying these potential drops is the charging of each capacitor. The overall charge on one side of the system or the other has the same magnitude as the charge on one side of either capacitor. The series connection thus corresponds to a capacitance $C = Q/V = Q/(V_1 + V_2)$, so that $C^{-1} = (C_1)^{-1} + (C_2)^{-1}$, the sum of the reciprocals of the capacitances.

The electrical potential at $\mathbf{r}$ resulting from a system of charges localized near the origin of coordinates can be written as a series in inverse powers of r, which is useful if r is large compared to distances between charges of the system. This occurs, for instance, when the system of charges is of molecular size and r is of macroscopic (or even microscopic) size. The first few terms in the expansion are

$$\phi = Qr^{-1} + (\mathbf{p}\cdot\mathbf{r})r^{-3} + \sum_{ij} Q_{ij}T_{ij}r^{-3} \qquad [69]$$

corresponding to the total charge Q, the dipole moment, and the quadrupole moment of the system. The dipole moment $\mathbf{p}$ is a vector given by

$$\mathbf{p} = \int \mathbf{r}'\rho(\mathbf{r}')\,d^3r'$$

if the system of charges is a continuous charge distribution, or given by $\Sigma \mathbf{r}_j q_j$ if it is a collection of discrete charges. The quantities Q_{ij} are components of the quadrupole moment tensor,

$$Q_{ij} = \int r'_i r'_j \rho(\mathbf{r}')\,d^3r', \qquad i,j = 1, 2, 3$$

and the T_{ij} are dimensionless functions of the components of r, for example, $T_{11} = T_{xx} = \frac{1}{2}[3(x^2/r^2) - 1]$.

For a neutral system, the second term in equation [69] is the leading term in the electrical potential. The simplest arrangement of charges which gives such a term is a pair of charges of equal size and opposite sign. The dipole moment of such a system is

$$\mathbf{p} = q\mathbf{r}_0$$

where q is the magnitude of the charges and $\mathbf{r}_0$ is the vector leading from the negative charge to the positive charge. The electric field at $\mathbf{r}$ resulting from a dipole at the origin approaches

$$\mathbf{E}(\mathbf{r}) = \frac{-\mathbf{p}}{r^3} + \frac{3(\mathbf{p} \cdot \mathbf{r})\mathbf{r}}{r^5} \qquad [70]$$

when r becomes large compared to r_0. In the limit of $q \to \infty$ and $r_0 \to 0$ such that p remains constant, one has a mathematical or point dipole, for which equation [70] gives the field for all nonzero $\mathbf{r}$. Neutral molecules (dimensions 10^{-8}–10^{-7} cm) appear as point dipoles when distances larger than 10^{-6} cm are considered. Thus it is often convenient to consider, analogously to the charge density, a dipole moment density called the polarization $\mathbf{P}$.

$\mathbf{P}$ is defined such that $\mathbf{P}(\mathbf{r}) \, d^3r$ is the dipole moment of the volume d^3r. Since the electrostatic potential at $\mathbf{r}$ resulting from a dipole $\mathbf{p}$ at the origin is $\mathbf{p} \cdot \mathbf{r}/r^3$, the electrostatic potential resulting from a polarized body is

$$\phi^P = \int \frac{\mathbf{P}(\mathbf{r}') \cdot (\mathbf{r} - \mathbf{r}')}{|\mathbf{r} - \mathbf{r}'|^3} \, d^3r' = \int \mathbf{P}(\mathbf{r}') \cdot \nabla_{r'} |\mathbf{r} - \mathbf{r}'|^{-1} \, d^3r'$$

One may write $\mathbf{P} \cdot \nabla_{r'} |\mathbf{r} - \mathbf{r}'|^{-1}$ as $\nabla_{r'} \cdot (\mathbf{P} |\mathbf{r} - \mathbf{r}'|^{-1}) - |\mathbf{r} - \mathbf{r}'|^{-1} \nabla_{r'} \cdot \mathbf{P}$ and use Gauss' theorem, $\int_V \nabla \cdot \mathbf{R} \, dv = \int R_n \, dS$, where S is the surface bounding V and R_n is the component of $\mathbf{R}$ along the outward normal to the surface. Now the potential resulting from the polarized body is

$$\phi^P = \int P_n(\mathbf{r}') |\mathbf{r} - \mathbf{r}'|^{-1} \, dS' - \int [\nabla_{r'} \cdot \mathbf{P}(\mathbf{r}')] |\mathbf{r} - \mathbf{r}'|^{-1} \, d^3r'$$

so that the polarization is equivalent to a surface charge density P_n and a volume charge density $-\nabla \cdot \mathbf{P}$, which we will call the polarization charge density.

For a parallel-plate capacitor with an insulating medium between the plates, the capacitance per unit area becomes $\epsilon/4\pi d$ (instead of $1/4\pi d$ with vacuum between the plates), where ϵ is slightly above 1 for gases, 5–8 for glass, and 80 for water. This is because the medium is polarized uniformly in the direction perpendicular to the plates, producing a surface charge density equal to $(\epsilon V/4\pi d) - (V/4\pi d)$, where V is the potential difference between the plates. Thus the polarization is

$$P = \frac{(\epsilon - 1)(V/d)}{4\pi} = \frac{(\epsilon - 1)E}{4\pi} \qquad [71]$$

where E is the electric field. The constant of proportionality between P and E is called the electric susceptibility χ:

$$\chi = \frac{\epsilon - 1}{4\pi}$$

In mks units, one writes $P = \epsilon_0(\epsilon - 1)E$ and $\epsilon - 1 = \chi$, the susceptibility χ being greater by a factor of 4π than that used with electrostatic units. In nonisotropic media, the polarization produced by an electric field may not be in the direction of field, and the susceptibility must be taken as a tensor.

For an isolated molecule, the dipole moment vanishes in the absence of a field, since any permanent dipole the molecule may have is randomly oriented. An electric field orients the permanent dipole and deforms the charge distribution of the molecule; both effects are proportional to the field for small fields. If the induced moment on a molecule is αE and the medium has n molecules per unit volume, the polarization will be $n\alpha E$ and $\chi = n\alpha$. Usually, however, the electric field that polarizes an atom or molecule differs from the macroscopic electric field, the former not including the field of the atom or molecule itself, and the latter involving a time-average over the positions of charges in the system. (See Chapter 3, Section D for more details.)

H. DOUBLE LAYERS

Within a dielectric medium, one still has $\mathbf{E} = -\nabla\phi$, but instead of $\nabla \cdot \mathbf{E}$ being equal to $4\pi\rho$, we have

$$\nabla \cdot \mathbf{E} = 4\pi(\rho - \nabla \cdot \mathbf{P}) \qquad [72]$$

where the polarization charge density $-\nabla \cdot \mathbf{P}$ has been added to the free charge density. At the boundary between two dielectric media, the tangential component of $\mathbf{E}$ is continuous, since otherwise a finite displacement of a charge along the boundary would involve different amounts of work depending on whether one was on one side of the boundary or the other. However, the normal component of $\mathbf{E}$ is not continuous, since the polarization behaves like a surface charge density. If there is no true surface charge density, we have

$$(E_n)_1 - (E_n)_2 = 4\pi[(P_n)_2 - (P_n)_1] \qquad [73]$$

Equations [72] and [73] suggest the definition of the electric displacement

$$\mathbf{D} = \mathbf{E} + 4\pi\mathbf{P} \qquad [74]$$

such that

$$\nabla \cdot \mathbf{D} = 4\pi\rho \qquad [75]$$

within an insulating medium and such that the normal component of D is continuous across an uncharged boundary between two such media. Within a medium, $\mathbf{D} = \epsilon\mathbf{E}$ according to equations [71] and [74].

The definition of $\mathbf{D}$ is chosen so that $\mathbf{D}$ in a dielectric medium behaves like $\mathbf{E}$ in a vacuum ($\epsilon = 1$), that is, it changes from point to point only by virtue of true charges. For example, across a charged surface the normal component of $\mathbf{D}$ changes by the true surface charge density,

$$(D_n)_1 - (D_n)_2 = 4\pi w$$

so that, inside one plate of a parallel-plate capacitor with $D = 0$ outside, $D = 4\pi w$ and $E = 4\pi w/\epsilon$. Integrating E to the other plate, one finds $4\pi wd/\epsilon$ for the potential difference across the system (and $4\pi d/\epsilon$ for the capacitance per unit area). If d is small the situation corresponds to a dipole layer with a surface dipole density wd. The electrochemical interface, having negligible thickness on a macroscopic scale and a finite nonzero potential drop across it, appears as such a dipole layer or as two oppositely charged layers infinitesimally separated. Hence we have the common term *double layer* (Barlow 1970).

In charging a parallel-plate capacitor by moving charge from the negative plate to the positive plate, work is required. When the voltage across the capacitor is V, the work to move an amount of charge dQ is $V\,dQ$. Since $V = Ed = 4\pi wd/\epsilon$, the work of charging is

$$\int_0^V V\,dQ = \frac{4\pi d}{\epsilon}\int_0^w wA\,dw = \left(\frac{2\pi Ad}{\epsilon}\right)w^2 = (8\pi)^{-1}(Ad)\,ED$$

where A is the surface area of a plate. Since Ad is the volume of the capacitor, $ED/8\pi$ may be considered the energy density of the field. For an arbitrary arrangement of conductors and insulators a similar expression for the work of charging can be established. To bring dQ from the surface of a conductor at potential V_1 to the surface of a conductor at potential V_2 requires work equal to $dQ(V_2 - V_1)$. A consequence of the Coulomb law (equation [61]) and the definition of $\mathbf{D}$ is that the integral of the normal component of $\mathbf{D}$ over a surface enclosing a conductor is 4π times the total charge, so dQ may be related to integrals of $\mathbf{D}$ and (Becker and Sauter 1964, Sections 19 and 32) one can show the work is

$$dW_T = (4\pi)^{-1}\int (\mathbf{E}\cdot d\mathbf{D})\,d\tau$$

Here, $d\mathbf{D}$ is the change in $\mathbf{D}$ resulting from the movement of dQ, and the integral extends over all space. The entire existing field can be imagined as having been built up by small charge transfers of this type. Therefore the energy density (energy per unit volume) is $(4\pi)^{-1}\int_0^D \mathbf{E}\cdot d\mathbf{D}$, which, if $\mathbf{D} = \epsilon\mathbf{E}$, can be integrated to $(8\pi)^{-1}\mathbf{E}\cdot\mathbf{D}$.

H. DOUBLE LAYERS

Consider a spatial charge distribution with charge density independent of the x and y positions, described by $\rho(z)$, in a medium of uniform dielectric constant ϵ. Then **D** and **E** are in the z direction and the Poisson equation (equation [75]) is

$$\epsilon \frac{dE}{dz} = 4\pi\rho(z)$$

Integrating from a to z, we have

$$E(z) - E(a) = 4\pi\epsilon^{-1} \int_a^z \rho(z')\, dz'$$

and, integrating again, we have

$$-\phi(z) + \phi(a) - (z - a)E(a) = 4\pi\epsilon^{-1} \int_a^z dz' \int_a^z dz''\, \rho(z'') \quad [76]$$

Suppose $\rho(z) = 0$ for $z < a$ and for $z > b$. Then, if $E(a) = 0$, $E(b)$ will vanish if $\int_a^b \rho(z')\, dz' = 0$. This electroneutrality corresponds to two or more layers of charge such that the total surface charge density vanishes. The potential difference across the region of nonzero $\rho(z)$ is, from equation [76],

$$\phi(b) - \phi(a) = -4\pi\epsilon^{-1} \int_a^b dz' \int_a^{z'} dz''\, \rho(z'')$$

$$= -4\pi\epsilon^{-1} \left[-\int_a^b \rho(z')\, z'\, dz' \right] \quad [77]$$

We have integrated by parts and used the electroneutrality condition to drop the boundary term. The integral $\int_a^b dz\, z\rho(z)$ is the dipole moment per unit area of the charge distribution. Although it need not consist simply of two oppositely charged layers, this situation is generally referred to in electrochemistry as a double layer.

The potential difference $\phi_b - \phi_a$ depends only on the dipole moment per unit area and the dielectric constant that is assumed to permeate the region between $z = a$ and $z = b$. If, instead of free charges, one has N dipole moments per unit area in this region, each of magnitude m in the z direction, the dipole moment per unit area is Nm, and

$$\phi_b - \phi_a = 4\pi\epsilon^{-1} Nm \quad [78]$$

This kind of double layer could correspond to a layer of dipolar molecules adsorbed on a surface.

As another example of a double layer, consider the interface formed by placing in contact two homogeneous media of dielectric constants ϵ_1 and ϵ_2, both containing a mobile species of charge q which can flow across the interface. Let the initial concentrations of this species in the two phases be n_1 and n_2. In each phase, there must be a compensating charge density that will be assumed immobile. We seek the concentration and potential as functions of position when equilibrium is reached. The concentrations of the mobile species in the bulk regions of the two phases, far from the interface, will remain n_1 and n_2. Assuming that the boundary between the media is at $z = 0$ and that the system is homogeneous in the x and y directions, we have the Poisson equation

$$\frac{d^2\phi}{dz^2} = -4\pi q \epsilon_1^{-1}(n - n_1), \quad z < 0 \qquad [79]$$

for medium 1 and

$$\frac{d^2\phi}{dz^2} = -4\pi q \epsilon_2^{-1}(n - n_2), \quad z > 0$$

for medium 2, where $-qn_1$ and $-qn_2$ are the charge densities of the immobile species.

Suppose the density of the mobile species in each medium is governed by the Boltzmann law,

$$\frac{n}{n_1} = e^{-q(\phi - \phi_1)/kT}, \quad z < 0$$

$$\frac{n}{n_2} = e^{-q(\phi - \phi_2)/kT}, \quad z > 0 \qquad [80]$$

where ϕ_1 and ϕ_2 are the values of the potential for $z \to -\infty$ and $z \to \infty$. Substituting in the Poisson equation we have, for $z < 0$,

$$\frac{d^2\phi}{dz^2} = \frac{-4\pi q n_1}{\epsilon_1}(e^{-q(\phi - \phi_1)/kT} - 1)$$

and for $z > 0$,

$$\frac{d^2\phi}{dz^2} = \frac{-4\pi q n_2}{\epsilon_2}(e^{-q(\phi - \phi_2)/kT} - 1)$$

If $q(\phi - \phi_1)/kT$ and $q(\phi - \phi_2)/kT$ are small, one can expand the exponentials and keep only the first two terms, which gives

$$\frac{d^2\phi}{dz^2} = \frac{4\pi q n_1}{\epsilon_1} \frac{q(\phi - \phi_1)}{kT} = \frac{\phi - \phi_1}{(l_1)^2}$$

$$\frac{d^2\phi}{dz^2} = \frac{4\pi q n_2}{\epsilon_2} \frac{q(\phi - \phi_2)}{kT} = \frac{\phi - \phi_2}{(l_2)^2}$$

where $l_1 = (\epsilon_1 kT/4\pi n_1 q^2)$ and $l_2 = (\epsilon_2 kT/4\pi n_2 q^2)^{1/2}$. The quantities l_1 and l_2 are sometimes called *Debye lengths*. If $\epsilon_1 = 10$, $T = 300$ K, $n_1 = 0.01$ mol/liter $= 6 \times 10^{18}$ cm^{-3}, and $q = e = 4.8 \times 10^{-10}$ esu, l_1 is about 1.5×10^{-7} cm or 1.5 mμm.

The solution to the equation for $z < 0$ must be taken as $\phi = \phi_1 + b \exp(z/l_1)$ so that $\phi \to \phi_1$ and $d\phi/dz \to 0$ as $z \to -\infty$, with b being a constant to be determined; for $z > 0$, $\phi = \phi_2 + b' \exp(-z/l_2)$. Since ϕ must be continuous at $z = 0$, we have $\phi_1 + b = \phi_2 + b'$. The continuity of the normal component of the electric displacement, $-\epsilon \, d\phi/dz$, requires $\epsilon_1 b/l_1 = -\epsilon_2 b'/l_2$. Solving, we find

$$b = \frac{\phi_2 - \phi_1}{1 + \epsilon_1 l_2/\epsilon_2 l_1}$$

$$b' = \frac{\phi_1 - \phi_2}{1 + \epsilon_2 l_1/\epsilon_1 l_2}$$

The overall potential drop across the interface is $\phi_2 - \phi_1$ and the potential at $z = 0$ is $\phi_1 + b$, so the fraction of the overall potential drop that occurs in phase 1 is $l_1 \epsilon_2 / (l_1 \epsilon_2 + l_2 \epsilon_1)$. Thus, the longer the Debye length and the smaller the dielectric constant for a phase, the more of the potential drop occurs within it. We can also calculate the total charge transferred from one phase to another in attaining equilibrium. The net charge (per unit area) for phase 1 is

$$q^{(1)} = \int_{-\infty}^{0} dz\, \rho = \int_{-\infty}^{0} dz\, \frac{-\epsilon_1 \phi''}{4\pi} = \frac{\epsilon_1 E(0)}{4\pi} \qquad [81]$$

For the present situation, $q^{(1)}$ is

$$\frac{-\epsilon_1 b}{4\pi l_1} = -(4\pi)^{-1}(\phi_2 - \phi_1)\left(\frac{l_1}{\epsilon_1} + \frac{l_2}{\epsilon_2}\right)^{-1}$$

which is also equal to $\epsilon_1 b'/4\pi l_2$ or $-q^{(2)}$. The capacitance per unit area is

$$\frac{dq^{(1)}}{d(\phi_1 - \phi_2)} = \frac{dq^{(2)}}{d(\phi_2 - \phi_1)} = (4\pi)^{-1}\left(\frac{l_1}{\epsilon_1} + \frac{l_2}{\epsilon_2}\right)^{-1}$$

that is, $(4\pi C)^{-1} = l_1/\epsilon_1 + l_2/\epsilon_2$, corresponding to two capacitors in series, one with capacitance $\epsilon_1/4\pi l_1$ per unit area and one with capacitance $\epsilon_2/4\pi l_2$. These obviously are the two phases. If $\epsilon_1 = 10$ and $l_1 = 1.5 \times 10^{-7}$ cm, $\epsilon_1/l_1 = 0.65 \times 10^8$ cm^{-1} = 7.2×10^{-5} C/(V · cm^2) or 72 μF/cm^2.

Another way of integrating Poisson's equation,

$$\frac{d^2\phi}{dz^2} = \frac{-4\pi\rho}{\epsilon}$$

is obtained by multiplying by $d\phi/dz$ before integrating. Then

$$\left[\left(\frac{-\epsilon}{4\pi}\right)\frac{1}{2}\left(\frac{d\phi}{dz}\right)^2\right]_{-\infty}^{z} = \int_{-\infty}^{z} \rho(z)\,\phi'(z)\,dz = \int_{\phi(-\infty)}^{\phi(-z)} \rho(z)\,d\phi$$

In particular, for $z = 0$ in the present problem,

$$\left(\frac{-\epsilon}{8\pi}\right)[\phi'(0)]^2 = \frac{-2\pi q^{(1)2}}{\epsilon_1} = \int^{\phi(0)} \rho\,d\phi \qquad [82]$$

where equation [81] has been used. The lower limit on the last integral is $\phi(-\infty)$ and ρ is to be considered as a function of ϕ. The differential capacity is

$$\frac{dq^{(1)}}{d[\phi(0) - \phi(-\infty)]} = [q^{(1)}]^{-1/2}\frac{-\epsilon_1}{4\pi}\rho(0)\,\phi(0)$$

since $\rho(-\infty)$, far from the surface, necessarily vanishes.

I. ELECTROCHEMICAL POTENTIALS

In thermodynamic terms, the equilibrium situation just described must correspond to constancy of the electrochemical potentials of the mobile species through the system. In phase i we write the electrochemical potential as

$$\tilde{\mu} = \mu^{i0} + RT \ln n + qN_A\phi \qquad [83]$$

(N_A = Avogadro's number), which separates the electrical part $q\phi$ and uses the dilute-solution form for the chemical part; μ^{10} of course differs from μ^{20}. The potential ϕ and the concentration n are functions of z. According to equation [80],

$$q(\phi - \phi_1) = -kT \ln \frac{n}{n_1}$$

in phase 1 so that, for $z < 0$,

$$\tilde{\mu} = \mu^{10} + RT \ln n + qN_A\phi_1 - N_A kT \ln \frac{n}{n_1}$$

$$= \mu^{10} + qN_A\phi_1 + RT \ln n_1 \qquad [84]$$

and similarly for $z > 0$ (phase 2). Expression [84] is clearly independent of position within phase 1 and equal to $\tilde{\mu}$ (equation [83]) at $z \to -\infty$ where $n = n_1$ and $\phi = \phi_1$. The equality of chemical potentials for $z \to -\infty$ and $z \to \infty$ is

$$\mu^{10} + RT \ln n_1 - (\mu^{20} + RT \ln n_2) = qN_A(\phi_2 - \phi_1) \qquad [85]$$

showing that the overall potential difference at equilibrium is determined by the properties of the bulk phases, μ^{i0} and n_i. The capacitance is then somewhat meaningless since one cannot vary $\phi_2 - \phi_1$.

If there is no possibility of direct transfer of charged particles from one phase to another, equilibrium between phases is not attained and there is no equation like [85] which determines the potential difference between the phases. Then $\phi_2 - \phi_1$ may be varied by charging the interface. This corresponds to a polarizable interface: The charging involves withdrawing charged species from one phase and introducing charged species into the other, perhaps through an external circuit, but not across the interface. Integrating the Poisson equation for phase 2 from $z = 0$ to $z = \infty$ gives

$$E(0_+) = -4\pi q \epsilon_2^{-1} \int_0^\infty (n - n_2)\, dz \qquad [86]$$

assuming no field at $z \to \infty$. In the negative z direction, $E(0_+)$ is the field of a sheet with surface charge density $w = q \int_0^\infty (n - n_2)\, dz$. Calculating $E(0_-)$ by integrating the Poisson equation across phase 1, where $E(-\infty)$ is assumed to vanish, we see that the continuity of the normal component of $\mathbf{D} = \epsilon \mathbf{E}$ requires that

$$4\pi q \int_{-\infty}^0 (n - n_1)\, dz = -4\pi q \int_0^\infty (n - n_2)\, dz \qquad [87]$$

which simply expresses the overall electrical neutrality of the interface as a whole.

If, say, $\int_0^\infty (n - n_2)\, dz > 0$ and q is negative, so that $E(0)$ is positive, the potential in phase 1 near $z = 0$ is lower than for $z \to -\infty$, that is, $-q(\phi - \phi_1) < 0$. According to equation [80], n is smaller than n_1 near $z = 0$. This is true even if the Boltzmann relation is replaced by something more suitable for electrons, such as

$$\frac{n}{n_1} = \left[1 + \frac{e(\phi - \phi_1)}{\epsilon_F}\right]^{3/2} \qquad [88]$$

which follows (Chapter 4, equation [54]) from the Sommerfeld free-electron model of a metal (the electrons have charge $q = -e$ and ϵ_F is the Fermi energy, a constant).

According to equation [80] or [88], positively charging the phase on the right ($z > 0$) will raise the potential near $z = 0$ and attract more electrons in the phase on the left to the region near the interface (accumulation), while negatively charging the phase on the right will decrease the electron concentration near the interface (depletion). If the negative charge is large enough, the quantity in square brackets in equation [88] may become negative, so that there will be a region of zero electron density (Schottky barrier) near the surface, which may prevent electron transport from the bulk to the surface.

3

STATISTICAL MECHANICS

A. INTRODUCTION

It is the goal of statistical mechanics to relate thermodynamic and other macroscopic properties to molecular properties. For a system of classical mechanical particles the formalism is well known and permits all properties of interest to be expressed in terms of the interparticle interactions (whose origin may well involve the laws of quantum mechanics). Reducing the formal expressions to manageable form or extracting numerical values for particular properties are other matters. We review some of the basic principles of classical statistical mechanics first.

The Hamiltonian function of a system of particles is the sum of the kinetic and potential energies:

$$H = T + \Phi \qquad [1]$$

The kinetic energy is written in terms of momenta, so that for N structureless particles it is

$$T = \sum_{i=1}^{N} \frac{(\mathbf{p}_i)^2}{2m_i} \qquad [2]$$

where $\mathbf{p}_i$ is the momentum of particle i and m_i is its mass. The potential energy is a function of particle coordinates (and, in general, particle orientations),

$$\Phi = \Phi(\mathbf{q}_1, \ldots, \mathbf{q}_N) \qquad [3]$$

Interactions that depend on particle momenta are excluded from consideration here. Then the positions and momenta of the particles are distributed according to

3. STATISTICAL MECHANICS

$$dP(\mathbf{q}_1, \ldots, \mathbf{p}_N) = \Omega(Z_N)^{-1} e^{-H/kT} d\mathbf{q}_1, \ldots, d\mathbf{q}_N d\mathbf{p}_1, \ldots, d\mathbf{p}_N h^{-3N} \quad [4]$$

where k is Boltzmann's constant, h is Planck's constant, Ω is a product of factors $1/N_s!$, where N_s is the number of indistinguishable particles of species s, and

$$Z_N = \Omega h^{-3N} \int e^{-H/kT} d\mathbf{q}_1, \ldots, d\mathbf{q}_N d\mathbf{p}_1, \ldots, d\mathbf{p}_N$$

The quantity dP is the probability that (1) the position of particle 1 is between $\mathbf{q}_1$ and $\mathbf{q}_1 + d\mathbf{q}_1$, (2) the position of particle 2 is between $\mathbf{q}_2$ and $\mathbf{q}_2 + d\mathbf{q}_2$, ..., (3) the momentum of particle 1 is between $\mathbf{p}_1$ and $\mathbf{p}_1 + d\mathbf{p}_1$, ..., and so on. The integration over momenta in Z_N is easily carried out when Φ involves positions only, giving

$$Z_N = \left[\frac{(2\pi m_1 kT/h^2)^{3N_1/2}}{N_1!}\right]\left[\frac{(2\pi m_2 kT/h^2)^{3N_2/2}}{N_2!}\right], \ldots, Q_N$$

where

$$Q_N = \int e^{-\Phi/kT} d\mathbf{q}_1, \ldots, d\mathbf{q}_N \quad [5]$$

is called the configurational integral. Each factor in square brackets in Z_N corresponds to a species of particles present.

If we are interested in the distribution of the particles in configuration space, we must integrate dP over momenta. This gives the N-particle distribution function (Hill 1956, Section 29)

$$\rho^{(N)}(\mathbf{q}_1, \ldots, \mathbf{q}_N) = Q_N^{-1} \exp\left(\frac{-\Phi}{kT}\right)$$

The probability that the position of particle 1 is within $d\mathbf{q}_1$ of $\mathbf{q}_1$, the position of particle 2 is within $d\mathbf{q}_2$ of $\mathbf{q}_2$, and so on, is $\rho^{(N)} d\mathbf{q}_1, \ldots, d\mathbf{q}_N$. By integrating $\rho^{(N)}$ over the positions of all the particles but the first, one obtains the distribution function for a single particle of species 1 which, when multiplied by the number of (indistinguishable) particles of this species, is the number density for species 1, represented by $n_i^{(1)}(\mathbf{r})$ or $n_i(\mathbf{r})$. The quantity $n_i(\mathbf{r}) d\mathbf{r}$ is the number of particles of species i in the infinitesimal volume $d\mathbf{r}$. When it is multiplied by Avogadro's number, it gives the density of species i in moles per unit volume. Two-particle distributions are obtained by integrating $\rho^{(N)}$ over all coordinates but those of two particles. Multiplication by numbers of particles of two different species then gives two-particle distribution functions for that pair of species, $n_{ij}^{(2)}(\mathbf{r}_1, \mathbf{r}_2)$ or $n_{ij}(\mathbf{r}_1, \mathbf{r}_2)$. For instance,

A. INTRODUCTION

$$n_{mn} = N_m N_n (Q_N)^{-1} \int d\mathbf{q}' \exp\left(\frac{-\Phi}{kT}\right) \quad [6]$$

where N_m is the number of particles of species m and $d\mathbf{q}'$ implies integration over coordinates of all particles except two, one from species m and one from species n. For two particles of the same species,

$$n_{mm} = N_m (N_m - 1)(Q_N)^{-1} \int d\mathbf{q}' \exp\left(\frac{-\Phi}{kT}\right) \quad [7]$$

since $N_m(N_m - 1)$ is the number of ordered pairs of different particles, where $d\mathbf{q}'$ now implies integration over coordinates of all particles except two from species m. According to the definition, $n_{ij}(\mathbf{r}_1, \mathbf{r}_2) d\mathbf{r}_1 d\mathbf{r}_2$ is the number of pairs of particles such that a particle of species i is in $d\mathbf{r}_1$ at $\mathbf{r}_1$ and a particle of species j is in $d\mathbf{r}_2$ at $\mathbf{r}_2$.

In a homogeneous and isotropic phase, $n_m^{(1)}$ is a constant, independent of position, while $n_{mn}^{(2)}$ depends only on r_{12}, the distance between the positions of the two particles. For a planar interface, assuming homogeneity in the x and y directions, $n_m^{(1)}(\mathbf{r})$ becomes $n_m^{(1)}(z)$.

The charge density is written as a sum of the distributions of the charged species:

$$\rho(\mathbf{r}) = \sum_i n_i(\mathbf{r}) q_i = \sum_i n_i(\mathbf{r}) z_i \mathcal{F}/N_A$$

where q_i is the charge on a particle of species i, $\mathcal{F}$ is the Faraday, and z_i is the number of electronic charges on a particle of species i. The dipole moment per unit volume or polarization is correspondingly

$$\mathbf{P}(\mathbf{r}) = \sum n_i(\mathbf{r}) \mu_i(\mathbf{r})$$

where $\mu_i(\mathbf{r})$ is the dipole moment of a particle of species i; it depends on $\mathbf{r}_i$ since local electric fields determine the magnitude and direction of a dipole.

Integration of a two-particle distribution over the coordinates of one particle yields a one-particle distribution:

$$\sum_j \int n_{ij}(\mathbf{r}_1, \mathbf{r}_2) d\mathbf{r}_2 = n_i(\mathbf{r}_1)$$

Another relation between the two-particle and one-particle distributions is associated with the names of Born, Green, Yvon, Bogulyubov, and others. By differentiating the definition of the one-particle distribution n_i (integral of $\rho^{(N)}$ over all coordinates but one) with respect to a spatial coordinate we obtain

$$\nabla_m n_i(\mathbf{r}_m) = \frac{N_i \left[\int e^{-\Phi/kT} (-\nabla_m \Phi/kT) \, d\mathbf{q} \, (d\mathbf{r}_m)^{-1} \right]}{\int e^{-\Phi/kT} \, d\mathbf{q}}$$

If Φ consists of pair interactions,

$$\Phi = \tfrac{1}{2} \sum_{kj}^{(j \neq k)} \phi_{jk}(\mathbf{r}_j, \mathbf{r}_k)$$

we have

$$\nabla_m \Phi = \sum_{j}^{(j \neq m)} \nabla_m \phi_{mj}(\mathbf{r}_m, \mathbf{r}_j)$$

so that, grouping together terms in the sum by the species of particle j, we get

$$kT \nabla_m n_i(\mathbf{r}_m) = -N_i \sum_j \int e^{-\Phi/kT} \nabla_m \phi_{mj}(\mathbf{r}_m, \mathbf{r}_j)$$
$$\cdot d\mathbf{q}(d\mathbf{r}_m \, d\mathbf{r}_j)^{-1} \, d\mathbf{r}_j \, (Q_N)^{-1}$$
$$= -\sum_k \int n_{ik}(\mathbf{r}_m, \mathbf{r}_j) \nabla_m u_{ik}(\mathbf{r}_m, \mathbf{r}_j) \, d\mathbf{r}_j \qquad [8]$$

where the last sum is now over species, not particles. We have grouped together terms in Σ_j belonging to particles j of the same species, all of which have the same interaction with particle m of species i; the interaction potential is written u_{ik}, representing the interaction of a particle of species i with one of species k. Equation [8] is that of Born, Green, Yvon, and others.

The quantity on the right-hand side of equation [8] is the average force exerted on the particles of species i in $d\mathbf{r}_m$ at $\mathbf{r}_m$ by all the particles of the system. Summing it over i would give the average force on the volume element $d\mathbf{r}_m$ at $\mathbf{r}_m$. On the other hand, dividing by $n_i(\mathbf{r}_m)$ would give the average force on a particle of species i located in the volume $d\mathbf{r}_m$ at $\mathbf{r}_m$. If we define the quantity U_i by

$$\nabla_m U_i = [n_i(\mathbf{r}_m)]^{-1} \sum_k \int n_{ik}(\mathbf{r}_m, \mathbf{r}_j) \nabla_m u_{ik}(\mathbf{r}_m, \mathbf{r}_j) \, d\mathbf{r}_j \qquad [9]$$

then U_i is the *potential of mean force* for species i, that is, the negative gradient of U_i gives the mean force on a particle of species i.

If there is an electrical field from sources external to the system, and some of the species are electrically charged, a term $\Sigma_j \, ez_j \phi$ would be added to Φ, where ϕ is the electrical potential and ez_j is the charge on an atom of species j. Then $\nabla_m \Phi$

would include a term $ez_m \nabla_m \phi$ and, on the right-hand side of equation [8], we would have the additional term

$$\frac{-ez_m \nabla_m \phi \, N_i \sum_j \int e^{-\Phi/kT} \, \mathbf{dq} (\mathbf{dr}_m)^{-1}}{\int e^{-\Phi/kT} \, \mathbf{dq}} = ez_m \mathbf{E} n_i(\mathbf{r}_m)$$

where $\mathbf{E}$ is the external electrical field. The electrostatic potential of the external field on the particles is simply added to the potential of mean force.

Now we may write equation [8] as

$$[n_i(\mathbf{r}_m)]^{-1} \nabla_m n_i(\mathbf{r}_m) = \frac{-\nabla_m U_i(\mathbf{r}_m)}{kT}$$

which may be integrated formally to

$$\ln n_i(\mathbf{r}_m) - \ln n_i(\mathbf{r}_0) = \frac{-[U_i(\mathbf{r}_m) - U_i(\mathbf{r}_0)]}{kT}$$

or

$$n_i(\mathbf{r}_m) = n_i(\mathbf{r}_0) \exp \left\{ \frac{-[U_i(\mathbf{r}_m) - U_i(\mathbf{r}_0)]}{kT} \right\} \qquad [10]$$

Equation [10] is exact. The density of any species varies in space according to a Boltzmann-like expression, but with the potential of mean force appearing as the potential energy. If the position of a particle of species i does not affect the positions of the other particles (including others of species i), as occurs in a rarefied system, $n_{ik}(\mathbf{r}_m, \mathbf{r}_j) = n_i(\mathbf{r}_m) n_k(\mathbf{r}_j)$. Then equation [9] becomes

$$\nabla_m U_i = \sum_k \nabla_m \int n_k(\mathbf{r}_j) u_{ik}(\mathbf{r}_m, \mathbf{r}_j) \, d\mathbf{r}_j$$

so that U_i is the mean potential energy of interaction between a particle of species i at $\mathbf{r}_m$ and the other particles.

The two-particle distribution n_{mn} can be treated similarly. Differentiating it with respect to the position of a particle of species m, we have

$$kT \nabla_h n_{mn}(\mathbf{r}_h \mathbf{r}_i) = N_m N_n \left[\int e^{-\Phi/kT} (-\nabla_h \Phi) \, \mathbf{dq} (\mathbf{dr}_h \, \mathbf{dr}_i)^{-1} \right] (Q_N)^{-1}$$

3. STATISTICAL MECHANICS

Here, $\Phi = \frac{1}{2}\Sigma_{kj}^{(k \neq j)} \phi_{jk}(\mathbf{r}_j, \mathbf{r}_k)$, so

$$\nabla_h \Phi = \sum_k^{(k \neq h)} \nabla_h \phi_{hk}(\mathbf{r}_h, \mathbf{r}_k) = \nabla_h \phi_{hi}(\mathbf{r}_h, \mathbf{r}_i) + \sum_k^{(k \neq h, i)} \nabla_h \phi_{hi}(\mathbf{r}_h, \mathbf{r}_k)$$

Grouping together terms in the sum by the species of particle k, we get

$$kT\nabla_h n_{mn}(\mathbf{r}_h \mathbf{r}_i) = -n_{mn}(\mathbf{r}_h \mathbf{r}_i) \nabla_h \phi_{mn}(\mathbf{r}_h \mathbf{r}_i)$$

$$- \int e^{-\Phi/kT} \sum N_m N_n N_p \nabla_h \phi_{mp}(\mathbf{r}_h, \mathbf{r}_k) \, \mathbf{dq} (\mathbf{dr}_h \, \mathbf{dr}_i)^{-1} (Q_N)^{-1}$$

The factor $N_m N_n N_p$ must be appropriately modified if two or three of the species are identical. One can now introduce the three-particle distribution, defined analogously to n_m and n_{mn} (equations [6–7]), and write

$$-kT\nabla_h n_{mn}(\mathbf{r}_h \mathbf{r}_i) = n_{mn}(\mathbf{r}_h \mathbf{r}_i) \nabla_h \phi_{mn}(\mathbf{r}_h \mathbf{r}_i)$$

$$+ \sum_p \int n_{mnp}(\mathbf{r}_h \mathbf{r}_i \mathbf{r}_k) \nabla_h \phi_{mp}(\mathbf{r}_h \mathbf{r}_k) \, \mathbf{dr}_k$$

$$\equiv n_{mn}(\mathbf{r}_h \mathbf{r}_i) \nabla_h U_{mn}(\mathbf{r}_h \mathbf{r}_i) \quad [11]$$

The last equality defines a potential of mean force for the two-particle interaction, U_{mn}. The two parts of U_{mn} are the direct force exerted by one particle on the other, which is the gradient of the interaction potential ϕ_{mn}, and the indirect force, exerted by one particle on the other via its influence on the positions of other, third, particles (correlations).

Formally, one has, on dividing both sides of equation [11] by $n_{mn}(\mathbf{r}_h \mathbf{r}_i)$ and integrating,

$$n_{mn}(\mathbf{r}_h \mathbf{r}_i) = n_{mn}(\mathbf{r}_{h0} \mathbf{r}_i) \exp\left\{\frac{-[U_{mn}(\mathbf{r}_h \mathbf{r}_i) - U_{mn}(\mathbf{r}_{h0} \mathbf{r}_i)]}{kT}\right\}$$

where $\mathbf{r}_{h0}$ is a reference value of $\mathbf{r}_h$. For a gas at low concentrations, the correlation term in U_{mn} may be neglected and U_{mn} is just ϕ_{mn}. One can take $\mathbf{r}_{h0}$ infinitely far from $\mathbf{r}_i$ and, since only interparticle distances, not directions, matter, we have

$$n_{mn}(|\mathbf{r}_h - \mathbf{r}_i|) = n_{mn}(\infty) \exp\left[\frac{-U_{mn}(|\mathbf{r}_h - \mathbf{r}_i|)}{kT}\right]$$

as U_{mn} vanishes at infinite interparticle distance. The two-particle distribution n_{mn} becomes the product of one-particle distributions (particle densities) at infinite distance, so that

$$\frac{n_{mn}(|\mathbf{r}_h - \mathbf{r}_i|)}{n_m n_n} = \exp\left[\frac{-\phi_{mn}(r_{hi})}{kT}\right]$$

The ratio of n_{mn} to $n_m n_n$ is often abbreviated as g_{mn}; the above equation shows that, for a sufficiently dilute gas, g_{mn} obeys a Boltzmann relationship with the interparticle potential ϕ_{mn}. If the correlations in equation [11] cannot be neglected, one can still, taking $\mathbf{r}_{h0}$ as infinity, write

$$\frac{n_{mn}(r_{hi})}{n_m n_n} = \exp\left[\frac{-U_{mn}(r_{hi})}{kT}\right] \qquad [12]$$

since U_{mn} vanishes at infinite distance. Equation [12] is exact: g_{mn} obeys a Boltzmann relationship with the potential of mean force. However, the potential of mean force may be a fairly complicated object.

Equation [8] may be interpreted in several ways. If it is divided by $n_i(\mathbf{r}_m)$ we obtain

$$\nabla_m kT \ln n_i(\mathbf{r}_m) = -\sum_k \int \left[\frac{n_{ik}(\mathbf{r}_m, \mathbf{r}_j)}{n_i(\mathbf{r}_m)}\right] \nabla_m u_{ik}(\mathbf{r}_m, \mathbf{r}_j) \, d\mathbf{r}_j \qquad [13]$$

which, if the right-hand side can be written as a gradient, say ∇Q, can be interpreted as the constancy of the electrochemical potential

$$\tilde{\mu}_i = \tilde{\mu}_i^0 + RT \ln n_i + RT \ln \gamma_i$$

where $\tilde{\mu}_i^0$ is the standard electrochemical potential, a constant, and the other two terms depend on position $\mathbf{r}_m$. Here, γ_i is an activity coefficient and n_i is a concentration. The term $RT \ln \gamma_i$ is, according to Equation [13], due to interparticle forces. Alternatively, one can sum equation [8] over species to obtain

$$\nabla\left[kT \sum_i n_i(\mathbf{r})\right] - \sum_{ik} \int n_{ik}(\mathbf{r}, \mathbf{r}') \nabla_m u_{ik}(\mathbf{r}, \mathbf{r}') \, d\mathbf{r}' = 0 \qquad [14]$$

and, if the second term can be written as a gradient, interpret this as $\nabla p = 0$, where the two terms are the kinetic and potential (interparticle force) contributions to the pressure p.

B. PRESSURE AND SURFACE TENSION

All the thermodynamic functions can be calculated in terms of Z_N (the problem will be the calculation of Z_N). The Helmholtz free energy is

$$A = -kT \ln Z_N$$

$$= -kT \ln Q_N - kT \sum_s N_s \left[\ln \left(\frac{2\pi m_s kT}{h^2} \right)^{3/2} - \ln N_s + 1 \right]$$

with the sum being over species. (Stirling's approximation, $\ln N! \sim N \ln N - N$, has been used.) Other thermodynamic functions can be obtained by differentiating A, that is,

$$p = -\left(\frac{\partial A}{\partial V} \right)_{T,N} \qquad S = -\left(\frac{\partial A}{\partial T} \right)_{V,N}$$

$$\mu_s = N_A \left(\frac{\partial A}{\partial N_s} \right)_{T,V} \qquad U = A - T \left(\frac{\partial A}{\partial T} \right)_{V,N}$$

where N_A is Avogadro's number. Thus if Φ is independent of momenta the energy is

$$U = kT^2 \left[\left(\frac{\partial \ln Q_N}{\partial T} \right)_{V,N} + \sum_s \frac{3}{2} N_s T^{-1} \right]$$

and the pressure is $kT(\partial \ln Q_N / \partial V)$.

The differentiation of Q_N with respect to V, required for the pressure p, may be carried out (Davidson 1962, Section 20.4) by a transformation to scaled variables, that is, the x coordinate of particle i is written

$$x_i = V^{1/3} x_i'$$

for a cubical box of volume V and similarly for y and z coordinates. Then the V dependence becomes explicit, the limits on the scaled (primed) variables become independent of V, and the differentiation is straightforward. Transforming back to the original coordinates after differentiation gives

$$\left(\frac{\partial \ln Q_N}{\partial V} \right)_{T,N} = \frac{\sum_s N_s}{V} - \frac{(Q_N)^{-1}}{3VkT} \int \sum_{i>j} \left\{ x_i \frac{\partial \phi_{ij}}{\partial x_i} + y_i \frac{\partial \phi_{ij}}{\partial y_i} + z_i \frac{\partial \phi_{ij}}{\partial z_i} \right.$$

$$\left. + x_j \frac{\partial \phi_{ij}}{\partial x_j} + y_j \frac{\partial \phi_{ij}}{\partial y_j} + z_j \frac{\partial \phi_{ij}}{\partial z_j} \right\} \exp \left(\frac{-\Phi}{kT} \right) d\mathbf{q}$$

It has been assumed that $\Phi = \frac{1}{2} \sum_{i \neq j} \phi_{ij}(\mathbf{r}_i, \mathbf{r}_j)$. If each two-particle potential ϕ_{ij} depends only on $r_{ij} = |\mathbf{r}_i - \mathbf{r}_j|$, the terms in curly brackets in the above equation reduce to $r_{ij}(d\phi_{ij}/dr_{ij})$. On grouping together integrals that are identical because they involve equivalent particles, we obtain

B. PRESSURE AND SURFACE TENSION

$$pV = -kTV\left(\frac{\partial \ln Z_n}{\partial V}\right)_{T,N} = kT \sum_s N_s$$

$$- \sum_{m,n} \frac{N_m N_n}{6} \int r_{ij} \frac{du_{mn}}{dr_{ij}} \exp\left(\frac{-\Phi}{kT}\right) d\mathbf{q}/Q_N \quad [15]$$

where the sum is now over species. If $m = n$ the factor $N_m N_n$ is replaced by $N_m(N_m - 1)$.

We may introduce the two-particle distribution functions of equation [6] into equation [15]. For a homogeneous system of identical particles of a single species one has (Davidson 1962, Section 20.4)

$$p = kT\frac{N}{V} - \frac{N(N-1)}{6Q_N V} \int r_{12} \frac{du}{dr_{12}} e^{-\Phi/kT} d\mathbf{q} \quad [16]$$

On carrying out the integration over all coordinates except 1 and 2 we recover the two-particle distribution function of equation [6]. Equation [16] is then the familiar pressure equation

$$pV = NkT - \frac{1}{6} \int r_{12} \frac{du}{dr_{12}} n^{(2)}(\mathbf{r}_1, \mathbf{r}_2) \, d\mathbf{r}_1 \, d\mathbf{r}_2$$

$$= NkT - \frac{V}{6} \int r_{12} \frac{du}{dr_{12}} n^{(2)}(r_{12}) \, 4\pi(r_{12})^2 \, dr_{12}$$

(Hill 1956, Section 39). The two terms correspond to kinetic and potential energy. For a system of several species of particles, we obtain

$$pV = kT \sum_s N_s - \frac{1}{6} \sum_{m,n} \int r_{12} \frac{du_{mn}}{dr_{12}} n_{mn}(\mathbf{r}_1, \mathbf{r}_2) \, d\mathbf{r}_1 \, d\mathbf{r}_2 \quad [17]$$

For a homogeneous system, each n_{mn} depends on r_{12} only.

As discussed in Chapter 2, Section B, the free energy of a system that is inhomogeneous in the z direction depends on the volume *and* on the cross-sectional area $\mathcal{A}$ normal to the z axis. This implies that the pressure may not be simply $-(\partial A/\partial V)$: The direction in which one deforms the system must be specified, since extension perpendicular to the interface increases the volume of bulk phases whereas extension parallel to the interface increases the interfacial area. This leads to the notion of normal and tangential pressures.

Alternatively, one can define the pressure as

$$-\left(\frac{\partial A}{\partial V}\right)_{T,N,\mathcal{A}} = kT\left(\frac{\partial \ln Q_N}{\partial V}\right)_{T,N,\mathcal{A}}$$

3. STATISTICAL MECHANICS

To carry out this differentiation one can introduce scaled coordinates as follows:

$$x_i = \alpha^{1/2} x'_i, \qquad y_i = \alpha^{1/2} y'_i, \qquad z_i = V\alpha^{-1} z'_i$$

Then

$$Q_N = V^N \int \exp\left(\frac{-\Phi}{kT}\right) dx'_1 \, dy'_1, \ldots, dz'_N$$

$$\left(\frac{\partial Q_N}{\partial V}\right)_{T,N,\alpha} = \frac{NQ_N}{V} - (kT)^{-1} \int \exp\left(\frac{-\Phi}{kT}\right) \left(\frac{\partial \Phi}{\partial V}\right)_{T,N,\alpha} dx'_1, \ldots, dz'_N$$

If $\Phi = \Sigma_{i<j} u_{ij}(\mathbf{r}_i, \mathbf{r}_j)$, we have

$$\left(\frac{\partial \Phi}{\partial V}\right)_{T,N,\alpha} = \sum_{i<j} \left(\frac{\partial u_{ij}}{\partial z_i} \alpha^{-1} z'_i + \frac{\partial u_{ij}}{\partial z_j} \alpha^{-1} z'_j\right)$$

$$= \sum_{i<j} \left(z_i \frac{\partial u_{ij}}{\partial z_i} + z_j \frac{\partial u_{ij}}{\partial z_j}\right) V^{-1}$$

Thus the pressure is

$$kT\left(\frac{\partial \ln Q_N}{\partial V}\right)_{T,N,\alpha} = \frac{NkT}{V} - V^{-1} \sum \int \exp\left(\frac{-\Phi}{kT}\right) z_{ij} \left(\frac{\partial u_{ij}}{\partial z_{ij}}\right) dq \qquad [18]$$

where $z_{ij} = z_j - z_i$. For an isotropic system, equation [18] is the same as equation [17], since one can substitute x or y for z, add together the three expressions, which are equivalent, and divide by 3. For an anisotropic system the x, y, and z terms are different and equation [18] should be identified with the pressure p_n normal to the interface.

We can use the same scaling to calculate the surface tension according to

$$\gamma = \left(\frac{\partial A}{\partial \alpha}\right)_{V,T,N} = -kT\left(\frac{\partial \ln Q_N}{\partial \alpha}\right)_{V,T,\alpha}$$

Then

$$\left(\frac{\partial Q_n}{\partial \alpha}\right)_{V,T,N} = -(kT)^{-1} \int e^{-\Phi/kT} \left(\frac{\partial \Phi}{\partial \alpha}\right)_{V,T,N} dx'_1, \ldots, dz'_N$$

where

$$\left(\frac{\partial \Phi}{\partial \alpha}\right)_{V,T,N} = \sum_{i<j} \left[\frac{\partial u_{ij}}{\partial x_i} \frac{x'_i}{2\alpha^{1/2}} + \frac{\partial u_{ij}}{\partial y_i} \frac{y'_i}{2\alpha^{1/2}} - \frac{\partial u_{ij}}{\partial z_i} \frac{Vz'_i}{\alpha^2} \right.$$
$$\left. + \frac{\partial u_{ij}}{\partial x_j} \frac{x'_j}{2\alpha^{1/2}} + \frac{\partial u_{ij}}{\partial y_j} \frac{y'_j}{2\alpha^{1/2}} - \frac{\partial u_{ij}}{\partial z_j} \frac{Vz'_j}{\alpha^2} \right]$$

Therefore the surface tension is

$$\gamma = (Q_N)^{-1} \int d\mathbf{q} \, \exp\left(\frac{-\Phi}{kT}\right) \sum_{i<j} \left[\frac{\partial u_{ij}}{\partial x_i} \frac{x_i}{2\alpha} + \frac{\partial u_{ij}}{\partial y_i} \frac{y_i}{2\alpha} - \frac{\partial u_{ij}}{\partial z_i} \frac{z_i}{\alpha} \right.$$
$$\left. + \frac{\partial u_{ij}}{\partial x_j} \frac{x_j}{2\alpha} + \frac{\partial u_{ij}}{\partial y_j} \frac{y_j}{2\alpha} - \frac{\partial u_{ij}}{\partial z_j} \frac{z_j}{\alpha} \right]$$

$$= (Q_N\alpha)^{-1} \int \exp\left(\frac{-\Phi}{kT}\right) \sum_{i<j} \left[\left(\frac{\partial u_{ij}}{\partial x_{ij}}\right) x_{ij} - \frac{\partial u_{ij}}{\partial z_{ij}} z_{ij} \right] d\mathbf{q}$$

where the equivalence of the x and y directions has been used. A comparison with equation [18] suggests that the expression in square brackets be identified with $p_n - p_t$, the difference between normal and tangential pressures. In Section C we confirm this.

We are concerned with systems in which electric fields are present and in which the interaction potentials u_{ij} may include electrostatic parts. It is convenient to separate out the electrostatic interactions, which are long-range, from the short-range interactions, which normally constitute the pressure (along with the kinetic contribution). To do this here we write $u_{ik} = u^e_{ik} + u^s_{ik}$ (electrostatic + short-range potentials) and

$$n_{ik}(\mathbf{r}_m, \mathbf{r}_j) = n_i(\mathbf{r}_m) n_k(\mathbf{r}_j) \left[1 + h_{ik}(\mathbf{r}_m, \mathbf{r}_j)\right]$$

where h_{ik} expresses the correlation between particles and is short range. The long-range part of the integrand in equation [14] is then $n_i(\mathbf{r}_m) n_k(\mathbf{r}_j) \nabla_m u^e_{ik}(\mathbf{r}_m, \mathbf{r}_j)$.

For example, suppose we have charged particles with no short-range forces, that is, $u^e_{ik}(\mathbf{r}_m, \mathbf{r}_j) = q_i q_k / r_{mj}$ and $u^s_{ik} = 0$, with no interparticle correlations, that is, $h_{ik} = 0$ or $n_{ik}(\mathbf{r}, \mathbf{r}') = n_i(\mathbf{r}) n_k(\mathbf{r}')$. Then equation [13] becomes

$$\nabla_m [kT \ln n_i(\mathbf{r}_m)] = -q_i \sum_k \int n_k(\mathbf{r}_j) q_k \nabla_m \frac{1}{r_{mj}} d\mathbf{r}_j$$
$$= -q_i \nabla_m \int \rho(\mathbf{r}_j) \frac{1}{r_{mj}} d\mathbf{r}_j$$

The last integral is the electrical potential $\phi(\mathbf{r}_m)$; thus we have

$$\nabla_m [\mu_i^0 + kT \ln n_i(\mathbf{r}_m) + q_i \phi(\mathbf{r}_m)] = 0$$

and the electrochemical potential appears as a sum of chemical and electrical ($q_i \phi$) parts. With the same assumptions, equation [14] becomes

$$\nabla \left[kT \sum n_i(\mathbf{r}) \right] = -\rho(\mathbf{r}) \int \rho(\mathbf{r}') \nabla(|\mathbf{r} - \mathbf{r}'|^{-1}) \, d\mathbf{r}' = -\rho(\mathbf{r}) \mathbf{E}(\mathbf{r})$$

where $\mathbf{E}$ is the electric field. If $\mathbf{E}$ is in the z direction, this is

$$\frac{d\left[kT \sum n_i(\mathbf{r}) - E^2/8\pi \right]}{dz} = 0$$

since $dE/dz = 4\pi\rho$; the electric field apparently adds a term $E^2/8\pi$ to the pressure.

If there are polarizable species as well as charged species present, this term becomes $(2\epsilon - 1) E^2/8\pi$, where ϵ is the dielectric constant, corresponding to the Maxwell stress (Becker and Sauter 1964, Section 36). For an interface perpendicular to the z direction, the expression for the surface tension, which includes a term $p_n - p_t$ or $p_z - p_x$, should then contain a term $(2\epsilon - 1) E^2/8\pi$, as discussed in Section C.

C. MECHANICAL EQUILIBRIUM

In the bulk-phase regions far from the interface, the pressure is isotropic and the same from point to point. Within the interface, the normal pressure p_n is independent of position, but the tangential pressure p_t varies. This follows from considerations of mechanical equilibrium. For the rectangular volume oriented along the x, y, and z axes, the condition of mechanical equilibrium is that the pressure–force on one face be equal and opposite to the pressure–force on the face opposite. This means that p_n must be independent of z:

$$\frac{dp_n}{dz} = 0$$

In the Gibbs thermodynamic description (Chapter 2, Section A) the interface has no thickness; its properties are defined by default. The model of the system as two homogeneous phases separated by a mathematical surface may be extended to the forces by imagining that a tension γ (the surface tension) acts in this surface, which will now be called the surface of tension. The tension makes the model

C. MECHANICAL EQUILIBRIUM

system mechanically equivalent to the real system, containing an interfacial region. The position of the surface of tension within the interface can be defined for a spherical interface (Sanfeld 1968), but not for a planar interface such as the one we deal with here.

Returning to the geometry of Chapter 2, Section A we have $p_z = p_n = p$, the pressure of the system, independent of z, whereas $p_x = p_y = p_t(z)$, with p_t becoming identical to p far from the interface. Now suppose we change the area $\mathcal{A}$ by $d\mathcal{A}$, keeping the volume $V = \mathcal{A}L$ constant: Retaining only first-order terms, this means

$$\mathcal{A} \to \mathcal{A} + d\mathcal{A}, \qquad L \to L - \frac{L}{\mathcal{A}} d\mathcal{A} \qquad [19]$$

Let the dimensions of the system in the x and y directions be X and Y; they increase by dX and dY such that $X\,dY + Y\,dX = d\mathcal{A}$ (see Figure 7). For the real system, the total force in the z direction is $\mathcal{A}p$ and does work $(\mathcal{A}p)(L/\mathcal{A})\,d\mathcal{A}$ on the system during the deformation; the forces in the x and y directions do work $dX[\int_{-b}^{a} p_t(z)\,Y\,dz]$ and $dY[\int_{-b}^{a} p_t(z)\,X\,dz]$, respectively. For the system containing the surface of tension, the work done on the system by forces in the z direction is again

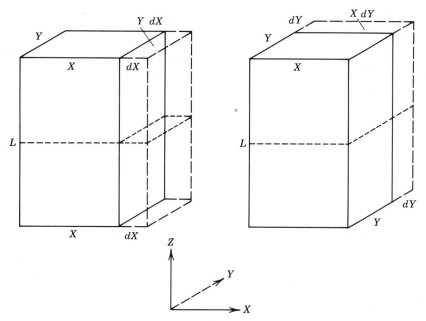

FIGURE 7. Lateral deformation of reference system containing two homogeneous phases joined at a surface of tension (dotted). The increase in area is $Y\,dX + X\,dY$ for simultaneous changes of X to $X + dX$ (left-hand diagram) and Y to $Y + dY$ (right-hand diagram).

$\alpha p(L/\alpha)\, d\alpha$, but $pLY\, dX - \gamma Y\, dX$ from forces in the x direction and $pLX\, dY - \gamma X\, dY$ from forces in the y direction. Equating the expressions for the two systems, we have

$$(Y\, dX + X\, dY) \int_{-b}^{a} p_t(z)\, dz = (pL - \gamma)(Y\, dX + X\, dY)$$

so that the surface tension must be

$$\gamma = \int_{-b}^{a} [p - p_t(z)]\, dz = \int_{-b}^{a} (p_n - p_t)\, dz \qquad [20]$$

The range of integration can be made $-\infty$ to ∞, since $p_t \to p_n = p$ far from $z = 0$.

In the presence of electrical interactions, one separates out the long-range (electrostatic) forces explicitly and assigns the remaining forces to the pressure, although there is some ambiguity in the way this is done (Sanfeld 1968; Hurwitz 1978b, Chapter II, Section 2). The electrostatic force on an element of volume $d\tau$ is given by (Sanfeld 1968; Becker and Sauter 1964, Sections 31–35)

$$\mathbf{F} = [\rho \mathbf{E} + (\mathbf{P} \cdot \nabla) \mathbf{E}]\, d\tau$$

where ρ is the charge density, $\mathbf{E}$ is the electric field, and $\mathbf{P}$ is the polarization (dipole moment per unit volume). The field $\mathbf{E}$ is the field resulting from charges outside $d\tau$; the net force resulting from Coulombic interactions between the charges inside $d\tau$ is zero. For the polarization a term $-2\pi \mathbf{P}^2$ must be added to the pressure (Hurwitz 1978b, Chapter II, Section 2) to represent the force on the dipoles inside $d\tau$ of the field of these same dipoles. (This is only one manifestation of the necessity, in the statistical mechanical interpretation of terms such as those in $\mathbf{F}$ which are quadratic in electrical charges and dipoles, for using the statistical average of a product and not the product of averages. Two-particle distributions n_{ij} are involved, just as in the pressure given in Section B.) Alternatively, a careful treatment of the force is necessary (Becker and Sauter 1964; Landau and Lifschitz 1960).

A force per unit volume can always be written as the derivative of a stress tensor σ such that $\Sigma_k\, \sigma_{ik}\, ds_k$ is the ith Cartesian component of the force on a surface area $\mathbf{ds}$; the components σ_{ii} are to be identified with pressure. The force density $\mathbf{F}$ is related to σ by

$$F_i = \sum_k \frac{\partial \sigma_{ik}}{\partial x_k}$$

Considering the work done in displacing unit area of surface within a medium, Landau and Lifschitz (1960, Section 15) show

C. MECHANICAL EQUILIBRIUM

$$\sigma_{ik} = \delta_{ik}\left[\Phi - d_m\left(\frac{\partial \Phi}{\partial d_m}\right)_{E,T}\right] + \frac{E_i D_k}{4\pi}$$

where **D** is the electric displacement

$$\mathbf{D} = \mathbf{E} + 4\pi \mathbf{P}$$

d_m is the mass density, $\delta_{ik} = 0$ for $i \neq k$ and 1 for $i = k$, and

$$\Phi = A_0 - \int_0^E \frac{\mathbf{D} \cdot d\mathbf{E}}{4\pi}$$

where A_0 is the free energy that would exist in the absence of electric field. One then finds

$$4\pi\sigma_{ik} = \delta_{ik}\left[-4\pi p^0 - \int_0^E \mathbf{D} \cdot d\mathbf{E} + d_m \int_0^E \left(\frac{\partial \mathbf{D}}{\partial d_m}\right) d\mathbf{E}\right] + E_i D_k \quad [21]$$

where p^0 is the pressure that would exist in the absence of electric field, that is, the kinetic contribution and that resulting from short-range forces. We will have to consider that p^0 also depends on direction.

For a planar interface, **E** and **P** are in the direction of inhomogeneity, which we take as the z direction. The nonvanishing components of the stress tensor are σ_{xx}, σ_{yy}, and σ_{zz}, with

$$\sigma_{zz} = -p_n - \frac{E^2}{8\pi} - \int_0^E P\, dE + d_m \int_0^E \frac{\partial P}{\partial d_m} dE + \frac{ED}{4\pi}$$

where p_n is the normal (z direction) component of p^0. Thus the total force in the z direction on $d\tau$ is

$$\left[-\frac{dp_n}{dz} - \frac{E(dE/dz)}{4\pi} - \frac{dQ}{dz} + E\rho + \frac{dE}{dz}\frac{D}{4\pi}\right] d\tau$$

$$= \left[\rho E + P\frac{dE}{dz} - \frac{dQ}{dz} - \frac{dp_n}{dz}\right] d\tau \quad [22]$$

where

$$Q = \int_0^E \mathbf{P} \cdot d\mathbf{E} - \int_0^E d_m \left(\frac{\partial \mathbf{P}}{\partial d_m}\right)_{E,T} \cdot d\mathbf{E}$$

The Poisson equation

$$\frac{dD}{dz} = 4\pi\rho$$

has been used. The mechanical equilibrium condition is no longer that p_n be independent of z, but that the term in square brackets in equation [22] vanish. If $D = \epsilon E$ and $\epsilon - 1$ is proportional to d_m, then Q vanishes and the equilibrium condition is

$$0 = (4\pi)^{-1}\left(\frac{dD}{dz}E + (D - E)\frac{dE}{dz}\right) - \frac{dp_n}{dz}$$

$$= \frac{d[(DE/4\pi) - (E^2/8\pi) - p_n]}{dz} \quad [23]$$

This means that at equilibrium $p_n - (DE/4\pi) + (E^2/8\pi)$ must be independent of z. In the homogeneous phase where D and E necessarily vanish, $p_n = p$ so that the condition may be written

$$p_n(z) - \frac{D(z)\,E(z)}{4\pi} + \frac{E(z)^2}{8\pi} = p$$

The term $DE/4\pi$, expressed as a pressure, can be appreciable: A surface charge density of 10 μC/cm^2 or 3×10^4 esu/cm^2 produces a field of 3×10^4 statvolt/cm; if $\epsilon = 10$ so $D = 3 \times 10^5$ statvolt/cm, $DE/4\pi$ becomes 7.2×10^8 esu^2/cm^4 = 7.2×10^8 dyne/cm^2. Compare a pressure of 1 atm, about 10^6 dyne/cm^2.

The mechanical equilibrium condition (equation [23]) is verified for a simple system: charged and polarizable particles with no short-range interactions or correlations. The pressure p_n is $kT \sum n_i$. The density of polarizable particles (species 0) follows the Boltzmann law

$$n_0 = n_{00}\exp\left[\frac{\int_0^E \mu\,dE}{kT}\right] \quad [24]$$

where n_{00} is the density at infinity where the field E vanishes and $-\int_0^E \mu\,dE$ is the energy of the (induced) dipole in a field (μ depends on E). The density of charged species i ($i > 0$) follows the Boltzmann law

$$n_i = n_{i0}\exp\left[-q_i(\phi - \phi_0)/kT\right] \quad [25]$$

where n_{i0} is the density at infinity (homogeneous region) where $\phi = \phi_0$, and q_i is the charge on a particle. Thus

C. MECHANICAL EQUILIBRIUM

$$\frac{dp_n}{dz} = kT \sum \frac{dn_i}{dz} = n_0 \left[\frac{d\left(\int_0^E \mu \, dE \right)}{dz} \right] + kT \sum n_i \frac{-q_i}{kT} \frac{d\phi}{dz}$$

Since $\sum n_i q_i = \rho$ and $n_0 [d(\int_0^E \mu \, dE)/dz] = n_0 \mu (dE/dz) = P(dE/dz)$, we have

$$\frac{dp_n}{dz} = -\rho \frac{d\phi}{dz} + P \frac{dE}{dz}$$

$$= -(4\pi)^{-1} \frac{dD}{dz}(-E) + (4\pi)^{-1}(D-E)\frac{dE}{dz}$$

$$= (4\pi)^{-1} \frac{d(DE)}{dz} - (8\pi)^{-1} \frac{d(E^2)}{dz}$$

so equation [23] is verified.

For the representation of a charged interface by a surface of tension, we again consider the work involved in deforming the system in the x direction by dX (see Figure 7). Using σ_{xx} for the force per unit area, we have

$$W_X = -Y \, dX \int_{-b}^{a} \left[p_t(z) + (8\pi)^{-1} E(z)^2 + Q \right] dz$$

where p_t involves kinetic and short-range contributions and E, P, and D are in the z direction. W_X must be equal to the work that must be done on the model system of isotropic homogeneous phases separated by a surface:

$$W_X = -Y \, dX \int_{-b}^{a} p \, dz + \gamma \, Y \, dX$$

Thus

$$\gamma = \int_{-b}^{a} \left[p - p_t(z) - (8\pi)^{-1} E(z)^2 - Q \right] dz$$

and the region of integration can be made $-\infty$ to ∞. If $Q = 0$ ($D = \epsilon E$, $\epsilon - 1$ proportional to d_m), we have

$$\gamma = \int_{-\infty}^{\infty} \left[p - p_t(z) - \frac{E^2}{8\pi} \right] dz \qquad [26]$$

Note that, according to equations [23] and [26], the electric field effects may be taken into account by replacing p_n by $p_n - (DE/4\pi) + (E^2/8\pi)$ and by replacing p_t by $p_t + (E^2/8\pi)$ (Badiali and Goodisman 1975).

108   3. STATISTICAL MECHANICS

Suppose the electrical conditions of the interface are changed, but p is maintained constant. Returning to the simple model with no correlations or short-range forces (equations [24] and [25]) we have for the change in tangential pressure

$$\Delta p_t = kT \sum \Delta n_i = -\sum n_i q_i \Delta(\phi - \phi_0) + n_0 \mu \Delta E \qquad [27]$$

where μ depends on E. If, as is usually the case, the interface consists of two phases in contact, there will be two homogeneous regions and the reference potential ϕ_0 will be different for different species.

Assuming each charged species can be assigned either to phase 1 ($z < 0$) with reference potential ϕ^1 or to phase 2 ($z > 0$) with reference potential ϕ^2 (this division of species corresponds to an ideally polarizable interface), we have from equations [26] and [27]

$$\Delta\gamma = \int_{-\infty}^{\infty} \left[\overset{1}{\sum} n_i q_i \Delta(\phi - \phi^1) + \overset{2}{\sum} n_i q_i \Delta(\phi - \phi^2) \right] dz$$

$$- \int_{-\infty}^{\infty} P\Delta E \, dz - \int_{-\infty}^{\infty} (8\pi)^{-1} \Delta(E^2) \, dz.$$

The sum $\sum^1 n_i q_i$, involving charged particles assigned to phase 1, gives ρ^1, the charge density of the phase-1 side of the interface. Since $\sum n_i q_i = \rho^1 + \rho^2 = \rho = (4\pi)^{-1}(dD/dz)$ and $4\pi P = D - E$, the change in surface tension is

$$\Delta\gamma = \int_{-\infty}^{\infty} \rho\Delta\phi \, dz - \left[\int_{-\infty}^{\infty} \rho^1 \, dz \right] \Delta\phi^1 - \left[\int_{-\infty}^{\infty} \rho^2 \, dz \right] \Delta\phi^2$$

$$- (4\pi)^{-1} \int_{-\infty}^{\infty} (D - E)\Delta E \, dz - (4\pi)^{-1} \int_{-\infty}^{\infty} E\Delta E \, dz$$

$$= (4\pi)^{-1} \int_{-\infty}^{\infty} D\Delta E \, dz - \left[\int_{-\infty}^{\infty} \rho^1 \, dz \right] \Delta\phi^1$$

$$- \left[\int_{-\infty}^{\infty} \rho^2 \, dz \right] \Delta\phi^2 - (4\pi)^{-1} \int_{-\infty}^{\infty} D\Delta E \, dz$$

Now $\int_{-\infty}^{\infty} \rho^1 \, dz = q^1$, the surface charge on the phase-1 side of the interface, and $\int_{-\infty}^{\infty} \rho^2 \, dz = q^2 = -q^1$, so that $\Delta\gamma = -q^1(\Delta\phi^1 - \Delta\phi^2)$. This is just the Lippman equation, which has here been given a derivation in terms of molecular and electrostatic concepts using statistical mechanics, albeit for a very simple model.

D. DIELECTRIC CONSTANTS

The relationship between D and E in a particular medium should be explicable in terms of the properties of the molecules. In the simplest case, isolated molecules

D. DIELECTRIC CONSTANTS

with polarizability α, an electric field **E** distorts the electronic cloud of a molecule and induces a dipole moment $\alpha \mathbf{E}$ in each. If there are n nonpolar molecules per unit volume, the polarization **P** is $n\alpha \mathbf{E}$, the electric displacement is $\mathbf{E} + 4\pi n\alpha \mathbf{E}$, and the dielectric constant is

$$\epsilon = \frac{D}{E} = 1 + 4\pi n\alpha$$

For a polar molecule free to rotate, there is no polarization in the absence of an external field since the orientations of the molecules are distributed randomly, and, in the presence of a field, the average dipole moment is in the field direction. If the field is not too large, the magnitude of this moment is $\mu^2/3kT$ times the field, as shown in the next paragraph. Thus the rotating dipolar molecule behaves like a nonpolar molecule with polarizability $\mu^2/3kT + \alpha_e$, where α_e is the electronic polarizability. In a fluid where the molecules are close together, interactions between the dipole moments of nearby molecules complicate the picture. The field that is effective in polarizing a molecule is not the applied field only, but includes the field of neighboring molecules, which are polarized.

Before turning to this problem, let us consider a molecule with permanent dipole moment μ acted on by an electric field **E**. The interaction energy is $-\mu E \cos \theta$, where θ is the angle between the dipole moment and the field. Thus $\bar{\mu}$, the average of the component of the dipole moment in the field direction $\mu \cos \theta$, is

$$\frac{\int \mu \cos \theta \exp\left[-\frac{-\mu E \cos \theta}{kT}\right] \sin \theta \, d\theta \, d\phi}{\int \exp\left[-\frac{-\mu E \cos \theta}{kT}\right] \sin \theta \, d\theta \, d\phi} = \frac{\mu \int_{-1}^{1} x e^{bx} \, dx}{\int_{-1}^{1} e^{bx} \, dx}$$

$$= \mu \left[(b-1) e^b - (-b-1) e^{-b}\right] b^{-1} \left[e^b - e^{-b}\right]^{-1} \qquad [28]$$

where $b = \mu E/kT$ and $x = \cos \theta$. Defining the Langevin function $\mathcal{L}(x)$ as

$$\mathcal{L}(x) = \coth x - x^{-1}$$

we have $\bar{\mu} = \mu \mathcal{L}(\mu E/kT)$. For small values of x, $\mathcal{L}(x)$ becomes

$$\frac{(1 + x + x^2/2 \ldots) + (1 - x + x^2/2 \ldots)}{(1 + x + \tfrac{1}{2}x^2 + \tfrac{1}{6}x^3) - (1 - x + \tfrac{1}{2}x^2 - \tfrac{1}{6}x^3)} - \frac{1}{x}$$

$$= \frac{1 + x^2/2}{x + x^3/6} - \frac{1}{x} = x^{-1} \frac{x^2/2 - x^2/6}{1 + x^2/6} = \frac{x}{3}$$

Thus the orientation of a permanent dipole moment corresponds to a polarizability (induced moment divided by electric field) of $\mu(\mu E/3kT)/E = \mu^2/3kT$.

For a water molecule α_e is several times 10^{-24} cm^3 and $\mu = 1.83$ D $= 1.83 \times 10^{-18}$ esu · cm. At $T = 300$ K, $\mu^2/3kT = 2.70 \times 10^{-23}$ cm^3, so it is much more important than the electronic polarizability. With a molecular density n of 3.34×10^{22} cm^{-3}, $1 + 4\pi n(\mu^2/3kT) = 12.3$, much lower than the actual dielectric constant of about 80. Although it is important to correct for the difference between average electric field and polarizing electric field, this result already indicates that the orienting unit is something bigger than an individual molecule. If it is m molecules, the value of n to use would be 3.34×10^{22} cm$^{-3}/m$ and the value of μ would be something less than $m(1.83$ Debye$)$, since the dipole moments on neighboring molecules are not parallel. A value of about 7 is thus estimated for m.

It is first necessary to elucidate the difference between the average field and the field polarizing a molecule. One way of doing this is to consider a sphere of radius r within a uniformly polarized dielectric and calculate the electric field at the center (Eyring et al. 1982, Chapter 7.3). This field includes the applied field $\mathbf{E}$, the field of the polarized molecules outside the sphere, and the field of the polarized molecules within it. The last field is zero for a symmetrical arrangement of molecules. If the radius r is large enough, the molecules outside can be represented by a continuous dielectric with polarization $\mathbf{P}$, which (Chapter 2, Section H) corresponds to a volume charge density $\nabla \cdot \mathbf{P}$ and a surface charge density P_n on the sphere. The result is that the field which polarizes a molecule is $\mathbf{E} + 4\pi\mathbf{P}/3$, so that the dipole moment on a molecule is $(\alpha_e + \mu^2/3kT)(\mathbf{E} + 4\pi\mathbf{P}/3)$. One then equates n times this dipole moment to $\mathbf{P}$ and solves for $\mathbf{P}$ and for the dielectric constant. The result (Clausius–Mossotti theory) is: $\epsilon = (1 + 8\pi n\alpha/3)/(1 - 4\pi n\alpha/3)$, where $\alpha = \alpha_e + \mu^2/3kT$. It appears that ϵ will be negative at temperatures low enough for the denominator to be negative and will be infinite when $\alpha = 3/4\pi n$. Neglecting α_e, we see this occurs at $T = 4\pi\mu^2 n/9k$, which, for water, is 1132 K. The theory is thus insufficient for polar liquids: The problem is that the derivation of the Lorentz field, $\mathbf{E} + 4\pi\mathbf{P}/3$, assumes the moments of all the molecules to be parallel and ignores instantaneous correlations between neighboring molecules; in the language of Section A, it equates $n_{ij}(\mathbf{r}_1, \mathbf{r}_2)$ to $n_i(\mathbf{r}_1) n_j(\mathbf{r}_2)$.

An approximate theory developed by Onsager to treat the dielectric constant of polar liquids is an improvement on Clausius–Mossotti theory. A molecule is supposed to occupy a spherical cavity of radius a in a dielectric; the molecules outside the sphere are treated as a continuum with dielectric constant ϵ. If there is an electric field $\mathbf{E}$ in the dielectric, the electrostatic potential is

$$\phi_{\text{in}} = \frac{-3\epsilon \mathbf{E} \cdot \mathbf{r}}{2\epsilon + 1}$$

inside the cavity and

$$\phi_{\text{out}} = -(\mathbf{E} \cdot \mathbf{r})\left[1 - \frac{(a^3/r^3)(1-\epsilon)}{2\epsilon + 1}\right]$$

D. DIELECTRIC CONSTANTS

outside. This potential satisfies the Poisson equation as well as the requirements of continuity of ϕ at $r = a$, continuity of the normal component of $\mathbf{D}$ at $r = a$ $(\partial\phi_{in}/\partial r = \epsilon(\partial\phi_{out}/\partial r)$ at $r = a)$, and $\phi \to -\mathbf{E} \cdot \mathbf{r}$ for $r \to \infty$. On the other hand, if there is a dipole moment $\mathbf{m}$ at the center of a cavity, the electric field is

$$\phi'_{in} = \frac{\mathbf{m} \cdot \mathbf{r}}{r^3} - g\mathbf{m} \cdot \mathbf{r}$$

inside and

$$\phi'_{out} = \frac{(1 - a^3 g) \mathbf{m} \cdot \mathbf{r}}{r^3}$$

outside, where $g = 2a^{-3}(\epsilon - 1)/(2\epsilon + 1)$. The potential in the cavity when there is both a moment $\mathbf{m}$ at the center and a polarized dielectric is the sum of ϕ_{in} and ϕ'_{in}, so that the electric field in the cavity is

$$-\nabla(\phi_{in} + \phi'_{in}) = \nabla\left[\frac{3\epsilon(\mathbf{E} \cdot \mathbf{r})}{2\epsilon + 1} - \mathbf{m} \cdot \mathbf{r}(r^{-3} - g)\right]$$

$$= \frac{3\epsilon \mathbf{E}}{2\epsilon + 1} + g\mathbf{m} - \nabla \frac{\mathbf{m} \cdot \mathbf{r}}{r^3} \quad [29]$$

The last term is just the field of the central dipole itself and should not be included in the local field $\mathbf{E}_{local}$ which acts to polarize this molecule.

The dipole moment $\mathbf{m}$ is the sum of the induced moment $\alpha\mathbf{E}_{local}$ and the permanent moment $\boldsymbol{\mu}$, whose value depends on $\mathbf{E}_{local}$:

$$\mathbf{m} = \boldsymbol{\mu} + \alpha\mathbf{E}_{local} = \boldsymbol{\mu} + \alpha\left(\frac{3\epsilon\mathbf{E}}{2\epsilon + 1} + g\mathbf{m}\right) \quad [30]$$

One can solve equation [30] for $\mathbf{m}$ to get

$$\mathbf{m} = (1 - \alpha g)^{-1}\left(\boldsymbol{\mu} + \frac{3\epsilon\alpha\mathbf{E}}{2\epsilon + 1}\right) \quad [31]$$

and then, according to equation [30],

$$\mathbf{E}_{local} = \frac{3\epsilon\mathbf{E}}{2\epsilon + 1} + g(1 - \alpha g)^{-1}\left(\boldsymbol{\mu} + \frac{3\epsilon\alpha\mathbf{E}}{2\epsilon + 1}\right)$$

Now equation [31] gives the average value of $\mathbf{m}$ in the direction of $\mathbf{E}$ as

$$\overline{m} = (1 - \alpha g)^{-1}\left(\overline{\mu} + \frac{3\epsilon\alpha E}{2\epsilon + 1}\right)$$

where

$$\bar{\mu} = \frac{\int \mu \cos\theta \exp(-\boldsymbol{\mu} \cdot \mathbf{E}_{local}/kT) \sin\theta \, d\theta}{\int \exp(-\boldsymbol{\mu} \cdot \mathbf{E}_{local}/kT) \sin\theta \, d\theta} \qquad [32]$$

where θ is the angle between $\boldsymbol{\mu}$ and $\mathbf{E}$. In the exponential,

$$\boldsymbol{\mu} \cdot \mathbf{E}_{local} = \frac{3\epsilon\mu E \cos\theta}{2\epsilon + 1} + g(1 - \alpha g)^{-1} \frac{3\epsilon\alpha\mu E \cos\theta}{2\epsilon + 1} + \mu^2 g(1 - \alpha g)^{-1}$$

and the last term yields the same θ-independent factor in numerator and denominator of equation [32], which cancels. This leaves

$$\bar{\mu} = \frac{\mu \int_0^\pi \cos\theta \exp(-\mu F \cos\theta/kT) \sin\theta \, d\theta}{\int_0^\pi \exp(-\mu F \cos\theta/kT) \sin\theta \, d\theta} = \mu \mathcal{L}\left(\frac{\mu F}{kT}\right) \qquad [33]$$

where $\mathcal{L}$ is the Langevin function and

$$F = [3\epsilon + g(1 - \alpha g)^{-1} 3\epsilon\alpha] (2\epsilon + 1)^{-1} E$$
$$= 3\epsilon(1 - \alpha g)^{-1} (2\epsilon + 1)^{-1} E \qquad [34]$$

Finally, one can calculate $\bar{m}$. Then the polarization or dipole moment per unit volume is in the direction of E with magnitude

$$P = \left(\frac{4\pi a^3}{3}\right)^{-1} (1 - \alpha g)^{-1} \left[\mu \mathcal{L}\left(\frac{\mu F}{kT}\right) + \frac{3\epsilon\alpha E}{2\epsilon + 1}\right]$$

since $4\pi a^3/3$, the volume assigned to a molecule, is the reciprocal of the number of molecules per unit volume.

The polarization will be proportional to the average field if $\mu F/kT$ is small enough to replace $\mathcal{L}(\mu F/kT)$ by $\mu F/3kT$. Assuming this is the case, we have for the polarization

$$P = \frac{9}{4\pi a^3} (1 - \alpha g)^{-1} (2\epsilon + 1)^{-1} \epsilon \left[\frac{\mu^2}{3kT} (1 - \alpha g)^{-1} + \alpha\right] E$$

and, since $P/E = (\epsilon - 1)/4\pi$, one has the following equation to solve for the dielectric constant:

$$\epsilon - 1 = \frac{9}{a^3} \frac{\epsilon}{(1 - \alpha g)(2\epsilon + 1)} \left(\frac{\mu^2/3kT}{1 - \alpha g} + \alpha \right) \quad [35]$$

If the direction of the electric field changes rapidly, the permanent moments will not be able to follow, and the dielectric constant, denoted here by ϵ^∞, will be due to α alone. With $\mu = 0$, equation [35] rearranges to

$$9\epsilon^\infty (\epsilon^\infty - 1)^{-1} \frac{\alpha}{a^3} = 2\epsilon^\infty + 1 - 2 \frac{\alpha}{a^3} (\epsilon^\infty - 1) \quad [36]$$

(note that g here is $2a^{-3}(\epsilon^\infty - 1)/(2\epsilon^\infty + 1)$). Equation [36] may be solved for α/a^3 to give

$$\frac{\alpha}{a^3} = \frac{\epsilon^\infty - 1}{\epsilon^\infty + 2} \quad [37]$$

This is the Clausius–Mossotti formula.

Substituting equation [37] for α into the full equation [35], with $g = 2a^3(\epsilon - 1)/(2\epsilon + 1)$, one gets some algebraic simplification on solving for $\mu^2/3kTa^3$:

$$\frac{\mu^2}{3kTa^3} = (2\epsilon + \epsilon^\infty)(\epsilon^\infty + 2)^{-2} \left(1 - \frac{\epsilon^\infty}{\epsilon} \right) \quad [38]$$

which is Onsager's formula. Since the number density is $n = (4\pi a^3/3)^{-1} = 3.34 \times 10^{22}$ and $\mu = 1.83$ D for water, $\mu^2/3kTa^3 = 3.77$ for $T = 300$ K. Then, putting $\epsilon^\infty = 2$, Onsager's formula gives $\epsilon = 31$; with $\epsilon^\infty = 6$, ϵ comes out to be 124. The correct high-frequency dielectric constant is probably in the range 2–6. Thus it is possible to get dielectric constants of the right size with this formula.

E. EVALUATING THE PARTITION FUNCTION

All properties of interest could be calculated from the partition function Z_N or the configurational integral Q_N (equation [5]). The configurational integral for a system of N particles interacting pairwise cannot be evaluated for the cases of interest to us, but approximate expressions are very useful. Cluster theories attempt to analyze the configurational integral into a sum of contributions and evaluate the most important ones. If all the particles are identical and ϕ_{hj} is the interaction energy between particles h and j (a function of $\mathbf{r}_h - \mathbf{r}_j$), we may write

$$Q_N = \int \exp\left(\frac{-\sum_{i<j} \phi_{ij}}{kT} \right) d\mathbf{q} = \int \prod_{i<j} (1 + \gamma_{ij}) \, d\mathbf{q} \quad [39]$$

114 3. STATISTICAL MECHANICS

where $d\mathbf{q}$ means integration over all particle coordinates and

$$\gamma_{ij} = \exp(-\phi_{ij}) - 1 \qquad [40]$$

The advantage of writing Q_N in this way is that γ_{ij} is appreciably different from 0 only when particles i and j are close to each other. The expansion of Q_N is a sum of terms, each being an integral over a product of up to $\frac{1}{2}N(N-1)$ γ_{ij}'s, and one can hope that terms with more γ_{ij}'s are smaller.

Many of the integrals are identical because of the equivalence of the particles, for example, $\int \gamma_{kl} \, d\mathbf{q}$ is independent of the choice of k and l. Some of the integrals, such as $\int \gamma_{kl} \gamma_{sr} \, d\mathbf{q}$, factor into products of simpler integrals over subsets of particle coordinates, and Q_N can be written (Eyring et al. 1982, Chapter 11) in terms of the "irreducible cluster integrals," the first of which are

$$2\beta_2 \equiv \int d\mathbf{q}_1 \, d\mathbf{q}_2 \gamma_{12} = V \int d\mathbf{q}_{12} \gamma_{12}$$

$$6\beta_3 \equiv \int d\mathbf{q}_1 \, d\mathbf{q}_2 \, d\mathbf{q}_3 \gamma_{12} \gamma_{13} \gamma_{23}$$

etc. Eventually, Q_N can be reduced to $V^N \exp\{\Sigma_2^\infty \beta_n c^n\}$, where c is the concentration V/N. Then all thermodynamic properties can be written in terms of the β_n.

For the electrolyte in the electrochemical interface, there are several complications: there are at least two different species of particles, and the particles interact with an external field as well as with each other. Furthermore, the long-range nature of the Coulombic field causes certain integrals to be divergent:

$$\beta_2 = \frac{1}{2} V \int r^2 \sin\theta \, dr \, d\theta \, d\phi \left[\exp\left(\frac{-e^2}{kTr}\right) - 1 \right] \to \infty \qquad [41]$$

In fact, this behavior is necessary (Barlow 1970, p. 215) because the contribution of the Coulombic interactions to the free energy is $-kT\Sigma_2^\infty c^n \beta_n$, which approaches $-kTc^2\beta_2$ for $c \to 0$, whereas the interaction energy in Debye–Hückel theory, which should be valid for $c \to 0$, turns out to be proportional to $c^{3/2}$. The solution to the problem involves the expansion of γ_{ij} in powers of $(r_{ij})^{-1}$ and combining graphs for powers (Barlow 1970, pp. 215–219). Such treatments show how to correct Gouy–Chapman and other simple theories.

It is possible to evaluate the partition function by producing a large number of molecular configurations that are chosen to be a representative sample of the infinite number of possible configurations, so that properties of the system may be calculated as an average over the generated configurations. The methods by which this is done are usually characterized as Monte Carlo or as molecular dynamics. The former techniques (Valleau and Whittington 1977; Valleau and Torrie 1977) generate sequences of molecular configurations, with criteria for rejecting certain sequences so that the set of configurations represents the equilibrium distribution. Molecular dynamics techniques (Kushnick and Berne 1977) generate the configurations by numerical solution of Newton's equations of motion, thus following the motion of the system. In addition to helping to determine equilibrium properties, these techniques can thus provide information related to time-dependent processes, such as correlation times.

The intermolecular potentials used in statistical mechanical studies such as Monte Carlo and molecular dynamics calculations are most often sums of spherically symmetric potential functions, each dependent only on the distance between a site on one molecule and a site on the other. These sites may be nuclear positions or may be auxiliary sites between nuclear positions. The most successful models of water use auxiliary sites for Coulombic interactions, so that the distribution of charge is represented by point charges at positions of atomic cores, bonds, and lone pair electrons. In addition to the Coulombic potentials, one uses short-range potentials representing the polarization and exchange interactions of atomic cores (for example, the 6-12 or exponential-6 forms).

The intermolecular interaction potentials are those required to reproduce liquid properties, so that they are effective potentials, not necessarily the actual interaction potentials between pairs of molecules. They must represent intermolecular forces, such as those associated with polarizability, which are not pairwise additive. Therefore the interaction potentials are not the true pair potentials and in fact may depend on the state of the system (Rossky 1985).

Of course, it is not the partition function itself that is of interest, but quantities derived from it, particularly the distribution functions introduced in Section A. Techniques for approximating these functions are of course of great interest. Integral equation methods are analytical techniques for deriving the spatial distribution and correlation functions and thermodynamic properties that are expressible in terms of these functions.

In connection with these methods, one must define still another correlation function. The direct correlation function c_{ij} for species i and j is defined by the Ornstein–Zernike equation

$$h_{ij}(\mathbf{r}_i, \mathbf{r}_j) = c_{ij}(\mathbf{r}_i, \mathbf{r}_j) + \sum_k \int c_{ik}(\mathbf{r}_i, \mathbf{r}_k) \rho_k h_{kj}(\mathbf{r}_k, \mathbf{r}_j) \, d\mathbf{r}_k$$

with ρ_k the density of species k. The direct correlation function may seem to represent the direct effect of one particle on another, in contrast to h_{ij}, which represents the correlation between two particles caused by their direct interaction and also by the mediation of other particles. Whether or not this interpretation is valid, it has been found that simple approximations for the c_{ij} can yield extremely useful representations of the more complex correlation functions h_{ij}.

To find the c_{ij} and h_{ij} one requires a second relation between them. Two approximate relations that have proven very useful are the hypernetted chain (HNC) and Percus–Yevick (PY) approximations. For a single species, the HNC is

$$e^{h-c-(U/kT)} = h + 1$$

and the PY is

$$[e^{-U/kT} - 1](1 + h - c) = c$$

where U is the interaction potential. The HNC is known to work well for dense ionic fluids (Rossky 1985), whereas the PY produces unphysical results for such systems.

In this section and Section B some connections were made between statistical mechanical or molecular concepts and thermodynamic quantities. In the following sections, we consider some others important in electrochemistry: electrochemical potentials in Section F, surface charge in Section G, and electrode potentials in Section H.

F. ELECTROCHEMICAL POTENTIALS

The electrochemical potential $\tilde{\mu}_{e^-}$ of a conduction electron in a metal, defined as the work required to bring an electron to the interior of the metal from vacuum infinitely far away (where its energy is defined as zero), is the energy of the highest filled electronic level (at $T = 0$); the electron goes into the lowest empty level, which, in a band of energies, is the same as the highest filled level. For $T > 0$, if there are $\rho(\epsilon) \, d\epsilon$ one-electron states with energies between ϵ and $\epsilon + d\epsilon$, the number of electrons with energies between ϵ and $\epsilon + d\epsilon$ is

$$N(\epsilon) \, d\epsilon = \rho(\epsilon) \left\{ \exp\left[\beta(\epsilon - \epsilon_F)\right] + 1 \right\}^{-1} \qquad [42]$$

with $\beta = 1/kT$. One can show (see Chapter 4, equation [62]) that the work required to introduce an electron, that is, the electrochemical potential, is $\tilde{\mu}_e = \epsilon_F$, so ϵ_F can be replaced by $\tilde{\mu}_e$ in equation [42].

It is also possible to define the electrochemical potential and Fermi energy for a system containing bound electrons, such as a semiconductor or a redox couple in an electrolyte (Lamy 1978, Chapter 11, Section 3). Consider first a dilute gas of N_e electrons in a volume V at temperature T. If the electrical potential vanishes and interelectronic interaction is negligible, the energy levels are $\hbar^2 k^2 / 2m$ and the state density $\rho(\epsilon)$ is $g_e V(\epsilon/2)^{1/2} m^{3/2} / \pi^2 \hbar^3$ (see Chapter 4, equation [52]), where g_e is the spin degeneracy, equal to 2 for electrons. The total electron density is

$$\frac{N_e}{V} = g_e m^{3/2} \pi^{-2} \hbar^{-3} \int_0^\infty \left(\frac{\epsilon}{2}\right)^{1/2} \left\{ \exp\left[\beta(\epsilon - \tilde{\mu}_e)\right] + 1 \right\}^{-1} d\epsilon \qquad [43]$$

If the quantity in curly brackets is small, as is usually the case, $\exp[\beta(\epsilon - \tilde{\mu}_e)] \gg 1$ and one has a Boltzmann distribution:

F. ELECTROCHEMICAL POTENTIALS

$$\frac{N_e}{V} = g_e m^{3/2} \pi^{-2} \hbar^{-3} \exp(\beta\tilde{\mu}_e) \int_0^\infty \left(\frac{\epsilon}{2}\right)^{1/2} \exp(-\beta\epsilon)\, d\epsilon$$

$$= g_e \left(\frac{mkT}{\pi\hbar^2}\right)^{3/2} (2^{-1/2}) \exp(\beta\tilde{\mu}_e) \qquad [44]$$

Then

$$\tilde{\mu}_e = kT \ln \frac{N_e}{Vd_e}, \qquad d_e = g_e \left(\frac{mkT}{\pi\hbar^2}\right)^{3/2} (2^{-1/2})$$

Similarly, for a gas of protons or H^+ ions, $\tilde{\mu}_{H^+} = kT \ln(N_+/Vd_+)$ where $d_+ = g_{H^+}(m_H kT/\pi\hbar^2)^{3/2}(2^{-1/2})$ with $g_+ = 2$. For a gas of H atoms, $\tilde{\mu}_H = kT \ln(N_H/Vd_0)$, where $d_0 = g_H(m_H kT/\pi\hbar^2)^{3/2}(2^{-1/2})$ with $g_H = 4$.

If one has a dilute plasma containing electrons, protons, and H atoms, and thermal equilibrium exists with respect to the reaction $H^+ + e^- \to H$, the electrochemical potentials must obey

$$\tilde{\mu}_H = \tilde{\mu}_{e^-} + \tilde{\mu}_{H^+} \qquad [45]$$

However, all energy levels must now be referred to the same zero, and a hydrogen atom at rest is lower in energy than a proton at rest plus an electron at rest, by the ionization potential I ($=13.6$ eV). Thus the energy levels of H are $(\hbar^2 k^2/2m_H) - I$. Then for hydrogen atoms we have

$$\rho(\epsilon) = g_H V[(\epsilon + I)/2]^{1/2} m^{3/2} \pi^{-2} \hbar^{-3}$$

and, instead of equation [44], we get

$$\frac{N_H}{V} = g_H m_H^{3/2} \pi^{-2} \hbar^{-3} \exp(\beta\tilde{\mu}_H) \int_{-I}^\infty \left(\frac{\epsilon + I}{2}\right)^{1/2} \exp(-\beta\epsilon)\, d\epsilon$$

or $\tilde{\mu}_H = kT \ln(N_H/Vd_0) - I$. The equilibrium condition (equation [45]) is now

$$\ln \frac{N_H}{Vd_0} - \frac{I}{kT} = \ln \frac{N_e}{Vd_e} + \ln \frac{N_+}{Vd_+} \qquad [46]$$

or, since $N_e = N_+$, we have $N_e = [N_H V d_e d_+/(d_0)^2]^{1/2} e^{-I/2kT}$. Then the electrochemical potential of the electron is

$$\tilde{\mu}_e = kT \ln \frac{N_e}{Vd_e} = kT \ln \left\{\left[\frac{N_H d_+}{Vd_e (d_0)^2}\right]^{1/2}\right\} - \frac{1}{2} I.$$

As suggested by treatments of semiconductors, one can introduce the notion of a hole p^+ with electrochemical potential $\tilde{\mu}_{p^+}$ such that $\tilde{\mu}_{e^-} + \tilde{\mu}_{p^+} = 0$, corresponding to the reaction $e^- + p^+ \to 0$. Then

$$\tilde{\mu}_{p^+} = -\tilde{\mu}_{e^-} = \frac{1}{2}I + kT \ln\left[d_0\left(\frac{Vd_e}{N_H d_+}\right)^{1/2}\right] \quad [47]$$

Similarly, one can consider a plasma of hydrogen atoms, electrons, and hydride ions. The electron affinity of hydrogen is $A = 0.75$ eV, so that the lowest energy level of H^- is 0.75 eV below that of a hydrogen atom and an electron at rest. Again taking the electrochemical potential of hydrogen atoms as $kT \ln(N_H/Vd_0) - I$, one must take the electrochemical potential of hydride ions as $kT \ln(N_{H^-}/Vd_-) - I - A$, where $d_- = g_-(m_H kT/\pi\hbar^2)^{3/2}(2^{-1/2})$. For the reaction $H + e^- \to H^-$ to be at equilibrium, $\tilde{\mu}_H + \tilde{\mu}_{e^-} = \tilde{\mu}_{H^-}$, or

$$\ln\frac{N_H}{Vd_0} - \frac{I}{kT} + \ln\frac{N_e}{Vd_e} = \ln\frac{N_{H^-}}{Vd_-} - \frac{I+A}{kT}$$

Finally suppose the species, H H^+ and H^-, are considered. The equilibrium condition

$$\tilde{\mu}_{H^+} + \tilde{\mu}_{H^-} = 2\tilde{\mu}_H$$

gives

$$kT\left(\ln\frac{N_{H^+}}{Vd_+} + \ln\frac{N_{H^-}}{Vd_-}\right) - (I + A) = 2\left(kT\ln\frac{N_H}{Vd_0} - I\right)$$

Assuming electroneutrality, $N_{H^+} = N_{H^-}$, and noting that $d_+ d_- = (d_0)^2$, we have

$$kT \ln(N_{H^+}/N_H)^2 = -(I - A) \quad [48]$$

so that, according to equation [45],

$$\tilde{\mu}_{e^-} = \tilde{\mu}_H - \tilde{\mu}_{H^+} = kT\ln\left(\frac{N_H d_+}{N_{H^+} d_0}\right) - I$$

$$= kT\ln\left(\frac{d_+}{d_0}\right) + \frac{1}{2}(I - A) - I \quad [49]$$

The electrochemical potential of electrons is defined even though the number of free electrons in this system is vanishingly small.

In fact, equation [49] defines the electrochemical potential of bound electrons.

F. ELECTROCHEMICAL POTENTIALS

Now consider the work function for a neutral system of positive and negative ions, neutral atoms, and electrons. It would be $\Phi = -\tilde{\mu}_{e^-}$, since the electronic energy levels form a continuum, so that the energy of the highest filled level is the same as that of the lowest empty level. Since the positive ions and the neutral atoms have the same mass, equation [49] gives

$$\Phi = -kT \ln \frac{g_+}{g_H} + \frac{1}{2}(I + A) \qquad [50]$$

It is interesting that one gets a reasonable value for the work function of a metal from equation [50], even though a metal is certainly not a gas of positive ions, negative ions, and neutral atoms. The first term in equation [50] is negligible compared to ionization potentials: At $T = 300$ K, $kT \ln 2$ is 2.9×10^{-14} erg or 1.8×10^{-2} eV. For Na, $I = 5.1$ eV, $A = 0.5$ eV, and $\frac{1}{2}(I + A) = 2.8$ eV; the work function is actually something over 2.4 eV.

To define $\tilde{\mu}_{e^-}$ for a redox system in solution, one can consider the reduction reaction

$$O_O + e^- \to R_R \qquad [51]$$

where e^- is an isolated electron infinitely far from the solution, and O and R, the oxidized and reduced species, are solvated. The subscripts on O and R indicate that O has a solvent sheath appropriate to O and that R has a solvent sheath appropriate to R. The free energy of the species O_O differs from that of the isolated species O in the gas phase, just as a solvated electron $e_e^-(aq)$ has a free energy different from that of an isolated electron e^-. The instantaneous transfer of an electron to O_O would produce R_O; subsequent rearrangement of solvent to produce R_R would involve some additional free energy. The energy levels of solvated species form continua rather than being sharp like those of isolated molecules. Thus one has a density of states for the solvated oxidized species, and similarly for the solvated reduced species.

Let A be the electron affinity of O_O, that is, A is the free-energy change for the reaction $R_O \to O_O + e^-$, the reaction occurring without change in solvation. Similarly, let I be the ionization potential of R_R, that is, the free-energy change for the reaction $R_R \to O_R + e^-$. Let L_R be the free-energy change associated with solvent rearrangement to produce R_O from R_R, and let L_O be the free energy required to produce O_R from O_O. If equation [51] is at equilibrium,

$$\tilde{\mu}_{e^-} = \tilde{\mu}_{R(R)} - \tilde{\mu}_{O(O)} = (\tilde{\mu}_{R(O)} - L_R) - (\tilde{\mu}_{R(O)} + A) = -A - L_R$$

Alternatively,

$$\tilde{\mu}_{e^-} = \tilde{\mu}_{R(R)} - \tilde{\mu}_{O(O)} = (\tilde{\mu}_{O(R)} - I) - (\tilde{\mu}_{O(R)} - L_O) = L_O - I$$

One can also take the average, obtaining

$$\tilde{\mu}_{e^-} = -\tfrac{1}{2}(A + I) + \tfrac{1}{2}(L_O - L_R) \qquad [52]$$

which differs from equation [49] by the term $\tfrac{1}{2}(L_O - L_R)$, due to solvent rearrangement. Equation [52], or the preceding equations, gives the electrochemical potential of a free electron in equilibrium with the species in the solution. In other words, this is the electronic Fermi level for a solution in which the redox couple O/R is at equilibrium.

If an inert electrode, acting as a source of electrons, dips into the solution, $\tilde{\mu}_{e^-}$ also represents the Fermi level of electrons in the inert electrode. Now the emf of the cell

$$L \mid H_2(a = 1) \mid H^+(a = 1) \parallel O, R(\text{solution}) \mid R$$

(where L and R are of the same metal) is

$$\mathcal{E}^{SHE} = \phi^R - \phi^L = -(\tilde{\mu}_e^R - \tilde{\mu}_e^L)/\mathcal{F} \qquad [53]$$

where $\tilde{\mu}_e^R$ is the electronic Fermi level for $O_O + e^- \rightarrow R_R$ and $\tilde{\mu}_e^L$ is the corresponding quantity for the H^+/H_2 couple. To obtain the latter, one needs to consider a thermodynamic cycle for

$$H^+_{H^+} + e^- \rightarrow \tfrac{1}{2} H_2(\text{aq})$$

since the Fermi level is now given by $\tilde{\mu}_{e^-} = \tfrac{1}{2}\mu^0_{H_2} - \tilde{\mu}^0_{H^+(H^+)}$. The electrochemical potential of $H^+_{H^+}$ is the negative of the free-energy change in bringing H^+ from the interior of the solution to infinity, estimated (Lamy 1978) as 11.4 eV/mol. In forming neutral H atoms from H^+ in the gas phase 13.6 eV/mol is liberated and an additional 2.1 eV is liberated in forming $\tfrac{1}{2}$ mole of H_2; it is estimated that 0.2 eV is required to bring the $\tfrac{1}{2}$ mole of H_2 from the interior of the solution to the gas phase. Thus

$$\tfrac{1}{2}\mu^0_{H_2} - \tilde{\mu}^0_{H^+(H^+)} = 11.4 - 13.6 - 2.1 - 0.2 \text{ eV} = -4.5 \text{ eV}$$

If $\tilde{\mu}_{e^-}$ for the standard hydrogen electrode is -4.5 eV, the Fermi level or electrochemical potential of the electron for any redox system is, according to equation [53],

$$\tilde{\mu}_{e^-} = \epsilon_F^{\text{redox}} = -\mathcal{F}\mathcal{E}^{SHE} - 4.5 \text{ eV}$$

where $\mathcal{E}^{SHE}$ is the emf of the cell above, with the redox electrode on the right and the standard hydrogen electrode on the left.

In foregoing treatment invoked the equality of electrochemical potentials or Fermi levels of electrons in two phases in contact when there is equilibrium with respect to exchange of electrons. The Fermi level in a metal is the negative of the experimentally measurable work function. For an oxidation–reduction couple in solution, the Fermi level is given by equation [52]. When the metal is in equilibrium with solution, the Fermi levels of metal and solution become equal, which implies that a change in Fermi level takes place in each phase. This means there is a difference in outer potentials between two phases in contact, associated with a surface charge at the metal–solution interface. We return to electrode potentials in Section H.

G. SURFACE CHARGE

The surface charge on an electrode or interface is defined unambiguously only for a polarizable interface, for which each charged component can be assigned to one phase or the other. In this case, Gibbs' equation is written (Chapter 2, equation [26])

$$d\gamma = -q^P \, dE - \sum^P \Gamma_i \, d\mu_i - \sum^Q \Gamma_j \, d\mu_j - \sum_k d\mu_k \qquad [54]$$

where $E = \phi^P - \phi^Q$, sums P and Q involve charged components on the P and Q sides of the interface, respectively, and the last sum is over electrically neutral components. The surface charge on phase P is $q^P = \Sigma^P \Gamma_i z_i \mathfrak{F}$ and is equal to the negative of $q^Q = \Sigma^Q \Gamma_i z_i \mathfrak{F}$. Now, keeping the compositions of both phases constant and varying E (which can be done if the electrode is connected in a cell with an electrode of invariable potential),

$$\left(\frac{\partial \gamma}{\partial E}\right)_{\text{comp}} = -q^P \qquad [55]$$

Of course, the actual values of the surface excesses Γ_i depend on where the Gibbs dividing surface is placed, and are not individually measurable, but a sum like $\Sigma^P \Gamma_i z_i$ is invariant to the location of this surface. A displacement of the surface by D will change the sum by $D \Sigma^P c_i z_i$, where c_i is the bulk concentration of species i, and bulk electroneutrality means $\Sigma^P c_i z_i = 0$. In a molecular-level description, q^P would be calculated as $\Sigma^P N_i z_i \mathfrak{F}$, where N_i, the number of moles of species i, is the integral of the density of this species over the system; since $\Sigma^P c_i z_i = 0$, only the region near the surface gives a nonvanishing contribution to q^P. Thus q^P represents an excess or deficiency of free charge on one side of the interface.

The thermodynamic definition of equation [54] also implies that q^P is the amount of electricity or free charge that must be supplied to one side of the interface when its surface is increased by unity, at constant potential and constant composition of the two phases. However, for nonpolarizable interfaces the two definitions are not equivalent, as discussed by Frumkin, Petrii, and Damaskin (1980, Chapter 6, Section 2). With their preferred definition of q^M as the amount of electricity to be supplied to the metal side of a metal–solution interface when its surface increases by unity with the bulk composition of the solution kept constant, q^M may no longer be the free-charge density on the surface. To understand this, one must remember that, for a nonpolarizable electrode, the potential difference across the interface is determined by the concentrations of the species in the homogeneous phases adjacent to the interface and cannot be varied independently of the chemical potentials of those species as in equation [55].

Consider the simple case of a metal Me, whose atoms can dissolve in the liquid metal electrode M (forming, e.g., an amalgam), in equilibrium with ions Me$^+$ in solution. The equilibrium relation between electrochemical potentials, $\tilde{\mu}_{Me^+}^{(S)} + \tilde{\mu}_{e^-}^{(M)} = \tilde{\mu}_{Me}^{(M)}$, fixes the electrode–solution potential difference according to

$$\mathfrak{F}E = \mathfrak{F}(\phi^M - \phi^S) = \mu_{Me^+}^{(S)} + \mu_{e^-}^{(M)} - \mu_{Me}^{(M)} \qquad [56]$$

Supposing that the electrode consists of atoms of M, atoms of Me, and electrons, one has $q^M = -\mathfrak{F}\Gamma_{e^-}$ and, for the Gibbs adsorption equation (equation [54]),

$$d\gamma = \mathfrak{F}\Gamma_{e^-}\, dE - \Gamma_{e^-}\, d\mu_{e^-} - \Gamma_M\, d\mu_M - \Gamma_{Me}\, d\mu_{Me}$$
$$- \Gamma_{Me^+}\, d\mu_{Me^+} - {\sum}'\, \Gamma_i\, d\mu_i \qquad [57]$$

where Σ' is a sum over charged and uncharged solution components other than Me$^+$. Because of equation [56], one cannot vary E while keeping all chemical potentials constant, but equation [56] may be used to eliminate one chemical potential from the Gibbs adsorption equation.

Thus, substituting for $\mu_{Me^+}^{(S)}$ yields

$$d\gamma = \mathfrak{F}(\Gamma_{e^-} - \Gamma_{Me^+})\, dE - (\Gamma_{e^-} - \Gamma_{Me^+})\, d\mu_{e^-} - \Gamma_M\, d\mu_M$$
$$+ (\Gamma_{Me^+} - \Gamma_{Me})\, d\mu_{Me} - {\sum}'\, \Gamma_i\, d\mu_i$$

This means that $(\partial\gamma/\partial E) = -\mathfrak{F}(\Gamma_{Me^+} - \Gamma_{e^-}) = -q^M - \mathfrak{F}\Gamma_{Me^+}$ if E is varied by changing $\mu_{Me^+}^{(S)}$, keeping the composition of the metal phase constant. Alternatively, equation [56] may be used to eliminate $d\mu_{Me}$ from equation [57], yielding

$$d\gamma = \mathfrak{F}(\Gamma_{e^-} + \Gamma_{Me})\, dE - (\Gamma_{e^-} + \Gamma_{Me})\, d\mu_{e^-} - \Gamma_M\, d\mu_M$$
$$- (\Gamma_{Me} + \Gamma_{Me^+})\, d\mu_{Me^+} - {\sum}'\, \Gamma_i\, d\mu_i$$

Thus, if E is varied by changing μ_{Me} keeping the solution composition constant, $(\partial\gamma/\partial E)$ determines, not $-q^M$, but $-q^M + \mathfrak{F}\Gamma_{Me}$. The quantities $q^M + \mathfrak{F}\Gamma_{Me^+}$ and $q^M - \mathfrak{F}\Gamma_{Me}$ are called *total electrode charges* by Frumkin et al. (1980) and are contrasted to free-charge densities such as q^M.

For the more general case of a reversible electrode whose potential is determined by the redox reaction $O + ne^- \to R$, one likewise defines two total electrode charges, for constant chemical potential of O and for constant chemical potential of R. One can then have two kinds of electrocapillary curves and two Lippmann equations. In the notation of Frumkin et al.,

$$\left(\frac{\partial \gamma}{\partial E}\right)_{\mu_O, \mu_i} = -Q'$$

$$\left(\frac{\partial \gamma}{\partial E}\right)_{\mu_R, \mu_i} = -Q'' \qquad [58]$$

where i refers to a substance not participating in the redox reaction: $Q'(Q'')$ is determined when $\mu_O (\mu_R)$ is held constant during the variation of E.

Suppose O and R are both in solution. If the area of the electrode is increased by unity, one can imagine adding O and R to the system in the amounts Γ_O and Γ_R, so that the solution composition is unchanged, and no electricity is supplied from outside to the interface. One could also produce the same result by adding only O, in the amount $\Gamma_R + \Gamma_O$, plus an amount of electricity $-n\mathscr{F}\Gamma_R$; or by adding only R in the amount $\Gamma_R + \Gamma_O$, plus an amount of electricity $n\mathscr{F}\Gamma_O$. If only R and charge are added, one keeps μ_O, as well as chemical potentials of other components not participating in the redox process, constant; since the amount of charge to be added to the solution side is $n\mathscr{F}\Gamma_O$, Q' in equation [58] is $-n\mathscr{F}\Gamma_O$. If only O is added, μ_R is constant and Q'' in equation [58] is $n\mathscr{F}\Gamma_R$. Again, one must note that, to relate these quantities to the actual free surface charge density, a model for the surface with adsorbed components is necessary. One might assume (Frumkin et al. 1980, p. 226) that the only charge transfer is between O and R and that the Gibbs surface is chosen such that the surface excesses of O and R are Γ_O and Γ_R. The surface charge on the solution side is then $q^S = n\mathscr{F}(\Gamma_R - \Gamma_O)$, so $q^M = -n\mathscr{F}(\Gamma_R - \Gamma_O)$. Thus $Q' = -n\mathscr{F}\Gamma_O = -q^M - n\mathscr{F}\Gamma_R$ and $Q'' = n\mathscr{F}\Gamma_R = -q^M + n\mathscr{F}\Gamma_O$.

H. ELECTRODE POTENTIALS

A number of different meanings can be given to the term *electrode potential*. The electrode–solution potential difference is a difference of inner potentials.

$$\Delta_S^M \phi = \phi^M - \phi^S$$

and is, in principle, not measurable because it is a difference in electrical potentials between points in chemically different phases. Measuring instruments for potential actually can measure only differences in electron energies, which can include differences in chemical as well as electrical potentials. Alternatively, one can say that, on connecting one lead of an instrument to a point in M and the other to a point in N, an electrochemical cell is formed: The measured potential is a cell potential, which is a combination of $\Delta_S^M \phi$ and other interfacial potential differences. The question of the relation of $\Delta_S^M \phi$ to the electrode potential then arises.

The emf of the cell $M^{(1)} | \text{Ref} | S | M^{(2)}$ with leads of the same metal M is $\mathcal{E} = \phi^{(2)} - \phi^{(1)} = \Delta_1^2 \phi$ and is obviously a sum of three potential drops:

$$\mathcal{E} = \Delta_S^{(2)} \phi + \Delta_{\text{Ref}}^S \phi + \Delta_1^{\text{Ref}} \phi \qquad [59]$$

Here $\Delta_S^{(2)} \phi$ and $\Delta_{\text{Ref}}^S \phi$ are actual metal–solution potential differences for the two electrodes, but this is not true for $\Delta_1^{\text{Ref}} \phi$. The conventional way of writing the emf as a difference of half-cell potentials,

$$\mathcal{E} = \mathcal{E}^M - \mathcal{E}^{\text{Ref}} \qquad [60]$$

implies a division of $\Delta_1^{\text{Ref}} \phi$ into parts assigned to the M and Ref half-cells, but this cannot be done unambiguously (Trasatti 1980, Section 2.4.2). Clearly, one must distinguish between $\mathcal{E}^M$, a thermodynamically defined electrode potential, and $\Delta_S^M \phi$, an actual electrode–solution potential difference.

Using the equilibrium condition for electrons at the interface between $M^{(1)}$ and the reference electrode, $\tilde{\mu}_{e^-}^{(1)} = \tilde{\mu}_{e^-}^{\text{Ref}}$, we may write equation [59] as

$$\mathcal{E} = \Delta_S^{(2)} \phi - \Delta_S^{\text{Ref}} \phi + \frac{\mu_{e^-}^{\text{Ref}} - \mu_{e^-}^{(1)}}{\mathcal{F}} \qquad [61]$$

This is the difference between

$$E^M = \Delta_S^M \phi - \frac{\mu_{e^-}^{(M)}}{\mathcal{F}}$$

which involves only quantities relevant to the M–S interface, and $E^{\text{Ref}} = \Delta_S^{\text{Ref}} \phi - \mu_{e^-}^{\text{Ref}}/\mathcal{F}$, which involves only quantities relevant to the reference electrode. E^M is a possible definition of the electrode potential, referred to as an *electrode potential on the vacuum scale* (Khan 1983). Note that one may have two electrodes with the same value of E^M and yet different potential differences $\Delta_S^M \phi$. For different faces of the same metal, on the other hand, $\Delta_S^M \phi$ is the same when E^M is the same (since $\mu_{e^-}^M$ is a bulk quantity), but the breakdown of $\Delta_S^M \phi$ into dipolar and free-charge contributions will probably differ.

If the potential difference between metal $M^{(2)}$ and solution is established by the reaction

$$M^{z+}(S) + e^-(M) \rightarrow M^{(z-1)+}(S)$$

(oxidation–reduction electrode) the equality of electrochemical potentials gives

$$\mu^{(S)}_{M^{z+}} + \mu^{(M)}_{e^-} = \mu^{(S)}_{M^{(z-1)+}} + \mathcal{F}(\phi^M - \phi^S) \qquad [62]$$

Therefore

$$E^M = \phi^M - \phi^S - \frac{\mu^{(M)}_{e^-}}{\mathcal{F}} = \frac{\mu^{(S)}_{M^{z+}} - \mu^{(S)}_{M^{(z-1)+}}}{\mathcal{F}} \qquad [63]$$

The electrochemical potential of an electron in solution may be defined, as in Section F, by the (hypothetical) equilibrium between $M^{z+} + e^-$ and $M^{(z-1)+}$, that is,

$$\tilde{\mu}^{(S)}_{e^-} = \mu^{(S)}_{M^{(z-1)+}} - \mu^{(S)}_{M^{z+}} - \mathcal{F}\phi^S \qquad [64]$$

Comparing equation [64] to equation [62], one sees that this definition is equivalent to defining $\tilde{\mu}^{(S)}_{e^-}$ as equal to $\tilde{\mu}^{(M)}_{e^-}$. As for any equilibrium for exchange of a charged species between phases, the chemical potential difference for the electron balances the electrical potential difference:

$$\mu^{(M)}_{e^-} - \mu^{(S)}_{e^-} = \mathcal{F}(\phi^M - \phi^S)$$

Thus $\phi^M - \phi^S - \mu^{(M)}_{e^-}/\mathcal{F}$ is $-\mu^{(S)}_{e^-}/\mathcal{F}$, and one meaning for the electrode potential on the vacuum scale is the *chemical* potential of an electron in the solution divided by $-\mathcal{F}$.

Another meaning of E^M is seen by noting that $\mu^{(M)}_{e^-} - \mathcal{F}\phi^M$ is the work required to bring a mole of electrons from vacuum into the metal M. This means that $\mathcal{F}E^M$ would be the work required to remove an electron from the metal to vacuum plus the work required to bring it to the Galvani potential of the solution (without any interaction with solvent).

Although the measured potential difference $\mathcal{E}$ is a difference of potentials on the vacuum scale, each such potential is not itself an electrical potential difference, but involves as well a chemical potential. Furthermore, the definition of E^M as a single electrode potential is not unambiguous since a constant could be added to the definitions of both E^M and E^{Ref}. One sometimes calls E^M the *reduced absolute electrode potential*, written $_rE^M$ (Trasatti 1980, Section 4.4.3).

For the hydrogen electrode, the reduced absolute electrode potential can be evaluated using the equilibrium condition

$$\mu^{(Pt)}_{e^-} - \mathcal{F}\phi^{(Pt)} + \mu^{(S)}_{H^+} + \mathcal{F}\phi^{(S)} = \tfrac{1}{2}\mu_{H_2}$$

and a Born–Haber cycle:

$$_rE^0(H^+/H_2) = \Delta_S^M\phi - \frac{\mu_{e^-}^{(Pt)}}{\mathcal{F}}$$

$$= \frac{\mu_{e^-}^{(Pt)} + \mu_{H^+}^{OS} - \frac{1}{2}\mu_{H_2}}{\mathcal{F}} - \frac{\mu_{e^-}^{(Pt)}}{\mathcal{F}}$$

$$= \frac{\mu_{H^+}^{OS} + \frac{1}{2}\Delta G_{diss}^0 + \Delta G_{ion}^0}{\mathcal{F}}$$

Here, the zero level of $\mu_{H^+}^{OS}$ corresponds to H^+ outside the solution: ΔG_{diss}^0 and ΔG_{ion}^0 are the free energies for dissociation of H_2 and the ionization of H. The value 4.31 eV is obtained for $_rE^0(H^+/H_2)$, replacing the less precise value of 4.5 eV of Section G. The value of $_rE^M$ for another electrode is given by adding $_rE^0(H^+/H_2)$ to the emf of a cell in which the normal hydrogen electrode (NHE) is the reference electrode. The emf of such a cell is

$$\mathcal{E}(NHE) = {_rE^M} - {_rE^0(H^+/H_2)} = \Delta_S^M\phi - \frac{\mu_{e^-}^M}{e} - {_rE^0(H^+/H_2)}$$

Thus the actual metal–solution potential difference $\Delta_S^M\phi$ can be determined when $\mu_{e^-}^M$ is calculable, which can be done reliably for s–p metals but not yet for d metals.

Now consider such a cell when it is at the pzc for the metal ($\sigma = 0$), so that $\Delta_S^M\phi$ is due to dipole layers only:

$$\mathcal{E}^{\sigma=0} = (\chi^M + \delta\chi_S^M) - (\chi^S + \delta\chi_M^S) - \frac{\mu_{e^-}^M}{e} - {_rE^0(H^+/H_2)}$$

$$= \frac{\Phi^M}{e} + \delta\chi_S^M - (\chi^S + \delta\chi_M^S) - {_rE^0(H^+/H_2)} \qquad [65]$$

Here, $\Delta_S^M\phi$ is written as the difference of the surface dipole potential of the metal, $\chi^M + \delta\chi_S^M$, and that of the (aqueous) electrolyte, $\chi^S + \delta\chi_M^S$; χ^M is the surface dipole potential of the free metal surface in the absence of electrolyte and χ^S is that of the electrolyte in the absence of metal, so that $\chi^M - \mu_{e^-}^{(M)}/\mathcal{F}$ is the work required to remove an electron from the metal, or Φ^M, the work function for electron emission to vacuum. A rearrangement of equation [65] provides another route to $_rE^0(H^+/H_2)$:

$$_rE^0(H^+/H_2) = \frac{\Phi^M}{e} - \mathcal{E}^{\sigma=0} + \Delta_S^M\psi^{\sigma=0} - \chi^S \qquad [66]$$

H. ELECTRODE POTENTIALS

Here, we have used the fact that $\delta\chi_S^M - \delta\chi_M^S$ is $(\phi^M - \phi^S) - (\chi^M - \chi^S)$, the metal–solution potential difference minus the difference of metal–vacuum and solution–vacuum potential differences; since $\phi - \chi = \psi$, this is just the difference in Volta potentials of metal and solution. In [66], only χ^S is not an experimentally measurable quantity.

To define an absolute electrode potential, free from the ambiguity of the additive constant associated with writing equation [61] as a difference, Trasatti (1980, Section 2.4.4) uses the fact that $\mathcal{E}$ is actually $-(\tilde{\mu}_{e^-}^{(2)} - \tilde{\mu}_{e^-}^{(1)})$. For nonpolarizable interfaces, one considers the cell without transport,

$$M^{(1)}|\text{Ref}|S(\text{Ref})|S(M)|M^{(2)} \qquad [67]$$

for which $\tilde{\mu}_{e^-}^{\text{Ref}} = \tilde{\mu}_{e^-}^{(1)}$ so that the emf $\mathcal{E}$ is the difference between $-\tilde{\mu}_{e^-}^{(2)}$ and $-\tilde{\mu}_{e^-}^{\text{Ref}}$. Then the absolute electrode potential is defined as e^{-1} times the work required to extract electrons from (1) the free surface of the solution S(M) in electronic equilibrium with the metal or (2) from free surface of the metal in electronic equilibrium with the solution. For polarizable interfaces (no electronic equilibrium between electrode and solution) one can use the cell $M^{(1)}|\text{Ref}|S|M^{(2)}$ and write

$$-(\tilde{\mu}_e^{(2)} - \tilde{\mu}_e^{\text{Ref}}) = -(\tilde{\mu}_e^{(2)} - \tilde{\mu}_e^{*S}) + (\tilde{\mu}_e^{\text{Ref}} - \hat{\mu}_e^{*S})$$

where $\tilde{\mu}_e^{*S}$ refers to an electron in solution, so that $-(\tilde{\mu}_e^M - \tilde{\mu}_e^{*S})/e$ is defined as the absolute electrode potential. This is just $\Phi^{M/S}$, the electron work function of metal M in solvent S, divided by e. Electrodes of different metals with the same electrode potentials then have the same Fermi level, although surface and bulk contributions may differ. This is true for nonpolarizable as well as polarizable interfaces, and is experimentally verified by measuring the photoelectric threshold for electron emission into the solution. In an electrochemical cell, the difference of absolute potentials $-\tilde{\mu}_e^M/e$ for nonpolarizable electrodes is the same as the difference of $_rE^M$ values because the values of ϕ^S cancel off. Similarly, cancellation of μ_e^{*S} occurs in the difference of $\Phi^{M/S}/e$ values for polarizable electrodes.

If M is electronically polarizable and the reference electrode is nonpolarizable,

$$e\mathcal{E} = -(\tilde{\mu}_e^M - \tilde{\mu}_e^{\text{Ref}}) = -(\tilde{\mu}_e^M - \tilde{\mu}_e^{*S}) - (\tilde{\mu}^{*S} - \tilde{\mu}_e^{\text{Ref}})$$

A cell without transport like [67] is assumed but, if there is no potential difference across the liquid junction, $\tilde{\mu}_e^{*S}$ is the same in both solutions. Because each of the terms in parentheses is a work function, the cell potential may be considered as a difference of the work functions, for emission to solution, of the metal M and the metal of the reference electrode. For the NHE, $_rE^0(H^+/H_2) = \Delta_S^M\phi - \mu_e^M/e$ so that the work function is

$$\Phi^{M/S} = \tilde{\mu}_e^{*S} - \tilde{\mu}_e^M = \mu_e^{*S} - \mu_e^M + e\Delta_S^M\phi$$

$$= \frac{{}_rE^0(H^+/H_2)}{e} + \mu^{*H_2O} = 4.31 \text{ V} + (-1.34 \text{ V})$$

The quantity 2.97 V is then the work required to transfer electrons from the Fermi level of the metal to the solvent.

Half-cell potentials are conventionally given on a scale for which the half-cell potential of the NHE is taken as zero. The absolute potential of an electrode (equal to $\Phi^{M/S}$) is now obtained by adding 2.97 V to the cell potential. Values of Φ^{M/H_2O} at the pzc obtained in this way are given and discussed by Trasatti (1980, Section 4.4.3). Disagreements with measured work functions are ascribed to different final states of the emitted electron. The discrepancy between a measured work function and one calculated from cell potentials then indicates the difference in energy between a delocalized and a localized electron in water.

The relation between $\Delta_S^M\phi$ and the various electrode potentials can be discussed by considering the de-electronation reaction at the electrode. For

$$M(M) \rightarrow M^+(S) + e\ (M)$$

we may write $\tilde{\mu}_{M^+}^{(S)} + \tilde{\mu}_{e^-}^{(M)} = \mu_M^{(M)}$ and, separating each electrochemical potential into electrical and chemical parts, we have

$$\mathcal{F}\Delta_S^M\phi = \mu_{e^-}^{(M)} + \mu_{M^+}^{(S)} - \mu_M^{(M)}$$

Since the sum of the free energies of evaporation and ionization of M, $\Delta G_{ev}^0 + \Delta G_{ion}^0$, is equal to $\mu_{M^+}^{(V)} + \mu_{e^-}^{(V)} - \mu_M^{(M)}$ (V = vacuum) and $\mu_{M^+}^{(V)}$ and $\mu_{e^-}^{(V)}$ are zero,

$$\mathcal{F}\Delta_S^M\phi = \mu_{e^-}^{(M)} + \mu_{M^+}^{(S)} + \Delta G_{ev}^0 + \Delta G_{ion}^0$$

The reduced absolute electrode potential ${}_rE^M$ is

$$\Delta_S^M\phi - \frac{\mu_{e^-}^{(M)}}{\mathcal{F}} = \frac{\mu_{M^+}^{(S)}}{\mathcal{F}} + \Delta G_{ev}^0 + \Delta G_{ion}^0 \qquad [68]$$

and corresponds to the reaction

$$M(M) \rightarrow M^+(S) + e^-(V)$$

The absolute electrode potential $\Delta^{M/S}/e$ corresponds to

$$M(M) \rightarrow M^+(S) + e^-(S)$$

since ΔG for this reaction differs from ΔG for the preceding one by $\mu_{e^-}^{(S)}$ and (see equation [68])

$$\Delta_S^M \phi - \frac{\mu_{e^-}^{(M)}}{\mathscr{F}} + \frac{\mu_{e^-}^{(S)}}{\mathscr{F}} = \frac{\tilde{\mu}_{e^-}^{(S)} - \tilde{\mu}_{e^-}^{(M)}}{\mathscr{F}} = \frac{\Phi^{M/S}}{e}$$

Thus the various electrode potentials differ in the location at which electrons are stored. Since the electrons cancel when electrodes or half-cells are combined to form a cell, the cell potential is independent of the electrons' location.

The chemical potential of an electron in solution is sometimes referred to as the Fermi level of the redox couple in solution, since it is the work that would be required to bring an electron from vacuum into the solution, aside from the electrostatic work $-e\phi^S$. According to equation [64], $\tilde{\mu}_{e^-}^{(S)} + \mathscr{F}\phi^S = \mu_{M^{(z-1)+}} - \mu_{M^{z+}}$, the difference of chemical potentials between the reduced and oxidized species of the redox couple. Compare the discussion of Section F. If $\phi^M - \phi^S$ is positive, $\mu_{e^-}^{(M)}$ is greater than $\mu_{e^-}^{(S)}$, so that the Fermi level in the metal is at higher energy than the Fermi level in solution.

In considering electron transfer between metal and solution, one has to consider energy states of the solvated ions in solution. Each of the states of the oxidized species is actually a state of the ion–solvent complex, its energy including vibration, rotation, and so on. The ground state, with energy E_{ox}^0, corresponds to the lowest state of vibration, rotation, and so forth. The probability of occupation of states of energy E is assumed to be given by the Boltzmann distribution:

$$W_{ox}(E) = \frac{\exp\left[-(E - E_{ox}^0)/kT\right]}{\int_{E_{ox,0}}^{\infty} \exp\left[-(E - E_{ox}^0)/kT\right] dE}$$

The integral in the denominator is just $\int_0^\infty e^{-x/kT} dx = kT$. A similar distribution function exists for the states of the reduced species, which involve a band of energies starting from a ground state, with energy E_{red}^0. These states involve vibration, rotation, and so on, of the complex formed from the reduced species and associated solvent.

Looked at from the point of view of the electron, the states of the reduced species are filled, whereas those of the oxidized species are empty. When an electrode acts as a cathode, electrons in states localized on the electrode (mostly with energies at the Fermi level) undergo transitions to states localized on the ions. For this to occur, there must be empty ionic states at the electrode's Fermi energy, so $W_{ox}(\epsilon_F)$ becomes the density of final states, which determines the transition probability according to the Fermi golden rule (cf. Chapter 9, Section C).

4

STRUCTURE OF SURFACES

In the case of a polarizable interface or other interface between phases with no common components, one can identify the components of the interface with one phase or the other. This means that, for instance, we assign the charges of certain substances to the surface charge on one side of the interface and assign the charges of other substances to the surface charge on the other side. The charge density at any point is also divided into two parts, corresponding to substances found in one phase or the other, so that the electric field and potential within the interface (see Chapter 2, equations [23] and [24]) likewise can be written as a sum of two contributions. This suggests it is profitable to study the surfaces of single-phase systems, that is, to consider the interfacial region as a sum of two parts and treat each separately. Of course, the properties of the interface between two phases are sums of surface properties of the separate phases only if the influence of one phase on the other is minimal, which is most likely to occur when there is no penetration of one phase by the other. This may be appropriate for the metal–electrolyte interface. In this case, in fact, it is often supposed that, because the region of nonzero surface charge density of the metal is of much smaller thickness than that of the electrolyte, the potential difference and capacitance of the interface is essentially that of the electrolyte; for capacitance calculations, the metal is replaced by a charged sheet of zero thickness.

We will consider the electrolyte surface by itself in Sections A–D, the metal surface in Sections E and F, and the semiconductor surface in Sections G and H. For each, we will consider bulk properties and then discuss the surface. In Chapter 5 we will consider, in more detail, electrolyte and metal as they interact.

A. LIQUID WATER SURFACE

Water is probably the most common constituent of electrochemical systems. It is also one of the more difficult substances to treat theoretically because the molecules are electrically polar and hydrogen bonds complicate the structure. It appears further that, to understand the properties of water one must take into account the quadrupole or other higher electrical moments of the molecule. For example, it appears that the surface entropy of water, from $d\gamma/dT$, is less than half that of nonpolar liquids, indicating some ordering at the surface, as compared to the bulk. This implies preferred orientation of the molecules. Considering the dipoles alone, one has no reason to prefer one orientation compared with its opposite. Including the quadrupole moment, however, one may argue that the molecules tend to orient themselves so as to place the highest electrical fields (where dipole and quadrupole fields reinforce) in the direction of the liquid, of high dielectric constant, rather than the vapor.

The isolated molecule of H_2O involves an H–O–H angle of 104.5° and an H–O distance of 0.96 Å. Since the van der Waals radius of the molecule is about 1.4 Å, the hydrogen nuclei are shielded, at least partly, by the electronic charge. The four valence electrons not involved in H–O bonding must be placed in two oxygen lone-pair orbitals, protruding oppositely to the O–H bonds. The arrangement of electron pairs is thus roughly tetrahedral, with the angle between the lone pairs something above the tetrahedral angle of 109.5°, and the H–O–H plane is perpendicular to the plane formed by the axes of the lone-pair orbitals. The H–O bonding electrons, which are attracted by the protons as well as by the oxygen core, are less polarizable than the lone-pair nonbonding electrons.

Let the z axis be chosen to bisect the H–O–H angle, with the x axis perpendicular to the H–O–H plane. The lone pairs must now lie in the xz plane, and the bonds must lie in the yz plane. Clearly, the symmetry of the molecule means that the dipole moment is along the z axis. Its value is 1.83 D (1 Debye = 1×10^{-18} esu · cm) in the $+z$ direction. The bonds probably have a net positive charge at the H end and the lone pairs, pointing in the $-z$ direction, involve negative charges. These contribute to the quadrupole moment.

The dipole moment of 1.83 D for the isolated molecule represents only part of the dipole moment of a water molecule in the liquid phase, where the dipole induced in a molecule by its neighbors is substantial (see Chapter 3, Section D). The electronic polarizability of a water molecule is about 2.4×10^{-24} cm^3 and makes only a very small contribution to the high dielectric constant. Interactions between dipoles, which are sometimes thought of as the formation of molecular clusters, are extremely important in the surface region as well as the bulk.

Two important kinds of model for liquid water exist: structural and continuous. In structural models, one considers groups or clusters of water molecules, which may be short-lived, as well as individual water molecules. The number and properties of different structures, and different states of individual water molecules, vary from theory to theory. For reviews, see Frank (1973) and Hurwitz (1978a). In continuous models of water, one considers distortion (1) from a regular structure formed from four-coordinated oxygens or (2) a three-dimensional network of hydrogen-bonded molecules. In structural models, one has either (1) a mixture of individual water molecules and clusters or (2) individual molecules in interstices in the structure. Each of the models accounts for some, but not all, of what is known about bulk water. However, there seems to be increasing evidence for the existence of individual water molecules and clusters, as well as for their location in structural interstices.

As discussed in Chapter 3, Section E, it is convenient to use, for computer simulations of water structure, intermolecular interaction potentials that are superpositions of spherically symmetric interaction potentials between sites on the molecules. For this and other reasons, the intermolecular interaction potential suitable for computer simulation may differ from the actual interaction potential between two isolated water molecules. The simplest model for a polar fluid, namely, hard spheres, each with a single ideal dipole at the center, has been much studied, but seems not to be a good model for real fluids such as water. It is important to introduce a nonspherical shape, which, by limiting the orientational effects of the dipoles, leads to a large decrease in calculated dielectric constants as well as great changes in short-range structure. For water, the deviation from spherical shape is less important than for other polar molecules, but it *is* necessary, as remarked above, to have a quadrupole moment as well as a dipole to get the important tetrahedral arrangements associated with hydrogen bonding (Rossky 1985). Quadrupole and dipole moments may be obtained, in a site-interaction model, from a charge of $-q$ on the oxygen and charges of $q/2$ on each hydrogen.

The computer simulations show that, around apolar solutes, the structure of a polar solvent is dominated by a tendency to maintain the hydrogen bonding available in the bulk, since the solute–solvent interaction is weak. To do this, the hydrogen-bonding groups tend to straddle the nonpolar solute. Polar solutes interact with water molecules similarly to the way water molecules interact with each other. For ionic solutes, the solute–solvent interaction is stronger and longer-ranged than the interaction between solvent molecules, so that the hydrogen-bonded structure is broken for ionic charges of e or greater. The orientational disorder of the solvent is maximized for charges of about e; for larger charges, the electric field of the ion, while still disrupting the solvent–solvent interactions, enhances the ordering. What happens at a solid interface appears to be governed by similar effects. At a nonpolar surface, packing forces and the tendency to maximize hydrogen bonding dominate the structure of water. Highly charged and polar surfaces lead to strong orientation and disruption of the hydrogen-bonding pattern.

At the surface of any polar phase, there will be a preferential orientation of molecular dipoles, leading to a surface potential χ. The value of χ enters the measured free energy of hydration of ions, since the hydration process entails bringing an ion through the surface into bulk solvent as well as surrounding it with solvent molecules. The value of χ is not directly measurable, but is inferred from indirect observations. For water, it is accepted that χ has a small positive value; 0.13 ± 0.02 V, decreasing with increasing temperature, has been suggested (Trasatti 1980). The positive value implies that water molecules at the interface have their oxygen ends pointing away from the bulk. It appears that, with the hydrogen end of water molecules oriented toward bulk liquid, the formation of hydrogen bonds is facilitated as compared with the opposite orientation.

According to equation [17] of Chapter 1, χ^{H_2O} is equal to the difference between the real potential and the chemical potential of an ion in water, of which only the former is measurable:

$$\chi^{H_2O} = (z\mathcal{F})^{-1}(\alpha_{Mz+}^{(H_2O)} - \mu_{Mz+}^{(H_2O)})$$

If $\mu_{Mz+}^{(H_2O)}$ is estimated from models, negative values of χ^{H_2O} are generally obtained (Trasatti 1979, pp. 105–110), although χ^{H_2O} appears as a small difference between two large quantities. The fact that the temperature coefficient of χ^{H_2O} is known to be negative, about -0.40 mV/K, is evidence that χ^{H_2O} itself is positive: Higher temperature lowers the orientation of dipole moments, so it should decrease the magnitude of the surface potential. Other evidence in favor of a positive value of about 0.13 V for χ^{H_2O} is reviewed by Trasatti (1979, pp. 105–107).

For the electrode–water surface at the point of zero charge, water is adsorbed on a metal electrode with the oxygen end toward the metal. This does not necessarily imply an electronic interaction of water with the metal, since the same orientation is found for the water–air interface. In the presence of a hydrophobic surface, which could be macroscopic or the periphery of a large molecule, there should be a tendency for water molecules to turn their hydrogen bonds toward the liquid, away from the surface. This should stabilize clusters of water molecules, increase hydrogen bonding, and increase the viscosity. Small or highly charged ions can orient water dipoles, forming a hydration shell. For smaller ions, the hydration shell may fit well with the bulk hydrogen-bond structure, which is thus stabilized. It has been proposed that water molecules hydrating a cation can still form three hydrogen bonds (instead of four) with neighboring water molecules, whereas water molecules hydrating an anion can form only two. Thus the molecules in the hydration shell for a cation are energetically more stable than for an anion of the same size, which may contribute to the preferential specific adsorption of anions over cations on a surface.

Trasatti (1979, Chapter IV, Section 3) reviews work relevant to the structure of adsorbed water on a metal surface. Experimental information about water in the compact layer on a metal M may be obtained from a cell formed from an NHE on the left and an electrode of M on the right:

$$\text{Pt}\,|\,\text{H}_2(a=1)\,|\,\text{H}^+(a=1,\text{aq})\,|\,\text{M}\,|\,\text{Pt}'$$

The cell emf may be written

$$\mathcal{E} = \mathcal{F}^{-1}(\mu_{e^-}^{(Pt)} - \mu_{e^-}^{(M)}) + (\phi^{(M)} - \phi^{(S)}) + (\phi^{(S)} - \phi^{(Pt)}) \qquad [1]$$

where we have substituted for $\phi^{(Pt')} - \phi^{(M)}$ using the contact equilibrium condition, $\mu_{e^-}^{(Pt)} - \mathcal{F}\phi^{(Pt')} = \mu_{e^-}^{(M)} - \mathcal{F}\phi^{(M)}$. The potential difference $\phi^{(S)} - \phi^{(Pt)}$ is fixed by the H_2/H^+ equilibrium. The potential difference $\phi^{(M)} - \phi^{(S)}$ is the sum of the ionic or free-charge contribution $g(\text{ion})$ and the contributions of the metal and the water dipoles on the metal surface. The metal contribution is written χ^M

$+ \delta\chi_S^M$, where χ^M is the value of this quantity for the free metal surface, that is, in the absence of solution; the water contribution is $-g_w^M$ (g_w^M is potential in water minus potential at metal). Therefore

$$\mathcal{E} = \mathcal{F}^{-1}(\mu_{e^-}^{(Pt)} - \mu_{e^-}^{(M)}) + g(\text{ion}) + \chi^M + \delta\chi_S^M - g_w^M + (\phi^{(S)} - \phi^{(Pt)}) \quad [2]$$

The work function of metal M, Φ^M, is $-\mathcal{F}^{-1}\mu_{e^-}^{(M)} + \chi^M$. Now suppose $\mathcal{E} = \mathcal{E}_{pzc}$, the potential that puts M at the point of zero charge so $g(\text{ion}) = 0$. If $\delta\chi_S^M$ is small, a plot of $\mathcal{E}_{pzc}$ vs. Φ^M for various metals should reflect the variation of g_w^M, since $\phi^{(S)} - \phi^{(Pt)}$ and $\mu_{e^-}^{(Pt)}$ are independent of M.

The points for various metals on such a plot are joined by lines of slope unity, with metals on the same lines having the same g_w^M and thus the same degree of orientation of water on their surfaces. The difference between lines is the difference between g_w^M. Given a value for g_w^M for one metal (Hg), the value of g_w^M for any metal can thus be extracted, for example, it is ~ 0 for Au, ~ 80 mV for Hg or Sb, and 330 mV for Ga or Zn. These values, in turn, can be interpreted in terms of the average angle between the water dipoles and the normal to the surface.

If $\mathcal{E}$ becomes sufficiently negative, the surface charge q^M for Ga becomes a linear function of $\mathcal{E}$ and independent of solute in many cases, suggesting there is no specific adsorption. This is shown in Figure 8. Furthermore, for these values of $\mathcal{E}$ the plot of q^M vs. $\mathcal{E}$ for Hg becomes linear and parallel to that for Ga. In equation [2], $g(\text{ion})$ is due to the diffuse-layer charge (see Section B), and we have

$$\mathcal{E} = \mathcal{F}^{-1}\mu_{e^-}^{(Pt)} + \Phi^{(M)} + g^d(q^M) - g_w^M + (\phi^{(S)} - \phi^{(Pt)}) \quad [3]$$

where $\delta\chi_S^M$ is neglected, and the diffuse layer potential g^d depends only on q^M. If plots of q^M vs. $\mathcal{E}$ for Hg and Ga are parallel, it is suggested by equation [3] that their difference is just the difference in work functions, with g^d and the water dipole potential g_w^M being identical. For less negative values of $\mathcal{E}$, the plots of q^M vs. $\mathcal{E}$ for Hg and Ga differ more and more, indicating a large difference between water structures at the two metal surfaces.

The reciprocal capacitance $d\mathcal{E}/dq^M$ of the metal–solution interface, again neglecting $\delta\chi_S^M$ in equation [2], becomes $dg(\text{ion})/dq^M - dg_w^M/dq^M$ so that

$$C^{-1} = (C_{\text{ion}})^{-1} - (C_{\text{dip}})^{-1} + (C_{\text{GCS}})^{-1}$$

where C_{ion} refers to the ions in the compact layer and C_{GCS} refers to the ions in the diffuse layer. $(C_{\text{ion}})^{-1} - (C_{\text{dip}})^{-1}$ is $(C_i)^{-1}$, where C_i is the inner-layer capacity. For the mercury–NaF interface, there is no charge in the inner layer (no specific adsorption, see Section D), and one can use Gouy–Chapman theory (Section B) to calculate C_i (which is due to adsorbed water) from experimental capacities. Such results are shown in Figure 9. Attempts to formulate models that can explain these capacitances are discussed in Chapter 5.

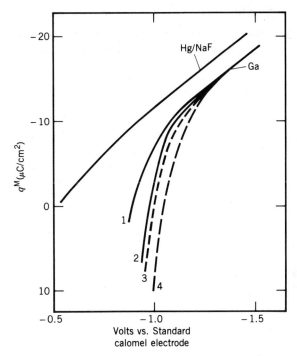

FIGURE 8. Surface charge vs. electrode potential for Hg/NaF and Ga with different electrolytes. Key: 1, NaClO$_4$; 2, KBr; 3, KI; 4, KCNS (cf. Hurwitz 1978a).

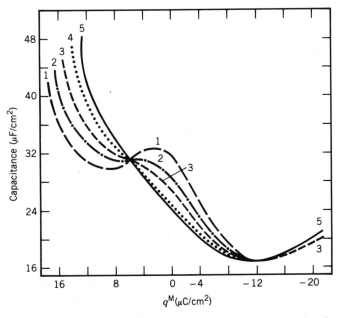

FIGURE 9. Capacitance of Hg/NaF interface at various temperatures and surface changes Key: 1, $T = 0°C$; 2, $T = 25°C$; 3, $T = 45°C$; 4, $T = 65°C$; 5, $T = 85°C$. (cf. Reeves 1980).

A measurable thermodynamic property that gives information on the solvent dipoles at the interface is the surface excess entropy, defined by

$$\Gamma_S = -\left(\frac{\partial \gamma}{\partial T}\right)_{p,\mu,E_\pm}$$

where μ is the chemical potential of the electrolyte and E_+ (E_-) is the potential of the electrode relative to an electrode reversible to cations (anions) (Habib 1977). Since γ is the derivative of the free energy of the system with surface area (Chapter 2, equations [7]–[9]), Γ_S is the derivative of the entropy with surface area. It differs from the interfacial entropy s^σ (Chapter 2, equation [6]), which is not measurable. To obtain the solvent contribution to Γ_S, one subtracts from Γ_S an ionic contribution, $\Gamma_+ \bar{s}_+ + \Gamma_- \bar{s}_-$, where Γ_+ and Γ_- are ionic surface excesses, whereas $\bar{s}_+$ and $\bar{s}_-$ are partial molar ionic entropies, which are experimentally accessible. In the few cases where measurements have been made, the solvent contribution is found to be roughly parabolic as a function of electrode charge. Its maximum occurs at a surface charge of 4–6 $\mu C/cm^2$, to the negative side of the pzc. This may be regarded as the surface charge density corresponding to the maximum disorder or minimum orientation of water molecules.

Another measurable quantity is $\Delta S^{M/S}$, the entropy of formation of the mercury–solution interface, which differs from Γ_S. As a function of surface charge density, it also goes through a maximum at q^M equal to about 4–6 $\mu C/cm^2$, negative to the point of zero charge (Habib 1977, pp. 133–135). The variation of the entropy of formation with q^M, like that of $\Gamma_S - (\Gamma_+ \bar{S}_+ + \Gamma_- \bar{S}_-)$, is attributed to changes in water dipole orientation produced by the field in the inner layer.

In Chapter 5 we will return to the properties of the water surface in contact with metal. Now, we turn to a description of ions in the double layer.

B. DIFFUSE-LAYER THEORY

Most ions are solvated, giving them a size much larger than that of a bare ion. The minimum distance of approach of the center of a solvated ion to the metal surface which is permitted by the solvation sheath is called the *outer Helmholtz plane*. The region outside this plane is referred to as the *diffuse part of the double layer*; the region inside is referred to as the *compact part* or *inner part*. In fact, there should be slightly different outer Helmholtz planes for different species of ion. Some ions are strongly enough attracted to the metal to become adsorbed, losing some of their solvent sheaths, which means they can approach the metal more closely than the outer Helmholtz plane. This "specific adsorption" depends on the properties of individual ions as well as of the electrode. The plane of centers of specifically adsorbed ions is called the *inner Helmholtz plane* (again, one could consider different inner Helmholtz planes for different ions).

The compact layer includes specifically adsorbed ions and solvent molecules, which also may be adsorbed on the electrode. In this section we consider the ions in the diffuse layer, for which a simple theory is generally accepted as adequate. In Section C we show how the theoretical descriptions of the compact and diffuse layers are to be combined. Some ideas about the ions in the compact layer are discussed in Section D.

In the diffuse layer, the competition between electrostatic forces and thermal disorder leads to a distribution of ions which may have a very large extension in the direction perpendicular to the electrode surface. The simple Gouy–Chapman theory for the diffuse layer assumes point ions in a continuous dielectric medium, with only electrostatic forces between the ions and the electrode. For planar symmetry, one has to consider the Poisson equation, which is (in electrostatic units)

$$\frac{d(\epsilon\, d\phi/dz)}{dz} = -4\pi\rho(z) \qquad [4]$$

where

$$\rho(z) = \sum q_i n_i(z) \qquad [5]$$

where q_i is the charge on ions of species i and n_i is their number density. The dielectric constant ϵ is assumed to be the same everywhere and the n_i are assumed to follow a Boltzmann distribution,

$$n_i = n_i(\infty)\, e^{-q_i\phi'/kT} \qquad [6]$$

where $n_i(\infty)$ is the value of n_i infinitely far from the electrode. In this region there is electroneutrality

$$\sum n_i(\infty) q_i = 0$$

and the average electrostatic potential ϕ is taken to be zero.

In the Gouy–Chapman theory, ϕ' of equation [6], which in all rigor is the potential of mean force (involving interparticle correlations: Chapter 3, equations [11]–[14]), is identified with ϕ, and the solvent is represented by a continuous dielectric permeating the entire space. In reality, an ion in the diffuse layer perturbs its environment, so that the potential of mean force at the position of an ion differs from the mean electrostatic potential, and, since a solvent molecule cannot occupy the space of an ion, the ion represents a cavity in the dielectric. One can expect to have problems when the ion concentration becomes large, so that interionic correlations, involving short-range repulsions, become important. Another obvious problem is the use of a continuous dielectric characterized by ϵ to represent a medium composed of molecules no smaller than the ions themselves. Formalisms that attempt to improve on the Gouy–Chapman theory by using a more correct representation of ϕ' have been introduced by a number of workers, and are discussed by Hurwitz (1978a, Section V.2). Unfortunately, experimental verification of the corrections to Gouy–Chapman theory has not been possible; in the concen-

4. STRUCTURE OF SURFACES

tration regime habitually studied experimentally, it seems that these corrections are not important.

Combining equations [4]–[6] gives an equation for ϕ:

$$\frac{\epsilon d^2 \phi}{dz^2} = -4\pi \sum q_i n_i(\infty) e^{-q_i \phi/kT} \qquad [7]$$

This equation can be linearized for small ϕ (indeed, the entire theory gets worse for higher potentials because correlations become more important) and solved for the general electrolyte, or solved as is for the special case of a simple electrolyte with $q_+ = -q_- = q$. (A more general solution is given by Reeves 1980, Section 3.4.) For the simple electrolyte, equation [7] is

$$\frac{\epsilon d^2 \phi}{dz^2} = -4\pi q n(\infty) \left(e^{-q\phi/kT} - e^{q\phi/kT} \right) \qquad [8]$$

Multiplying by $2d\phi/dz$ makes this

$$\frac{\epsilon d(d\phi/dz)^2}{dz} = 16\pi q n(\infty) \sinh\left(\frac{q\phi}{kT}\right) \frac{d\phi}{dz}$$

which integrates to (note $d\phi/dz$ vanishes as $z \to \infty$)

$$\left(\frac{d\phi}{dz}\right)^2 = 16\pi k T n(\infty) \epsilon^{-1} \left[\cosh\left(\frac{q\phi}{kT}\right) - 1 \right]$$

or, using the definition of the hyperbolic functions,

$$\frac{d\phi}{dz} = \pm \left(\frac{32\pi k T n(\infty)}{\epsilon}\right)^{1/2} \sinh\left(\frac{q\phi}{2kT}\right) \qquad [9]$$

where $-d\phi/dz$ is the electric field. Dividing by the sinh and integrating again we have

$$\left(\frac{2kT}{q}\right) \ln \tanh\left(\frac{q\phi}{4kT}\right) = \pm \left(\frac{32\pi k T n(\infty)}{\epsilon}\right)^{1/2} z + C$$

where C is a constant of integration; this equation rearranges to

$$\tanh\left(\frac{q\phi}{4kT}\right) = \exp\left[\pm \left(\frac{8\pi q^2 n(\infty) \epsilon}{kT}\right)^{1/2} z + \frac{Cq}{2kT} \right]$$

To have $\phi = 0$ as $z \to \infty$ we need the negative sign. For $z = 0$, $\phi = p$, the potential drop across the diffuse layer from $z = 0$ to $z \to \infty$, so finally

$$\frac{\tanh(q\phi/4kT)}{\tanh(qp/4kT)} = e^{-\kappa z} \qquad [10]$$

where we have introduced κ, defined by

$$\kappa^2 = \frac{4\pi \sum_i (q_i)^2 n_i(\infty)}{\epsilon kT} = \frac{4\pi q^2 [n_+(\infty) + n_-(\infty)]}{\epsilon kT} \qquad [11]$$

The inverse of κ is the Debye screening length in the Debye–Hückel ion-atmosphere theory of ionic activity coefficients.

If p is small enough for $\tanh(qp/4kT)$ to be approximated by $qp/4kT$, equation [10] is simply $\phi = pe^{-\kappa z}$, so that κ is also the characteristic decay length for the diffuse layer. For $q = e$, $\epsilon = 79$, and $n = 6 \times 10^{18}$ cm^{-3} (0.01 M solution), $\kappa = 2.9 \times 10^7$ cm^{-1}.

Equation [9], with Gauss' theorem relating electric fields to surface charge, allows one to relate the surface charge density of the metal, q^M, to $p = \phi(0)$:

$$q^M = \epsilon \left(\frac{2kTn(\infty)}{\pi\epsilon}\right)^{1/2} \sinh\left(\frac{qp}{2kT}\right) \qquad [12]$$

Note that $(d\phi/dz)_{z=0} = -4\pi q^M/\epsilon$ so that a surface charge density of 10 μC/cm^2 or 3×10^4 esu/cm^2 (which is easily realizable experimentally) gives a surface field of 4.8×10^3 esu/cm^2 or 1.4×10^6 V/cm, which is extremely large. The surface charge density in the diffuse layer is, according to equation [4],

$$q^S = \int_0^\infty dz\, \rho(z) = -(4\pi)^{-1} \int_0^\infty \left[\frac{d(\epsilon d\phi/dz)}{dz}\right] dz$$

$$= (4\pi)^{-1} \left[\frac{\epsilon d\phi}{dz}\right]_{z=0} = -q^M$$

Note that q^M is monotonic in p, with q^M (and q^S) vanishing when p vanishes. According to equation [12], the potential drop corresponding to a given surface charge decreases as the concentration $n(\infty)$ increases, since higher concentration means more efficient screening. Equation [10] shows that all the plots of $\phi(z)$ vs. z for different values of p differ only by a constant factor. Because the right-hand side of equation [10] is an exponential in z, this means that the plots are superposable by a simple translation in z.

4. STRUCTURE OF SURFACES

In general, the linearization of equation [7] yields

$$\frac{d^2\phi}{dz^2} = -4\pi \sum q_i n_i(\infty)\left(\epsilon^{-1} - \frac{q_i\phi}{\epsilon kT}\right) = \kappa^2 \phi$$

since $\sum q_i n_i(\infty) = 0$. The solution for ϕ is easily seen to be a linear combination of $e^{\kappa z}$ and $e^{-\kappa z}$ but the former must be rejected since $\phi \to 0$ as $z \to \infty$. Thus

$$\phi = Ae^{-\kappa z}$$

The constant A is determined, like C mentioned earlier, in terms of the potential drop across the diffuse layer p, so finally

$$\phi(z) = pe^{-\kappa z} \qquad [13]$$

and

$$q^M = -\left(\frac{\epsilon}{4\pi}\right)\left(\frac{d\phi}{dz}\right)_{z=0} = \left(\frac{\epsilon}{4\pi}\right)\kappa p \qquad [14]$$

(This is the linear approximation to equation [12], which would be written $q^M = \epsilon(2kTn(\infty)/\pi\epsilon)^{1/2} qp/2kT$.)

The capacitance per unit area is independent of potential, according to the linearized theory; according to equation [14], this capacitance is $\epsilon\kappa/4\pi$. If $\epsilon = 79$ (appropriate for H_2O), $T = 300$ K, and $\sum (q_i)^2 n_i(\infty) = 0.1e^2$ mol/dm^3, we calculate a capacitance of 4.6×10^7 cm/cm^2 = 51 μF/cm^2. This is larger than typical measured capacitances because the Gouy–Chapman theory describes only the diffuse part of the double layer. From the nonlinear theory (see equation [12]) we obtain

$$\frac{dq^M}{dp} = \left(\frac{q^2 \epsilon n(\infty)}{2\pi kT}\right)^{1/2} \cosh\left(\frac{qp}{2kT}\right) \qquad [15]$$

so that for $q = e$ an excursion of 0.2 V or 1/1500 esu gives $qp/2kT$ a value of 3.9 and increases the capacitance by a factor of 24. Thus the diffuse layer makes no appreciable contribution to capacitance except near the electrocapillary maximum. Given a theory for the compact part of the double layer (see Section C), one can combine it with the Gouy–Chapman theory by putting the distance from the outer Helmholtz plane for z in the above formulas and interpreting p as the electrical potential at this location.

The surface excess of an ion is obtained in Gouy–Chapman theory by integrating equation [6] from 0 to ∞, assuming the electrolyte phase ends at $z = 0$, and subtracting the integral of $n_i(\infty)$, the constant density of the reference system:

$$\Gamma_i = (N_A)^{-1} \lim (L \to \infty) \int_0^L n_i(\infty)\left(e^{-q_i\phi/kT} - 1\right)dz \qquad [16]$$

dividing by Avogadro's number to convert to moles. This is not the relative surface excess of thermodynamic theory, unless the solvent density is considered constant for all positive z. For the simple symmetric electrolyte equation [10] may be rearranged to

$$e^{q\phi/2kT} = \frac{1 + Pe^{-\kappa z}}{1 - Pe^{-\kappa z}} \qquad [17]$$

where $P = \tanh(qp/4kT)$. Using this in equation [16],

$$\Gamma_i = c_i \int_0^\infty (-4P_i e^{-\kappa z})(1 + P_i e^{-\kappa z})^{-2} dz$$

where $P_i = \tanh(q_i p/4kT)$ and $c_i = n_i(\infty)/N_A$, the concentration of ion i in the homogeneous phase. Therefore

$$\Gamma_i = \frac{-4c_i \kappa^{-1} P_i}{1 + P_i} = -2c_i \kappa^{-1}(1 - e^{-q_i p/2kT}) \qquad [18]$$

with serious problems in this theory becoming evident when $P_i \to -1$. In the linear theory for the general electrolyte, we may write

$$\Gamma_i = (N_A)^{-1} \int_0^\infty n_i(\infty) \frac{-q_i \phi}{kT} dz = -\frac{c_i q_i p}{kT} \int_0^\infty e^{-\kappa z} dz$$

$$= \frac{-c_i q_i p}{kT\kappa} \qquad [19]$$

using equation [13]. (For small values of p and P_i, equation [18] reduces to $-2c_i \kappa^{-1}(q_i p/2kT)$, identical to equation [19]). For unsymmetrical electrolytes, numerical integration is required to get Γ_i (Reeves 1980, p. 111).

Thus the surface excess of ion i in the diffuse layer is positive if $q_i p$ is negative and vice versa. When $p = 0$ (pzc) all Γ_i vanish. For small p, Γ_i is proportional to the ionic charge and to the potential drop p across the layer. If the electrolyte is $C^{+q}A^{-q}$, each ion makes the same contribution, $q_i \Gamma_i = cq^2 p/kT\kappa$, to the surface charge of the diffuse layer.

It has been shown (Chapter 2, equations [35]–[36]) that for a polarizable electrode held at a potential corresponding to the electrocapillary maximum (ecm) and dipping into a solution of C^+A^-, we have

$$\left(\frac{\partial \gamma}{\partial \ln a_\pm}\right)_{ecm} = -RT(\Gamma_{A^-} + \Gamma_{C^+})$$

Although surface excesses of anion and cation are equal at the ecm, one ion or the other may be specifically adsorbed in the compact layer, making the surface concentrations in the diffuse layer unequal. However, if γ at the ecm is found to be independent of electrolyte activity, so that Γ_{A^-} and Γ_{C^+} vanish, one can infer that all the ions are in the diffuse layer, that is, there is no specific adsorption. This turns out to be the case for alkali and alkaline-earth fluorides, implying that the ions involved are not surface-active.

Typically, plots of Γ_- vs. E for different anions converge to the same almost horizontal curve for negative E, that is, Γ_- eventually becomes independent of E (the actual values of E depend on the reference electrode of the cell, but the potential of the reference electrode remains constant). For positive values of E, different anions give markedly different plots, and the surface excesses Γ_- are large and vary rapidly with E. For cations, surface excesses vary linearly with E for more negative values of E, with plots for different cations being almost the same, except for some influence of cation charge. For more positive E values, the curves of cation surface excesses diverge at the same point as do the curves of anion surface excesses. Even for the same cation with different anions, one gets different Γ_+ curves. Except when the anion is fluoride, Γ_+ passes through a minimum and increases with increasing E. This behavior is found with Hg electrodes in aqueous solution, but also with other metals and solvents.

If one obtains $(\partial\gamma/\partial E_-)_{Tp\mu}$ or $(\partial\gamma/\partial E_+)_{Tp\mu}$ (subscript on E indicates reference electrode), that is, $-q^M$, it appears that for very negative q^M (potentials several tenths of volts more negative than the pzc) q^M vs. E has the same slope regardless of the anion, but the slopes are quite different for different anions for positive q^M. Note $q^M = -q^S = -(z_+\mathfrak{F}\Gamma_+ + z_-\mathfrak{F}\Gamma_-)$. Large negative q^M means negative potentials, for which $\Gamma_+ \gg \Gamma_-$. At large positive q^M, or positive potentials, measured surface excesses are such that $\Gamma_- \gg \Gamma_+$.

Diffuse-layer theory may be used for a z-z electrolyte to calculate surface excesses and charges, and results for Γ_+, Γ_-, and q^M vs. E agree well with experimental results for simple salts when $q^M \ll 0$. Going to more positive q^M, theory predicts Γ_- to increase markedly and Γ_+ to decrease, but Γ_+ in fact (except for fluoride) goes through a minimum and increases again. Diffuse-layer theory describes the cations well for $q^M < 0$. One may theorize that for $q^M > 0$ the cations are still only in the diffuse layer, since here there is more repulsion, so it will be even harder for them to approach closely to the electrode. One can assume in this region that Γ_+ is equal to the cation surface excess as calculated by the diffuse-layer theory. Then one can use the theory to calculate ϕ_2 (potential at the outer Helmholtz plane) and from ϕ_2 one can get, from the same theory, the diffuse-layer contribution to Γ_-. Then the inner-layer contribution, corresponding to "specifically adsorbed ions," is calculated by subtracting this from the actual Γ_-.

There are a number of tests for detecting specific adsorption and for evaluating the validity of the Gouy–Chapman theory for the diffuse layer. Reeves (1980, Chapter 3, Section 2) discusses the advantages and disadvantages of different procedures using different kinds of data (thermodynamic surface excesses, capacitances, etc.). For example, the capacitance C is $(\partial q^M/\partial E_-)_\mu$ and $q^M = -q^S = -(z\mathfrak{F}\Gamma_+ - z\mathfrak{F}\Gamma_-)$ for a z-z electrolyte, where E_- means the reference electrode is reversible to anions. One can measure the derivative $(\partial C/\partial E_-)_\mu$ and, assuming a concentration-independent inner-layer capacitance (see Chapter 5, Sections D and E), derive a diffuse-layer contribution and hence values for $\partial^2\Gamma_\mp/\partial(E_-)^2$. For KF, there is good agreement of observed values of $\partial^2\Gamma_+/\partial(E_-)^2$ and $\partial^2\Gamma_-/\partial(E_-)^2$ with those calculated by Gouy–Chapman theory, but, for KCl and KAc, there is agreement in $\partial^2\Gamma_+/\partial(E_-)^2$ only for negative potentials. This confirms once more that K^+ and F^- are not specifically adsorbed, but that other anions are.

Another test involves the quantity $(\partial\Gamma_\pm/\partial q^M)_{T,\mu}$. It follows from equations [34]–[36] of Chapter 2 that

$$d(\gamma + q^M E_+) \equiv dX = E_+ \, dq^M - \Gamma_\mp \, d\mu$$

where the upper (lower) signs refer to a reference electrode reversible to cations (anions). Thus $E_\pm = (\partial X/\partial q^M)$ and $\Gamma_\mp = -(\partial X/\partial\mu)$ so that

$$\left(\frac{\partial E_\pm}{\partial \mu}\right)_{T,q^M} = -\left(\frac{\partial \Gamma_\mp}{\partial q^M}\right)_{T,\mu} \qquad [20]$$

For a symmetrical electrolyte, $\mu = \mu^0 + RT \ln (a_\pm)^2$ so that

$$\frac{\partial E_\pm}{\partial \ln (a_\pm)^2} = \mp \frac{RTcqe^{\pm qp/2kT}}{kT\kappa} \frac{\partial p}{\partial q^M}$$

where equation [18] has been used for the surface excess. The derivative dq^M/dp is given by equation [15], so that [$c = n(\infty)/N_A$, $q = ze$, $N_A e = \mathcal{F}$]

$$\frac{\partial E_\pm}{\partial \ln (a_\pm)^2} = \mp \frac{RT}{2N_A q} \frac{e^{\pm qp/2kT}}{\cosh(qp/2kT)} \qquad [21]$$

Finally, one can get $e^{qp/2kT}$ in terms of q^M by rewriting equation [12] as

$$e^{qp/kT} - 1 = e^{qp/2kT} \left(\frac{kT\epsilon n(\infty)}{2\pi}\right)^{-1/2} q^M \equiv \alpha e^{qp/2kT}$$

and solving for $e^{qp/2kT}$. Then, with the upper signs in equation [21],

$$\frac{\partial E_+}{\partial \ln (a_\pm)^2} = -\frac{RT}{z\mathcal{F}} \frac{\alpha^2 + \alpha(\alpha^2+4)^{1/2} + 2}{\alpha^2 + \alpha(\alpha^2+4)^{1/2} + 4}$$

According to diffuse-layer theory, then, a plot of E_+ vs. $\ln (a_\pm)^2$ at constant q^M (Esin–Markov plot) should be linear. The slope should decrease in size with increasing q^M: it should be $-RT/z\mathcal{F}$ for large positive values of q^M ($\alpha \to +\infty$), $-RT/2z\mathcal{F}$ for $q^M = 0$ ($\alpha = 0$), and something approaching 0 for large negative values of q^M ($\alpha \to -\infty$). The results for NaF solutions bear out these predictions, with ~ 20 $\mu C/cm^2$ being a large enough surface charge to show the limiting behavior.

Gouy–Chapman theory is expected to have difficulties for concentrations of 0.01 M and higher, like the Debye–Hückel theory for bulk electrolytes (Habib and Bockris 1980, Section 4.7). Not only does it ignore short-range forces between ions and between ions and molecules, and consider point charges in a structureless dielectric, but it does not consider ion pair formation. It is clearly ridiculous when κ^{-1}, the diffuse-layer thickness, is less than the average distance between ions, $(1000/2N_A c)^{1/3}$. Many tests of the diffuse-layer theory, starting with Grahame's calculation (Grahame 1957) of capacitance of Hg/NaF (up to 0.1 M, near the pzc) have been made through the years. The conclusion is that the theory has not really been quantitatively tested.

Discussing the evidence for the validity of diffuse-layer theory, Reeves (1980, p. 114) concludes that it is not possible to prove or disprove it unambiguously from experimental data. Of course, if one can perform calculations with models that go beyond this theory, one can see where the theory's assumptions lead to error. Such models are discussed in Chapter 6, Section A. It is, however, usual to use the theory for the diffuse part of the double layer, removing diffuse-layer contributions and developing more theories for the remainder.

C. COMPACT LAYER

The capacitance $c_2 = dq^M/dp$ from equation [14] or [15] does not agree with measured capacitances, even for low concentrations of salts like NaF (neither ion of which is surface-active) and even near the ecm. The problem is not the Gouy–Chapman theory for the diffuse layer, but the existence of the compact layer, consisting of water molecules, in series with it. For 0.01 NaF at 0°C, C is measured as 14.5 $\mu F/cm^2$, whereas C_2 is ($\epsilon = 78$, $n(\infty) = 6.02 \times 10^{18}$ cm^{-3}) 2.27 $\times$ 10^7 cm^{-1} or 25.2 $\mu F/cm^2$. If the measured capacitance is that of two capacitances in series, that is, $C^{-1} = (C_1)^{-1} + (C_2)^{-1}$, where $C_2 = (8\pi q^2 \epsilon n(\infty)/kT)^{1/2}$ and C_1 is due to the compact layer, C_1 should be 34 $\mu F/cm^2$ or 3.05 $\times 10^7$ cm^{-1}. If this represents $\epsilon/4\pi d$ (Chapter 2, Section I), then ϵ/d is 3.8 $\times 10^8$ cm^{-1}. The thickness of a monolayer of water is about 3 Å, corresponding to a dielectric constant of about 11 in the compact layer, which seems reasonable for oriented water molecules. Of course, taking C_1 as a constant cannot explain the variation of capacity with potential and more complex theories are needed for the adsorbed water.

Often, the compact layer includes adsorbed ions as well as water. For a mercury electrode in contact with an aqueous solution of a z–z electrolyte,

$$-\left(\frac{\partial \gamma}{\partial \mu}\right)_{E_\mp} = \Gamma_{\pm,w} = -(\nu_\pm RT)^{-1}\left(\frac{\partial \gamma}{\partial \ln a_\pm}\right)_{E_\mp}$$

(Γ = relative surface excess, $\pm$ refers to a reference electrode reversible to anion or cation). From the Lippmann equation, one obtains the surface charge: $(\partial \gamma / \partial E_\pm)_\mu = q^M$. Now $-q^M = q^S = z\mathcal{F}(\Gamma_+ - \Gamma_-)$, where $\Gamma_i = \int_0^\infty [c_i - c_i(\infty)]\, dz$. This includes ions adsorbed specifically and also "electrostatically." To sort out specific adsorption (in the compact layer), one can assume that cations are not specifically adsorbed, so Γ_+ is calculable from diffuse-layer theory. Knowing the cation's contribution to q^S, and q^S itself from the Lippmann equation, one can get Γ_- and compare with the diffuse-layer contribution. For more details and related methods, see Habib and Bockris (1980, Section 4.5.1.1). Direct methods for determining adsorption also exist (Habib and Bockris 1980, Section 4.5.3). These methods are based on radiotracers and on ellipsometry. Radiotracer methods measure the surface excess of particular ions; diffuse-layer theory is then used to extract specific adsorption. The ellipsometric methods are sensitive to changes in the (complex) refractive index of the adsorbed layer, as well as the indices of the solution and the substrate metal; a model for the adsorbed layer is required to analyze the experimental data in terms of amounts of ion adsorbed. Careful comparison of results for different measurements of specific adsorption are found in Habib and Bockris (1980, Section 4.1.2).

To consider the contributions of the diffuse and compact layers together, suppose the solution side of the double layer is divided into the compact layer, extending from $z = 0$ to the outer Helmholtz plane at $z = z_2$, and the diffuse layer, extending from $z = z_2$ to $z \to \infty$. The surface charge density is correspondingly

$$q^S = q^1 + q^2$$

where q^1 is the contribution of adsorbed ions at the inner Helmholtz plane and q^2 is the contribution of the diffuse layer, for which Gouy–Chapman theory (equations [12] or [14]) is generally considered sufficient. Let ϕ_M be the potential at $z = 0$ and ϕ_2 the potential at z_2, the potential at $z \to \infty$ being zero. Thus ϕ_2 may be identified with p in the Gouy–Chapman theory. If q^1 is due to anions only,

$$q^1 = -z\mathcal{F}\Gamma_{A^-,1} = -z\mathcal{F}(\Gamma_{A^-} - \Gamma_{A^-,2}) \qquad [22]$$

where the diffuse-layer contribution is, according to equation [18],

$$\Gamma_{A^-,2} = -2c_{A^-}\kappa^{-1}(1 - e^{z\mathcal{F}p/2RT})$$

for a symmetrical electrolyte with ions of charge $\pm z$. For cations, assumed not present in the compact layer,

$$\Gamma_{C^+} = \Gamma_{C^+,2} = -2c_{C^+}\kappa^{-1}(1 - e^{-z\mathcal{F}p/2RT})$$

Since $z\mathcal{F}(\Gamma_{C^+} - \Gamma_{A^-}) = q^S = -q^M$, we may use these equations to calculate p or ϕ_2 in terms of q^1:

$$q^1 = -z\mathcal{F}[\Gamma_{A^-} - 2c\kappa^{-1}(e^{z\mathcal{F}p/2RT} - 1)]$$

4. STRUCTURE OF SURFACES

The diffuse-layer charge is given by

$$q^1 + q^M = -z\mathcal{F}(2c\kappa^{-1})(e^{-z\mathcal{F}p/2RT} - e^{z\mathcal{F}p/2RT}) \qquad [23]$$

The potential across the diffuse layer, p, is now ϕ_2, so

$$\phi_2 = \frac{2RT}{z\mathcal{F}} \sinh^{-1}\left[(q^1 + q^M)\left(\frac{2\epsilon RTc}{\pi}\right)^{-1/2}\right] \qquad [24]$$

To calculate capacitances, we divide the potential drop across the electrolyte into potential drops across compact and diffuse layers, so that

$$C^{-1} = \frac{d(\phi_M - \phi_2)}{dq^M} + \frac{d(\phi_2)}{dq^M} = (C_1)^{-1} + (C_2)^{-1} \qquad [25]$$

All derivatives are for constant solution composition. The diffuse-layer capacitance C_2 is calculated from equation [14] or [15] with $q^M + q^1$ of equation [23] replacing q^M and ϕ_2 replacing p in those equations. Thus for a symmetrical electrolyte with $q = ze = z\mathcal{F}/N_A$,

$$C_2 = \left(\frac{z^2\mathcal{F}^2\epsilon c}{2\pi RT}\right)^{1/2} \cosh\frac{z\mathcal{F}\phi_2}{2RT}\left[\frac{d(q^1 + q^M)}{dq^M}\right]^{-1}$$

so that

$$C^{-1} = \frac{d(\phi_M - \phi_2)}{dq^M} + \left(\frac{2\pi RT}{z^2\mathcal{F}^2\epsilon c}\right)^{1/2} \text{sech}\left(\frac{z\mathcal{F}\phi_2}{2RT}\right)\left[1 + \frac{dq^1}{dq^M}\right] \qquad [26]$$

allows the calculation of the compact-layer capacity from measured C. For example, according to equation [37] below,

$$dq^1/dq^M = \epsilon_c(z_2 - z_1)^{-1}\frac{d\phi_M}{dq^M} = \epsilon_c(z_2 - z_1)^{-1} C^{-1}$$

which, when substituted into equation [26], gives an equation for $z_2 - z_1$, the thickness of the compact layer, in terms of the measured capacity.

It should be noted that equation [26] does not represent two capacitances C_1 and C_2 in series, unless $dq^1/dq^M = 0$. This occurs when there is no adsorption of ions in the compact layer, so that $q^1 = 0$. Then indeed equation [26] gives C^{-1} as the sum of the reciprocals of the compact-layer and diffuse-layer capacitances. Furthermore, using equation [9] for $z = 0$, with $(d\phi/dz)_{z=0} = -4\pi q^M/\epsilon$ and $\phi_{z=0} = \phi_2$,

$$\cosh^2 \frac{z\mathcal{F}\phi_2}{2RT} = 1 + \sinh^2 \frac{z\mathcal{F}\phi_2}{2RT} = 1 + \frac{\pi(q^M)^2}{2\epsilon kTn(\infty)}$$

Then equation [26] becomes

$$(C_1)^{-1} = C^{-1} - \left(\frac{2\pi RT}{z^2\mathcal{F}^2\epsilon c}\right)^{1/2} \left[1 + \frac{\pi(q^M)^2}{2\epsilon kTn(\infty)}\right]^{-1/2} \quad [27]$$

which shows how one obtains the compact-layer capacity C_1 from measured values of the capacity C. This was done by Grahame for NaF at series of temperatures and concentrations (Grahame 1947, 1957). The results, in the form of plots of inner-layer capacitance vs. surface charge, were shown in Figure 9. The capacitances depend very little on electrolyte concentration, as should be the case if there is no specific adsorption. Attempts to understand them in terms of models have been made since Grahame's early work; these attempts continue to be made (see Chapter 5).

D. IONS IN THE COMPACT LAYER

The region between the electrode surface and the outer Helmholtz plane is supposed to contain specifically adsorbed ions and a monomolecular layer of solvent. The solvent molecules here are oriented by the electrostatic and other forces of the electrode, so that the dielectric constant, which measures the ability of the molecules to screen electric fields by polarization and orientation of their dipoles, should have a value in this region considerably less than the bulk value. The high bulk dielectric constant for water, about 78 at room temperature, is largely due to the dipolar orientation, and a value of 6, corresponding to the molecular electronic polarizability alone, is often taken for water in the compact layer. At more positive potentials of electrode vs. solution, the negative (oxygen) ends of the water molecules are preferentially directed toward the electrode; at more negative potentials, the hydrogen ends are more toward the electrode. The orientation increases with increasing positive or negative electric field. Higher field strength should also compress the solvent monolayer and increase the capacitance.

The original theory of Stern considered that the entire metal–solution potential drop occurred in the first monolayer of solution, which included adsorbed ions and solvent. The potential felt by the adsorbed ions, ψ, would then be some fraction of the potential of the metal (the potential outside the compact layer is zero in this model), and the electrostatic interaction of the ions with the potential would be an additive term in the free energy of adsorption. The adsorption isotherm (cf. Chapter 2, Section F) would then be

$$\theta = a_i K(\theta) \exp\left(\frac{-z_i e\psi}{kT}\right) \quad [28]$$

where θ is the fractional coverage of surface by ions i, the function $K(\theta)$ includes the effect of interactions between adsorbed molecules, a_i is the activity of ions i, and z_i is the charge on ion i.

A necessary condition on an adsorption isotherm for the compact layer was derived by Esin and Markov (Barlow, 1970), considering a polarizable electrode forming a cell with, on the left-hand side, an electrode reversible to cations C. According to equation [20]

$$-\left(\frac{\partial E_+}{\partial \mu_{CA}}\right)_{q^N} = -\left(\frac{\partial \phi^M}{\partial \mu_{CA}}\right)_{q^N} = \left(\frac{\partial \Gamma_{A^-}}{\partial q^N}\right)_{\mu_{CA}}$$

$$= \left(\frac{\partial \Gamma_{A^-,1}}{\partial q^N}\right)_{\mu_{CA}} + \left(\frac{\partial \Gamma_{A^-,2}}{\partial q^N}\right)_{\mu_{CA}} \quad [29]$$

(1 and 2 refer to compact- and diffuse-layer contributions). Here, E_+ is the potential of the cell $M|C|S|N|M$, where S is the solution containing the electrolyte CA; q^N is the surface charge density on the metal of the polarizable interface; and $\mu_{CA} = \mu_{CA}^0 + RT \ln (a_+)^2$ is the chemical potential of CA. For experiments on a number of salts and a range of values of q^N and μ_{CA}, Esin and Markov (Hurwitz, 1978a) found the coefficient on the left-hand side of [29] to be the same. Since $\Gamma_{A^-,2}$ was negligible under the conditions of the experiments, this means that $(\partial \Gamma_{A^-,1} / \partial q^N)_{\mu_{CA}}$ is constant.

Isotherms based on the Stern theory cannot explain this. For instance, if $\theta = a_{A^-} \exp(-z_{A^-} \phi_a / kT)$ and $\Gamma_{A^-,1} = C\theta$ with C a constant, $(\partial \Gamma_{A^-,1} / \partial q^N)_{\mu_{CA}} = \Gamma_{A^-,1}(-z_{A^-}/kT)(\partial \phi_a / \partial q^N)_{\mu_{CA}}$ and the last factor is constant but $\Gamma_{A^-,1}$ is not. Theories that replace the average potential ψ by a "micropotential," differing from point to point in the compact layer, are required. The calculation of the micropotential must take into account the effect of each ion on the distribution of ions around it, as well as the fact that each ion is a cavity in the dielectric. An average electrostatic potential can be calculated afterward by averaging laterally over the micropotential. Hurwitz (1978a, Section V.3) give details and references to some calculations of the micropotential.

For low occupation of the compact layer by ions, $K(\theta)$ becomes independent of θ and θ becomes proportional to a_{A^-} (Henry's law), so that from equation [28] we get

$$q^1 = Aa_{A^-} e^{-z\mathcal{F}\psi/RT} \quad [30]$$

It can be argued, making a series of approximations (Barlow 1970, pp. 234–235) that $(\phi_M - \phi_2)$ is proportional to q^1. Then, representing $\phi_M - \phi_2$ by V_1,

$$\frac{d \ln V_1}{d \ln a_-} = \frac{d \ln q^1}{d \ln a_-} = 1 - \frac{z\mathcal{F}}{RT} \frac{d\psi}{d \ln a_-}$$

$$= 1 - \frac{z\mathcal{F}}{RT} \frac{d\psi}{dV_1} \frac{dV_1}{d \ln a_-}$$

Rearranging, we obtain

$$\frac{dV_1}{d \ln a_-} \left[(V_1)^{-1} + \frac{z\mathcal{F}}{RT} \frac{d\psi}{dV_1} \right] = 1$$

For increasing a_-, the first term in square brackets becomes smaller than the second, so $dV_1/d \ln a_-$ approaches $(RT/z\mathcal{F})(dV_1/d\psi_1)$. This is an important part of the Esin–Markov coefficient $d\phi_M/d \ln a_{A^-}$. The Stern theory equates V_1 and ψ and, replacing $dV_1/d\psi$ by 1, it predicts too low a value for $dV_1/d \ln a_-$.

The difference between the average potential in the compact layer and the micropotential, due to the correlations between a particular ion and its neighbors, can be partly taken into account by a simple model for specific adsorption; this assumes that the ions are discrete charges, arranged in a hexagonal array with their centers on the inner Helmholtz plane at $z = z_1$. The value of ϕ_1, the average electrical potential at z_1, is between ϕ_M, the potential at the electrode boundary at $z = 0$, and ϕ_2, the potential at the outer Helmholtz plane at $z = z_2$. Interpolating linearly, we get

$$\phi_2 - \phi_1 = \frac{(z_2 - z_1)(\phi_2 - \phi_M)}{z_2} \qquad [31]$$

For the hexagonal arrangement of ions, it can be shown (Barlow 1970, p. 235) that

$$\phi_M - \phi_2 = \frac{q^1(z_2 - z_1)}{\epsilon_c}$$

where q^1 is the surface charge density of the specifically adsorbed ions and ϵ_c is the dielectric constant in this region. Combining with equation [31], we obtain

$$\phi_1 = \phi_2 + \frac{(z_2 - z_1)^2 q_1}{z_2 \epsilon_c} \qquad [32]$$

The potential ϕ_1 affects the electrochemical potential of adsorbed ions:

$$\tilde{\mu}_{i,1} = \mu_{i,1}^0 + RT \ln \Gamma_{i,1} + z_i \mathcal{F} \phi_1 \qquad [33]$$

where the conventional form for electrochemical potentials is used with concentration replaced by surface concentration. Since $\tilde{\mu}_{i,1} = \tilde{\mu}_i = \mu_i^0 + RT \ln a_i$, where $\tilde{\mu}_i$ is the electrochemical potential of ion i in bulk electrolyte, and $\phi = 0$ in bulk, far from the electrode, equation [33] gives

$$RT \ln \frac{\Gamma_{i,1}}{a_i} = -(\mu_{i,1}^0 - \mu_i^0) - z_i \mathcal{F} \phi_1 \qquad [34]$$

4. STRUCTURE OF SURFACES

In this adsorption isotherm, the quantity $\mu_{i,1}^0 - \mu_i^0$ is the standard molar free energy change on adsorption; it should depend on the activity of adsorbed ions and on the charge of the electrode.

Because of their higher polarizabilities and the better solvation of cations, anions are much more likely to be specifically adsorbed than cations. Then, considering adsorbed anions of charge $-z$, $q^1 = -z\mathfrak{F}\Gamma_{-,1}$; thus, using equation [34], we obtain

$$RT \ln \frac{-q^1}{z\mathfrak{F}a_-} = -(\mu_{-,1}^0 - \mu_-^0) + z\mathfrak{F}\phi_1$$

Then, keeping q^M constant and ignoring the dependence of the free energy of adsorption on activity of adsorbed ions, we get

$$RT\left[\left(\frac{\partial \ln(-q^1)}{\partial \ln a_-}\right)_{q^M} - 1\right] = z\mathfrak{F}\frac{\partial \phi_1}{\partial \ln a_-} \quad [35]$$

where q^1 is calculated from the measured Γ_- and the calculated diffuse-layer contribution. This equation may be integrated to calculate ϕ_1,

$$\phi_1 = \left(\frac{RT}{z\mathfrak{F}}\right) \int \left[\left(\frac{\partial \ln(-q^1)}{\partial \ln a_-}\right)_{q^M} - 1\right] d \ln a_- \quad [36]$$

where the integration constant is determined by the fact that ϕ_1 becomes equal to ϕ_2 when $a_- \to 0$ because then $q^1 \to 0$ (see equation [32]). In particular, equation [36] allows one to determine ϕ_1 at the pzc ($q^M = 0$) for different anion activities.

Returning to the model, note that at high electrolyte concentrations, q^1 becomes large and ϕ_2 may be neglected. Then equations [32] and [31] give

$$q^1 = \frac{\epsilon_c z_2 \phi_1}{(z_2 - z_1)^2} = \frac{\epsilon_c \phi^M}{z_2 - z_1} \quad [37]$$

Substituting for q^1 and for ϕ^1 into equation [35] we get

$$\left(\frac{\partial \ln \phi_M}{\partial \ln a_-}\right)_{q^M} - 1 = \frac{z\mathfrak{F}}{RT}\frac{\partial \phi_M}{\partial \ln a_-}\frac{z_2 - z_1}{z_2}$$

and, writing the first derivative as $(\phi_M)^{-1}(\partial \phi^M/\partial \ln a_-)$, we have

$$\left(\frac{\partial \phi_M}{\partial \ln a_1}\right)_{q^M} = \left[\frac{1}{\phi_M} - \frac{z\mathfrak{F}}{RT}\frac{z_2 - z_1}{z_2}\right]^{-1} \quad [38]$$

The derivative of ϕ_M with $\log_{10} a_-$ is sometimes called the Esin–Markov coefficient. If $\phi_M \gg RT/z\mathfrak{F}$, equation [38] becomes

$$\left(\frac{\partial \phi_M}{\partial \log_{10} a_-}\right)_{q_M} = -2.303 \frac{RT}{z\mathcal{F}} \frac{z_2}{z_2 - z_1}$$

which is a constant, as found experimentally. For $q_M = 0$ this quantity is the shift in the pzc with activity of surface-active ions (anions that can be specifically adsorbed). For adsorption of iodide, Esin and Markov found a shift in the pzc of 120 mV. Since $2.303 RT/\mathcal{F} = 59$ mV, this implies that the inner Helmholtz plane is halfway between the metal and the outer Helmholtz plane, which fits the picture of specifically adsorbed ions with their centers on $z = z_1$.

Large cations may specifically adsorb, raising the problem of finding specific adsorptions when both cations and anions adsorb specifically. Cations and anions will adsorb at different distances from the electrode, say z_1^+ and z_1^-. With z_2 being the distance of the outer Helmholtz plane, one has to consider surface charge densities on the metal (q^M) and at both inner Helmholtz planes (q_1^+, q_1^-), as well as the (average) electrical potentials at M, z_1^+, z_1^-, and z_2. The potential differences, assuming $z_1^+ < z_1^-$ and a constant dielectric constant in the compact layer, are as follows:

$$\phi_M - \phi_1^+ = \left(\frac{4\pi}{\epsilon_0}\right) q^M z_1^+ \quad [39]$$

$$\phi_1^+ - \phi_1^- = \frac{4\pi}{\epsilon_0} (q^M + q_1^+)(z_1^- - z_1^+) \quad [40]$$

$$\phi_2 - \phi_1^- = \frac{4\pi}{\epsilon_0} (q^M + q_1^+ + q_1^-)(z_2 - z_1^-) \quad [41]$$

Let q_2 represent the diffuse-layer surface charge, consisting of electrostatically adsorbed cations and anions:

$$q_2 = q_2^+ + q_2^-$$

The Gouy–Chapman theory may be used for q_2. With a z–z electrolyte and $\phi = 0$ at $x \to \infty$,

$$q_2 = -2\left(\frac{RT\epsilon C}{2\pi}\right)^{1/2} \sinh \frac{\mathcal{F}\phi_2}{2RT} \quad [42]$$

Of course, $q^M + q_1^+ + q_1^- + q_2 = 0$.

From electrocapillary or capacitance measurements one can deduce surface excesses of cations and anions:

$$\mathcal{F}\Gamma_+ = q_1^+ + q_2^+ \quad [43]$$

$$\mathcal{F}\Gamma_- = q_1^- + q_2^- \quad [44]$$

For $TlNO_3$, Delahay and coworkers (Delahay 1966) assumed $z_1^+ = 2$ A, $z_1^- = 3.15$ A, $\epsilon_0 = 6$, and solved the equations for q_1^+ and q_1^- as functions of electrode charge and $TlNO_3$ concentration. A problem with this method is that the results are sensitive to the choices of z_1^+, z_1^-, and ϵ_0. Habib and Bockris (1980, Section 5.13.3) discuss methods for the determination of specific adsorptions of simultaneously adsorbing anions and cations.

The adsorption of ions on a metal (or on a nonmetal) involves the desorption of water. The free energy of adsorption derived from experimental isotherms thus includes a contribution of the free energy of adsorption of water. Indeed, one is really measuring the relative interaction of ions and water with the electrode, as well as the relative interaction of ions and water with the bulk solution. The relative strength of interactions between electrode and solvent or adsorbate also governs whether adsorption is specific or diffuse. Of course the total adsorption of an ion depends on the electrode charge. A theory of ionic adsorption in the compact layer must consider partial dehydration of the ion, as well as dispersion and image interactions with the metal. The image forces depend on the dielectric constant in the electrolyte and its variation with distance from the electrode (Bockris and Khan 1979, Sections 1.4, 1.5).

Several authors have suggested that, in the specific adsorption of ions on a metal electrode, a partly covalent bond is formed between ions and metal (Habib and Bockris 1980, Section 4.9). The adsorbed species then transfers part of its charge to the metal. If the fraction of charge transferred is known, one can obtain information about the specific surface excess from electrical measurements. Whether covalent bond formation in fact contributes importantly to adsorption of anions on a metal is not settled. The anions are generally closed-shell ions, and most workers have discussed the process of specific adsorption from the point of view of electrostatic models. Thus, ΔH and ΔS for adsorption are considered to consist of contributions from image and dispersion forces, changes in hydrogen bonding, rearrangement of solvation spheres, and so on (Habib and Bockris 1980, Section 4.10).

Isotherms for ionic adsorption are discussed by Habib and Bockris (1980, Section 4.11). Electrochemical isotherms must show the dependence of adsorption on electrode charge (or potential) as well as on bulk activity. Bockris and co-workers calculated the free energy of adsorption considering (1) dispersion and image interactions of the ions with the metal and (2) lateral repulsions between adsorbed ions. A smooth function describing the variation of the dielectric constant with distance from the metal was constructed: Its value is about 6 out to $2\frac{1}{2}$ A, then rises linearly to about 50 near the outer Helmholtz plane at 6.3 A, then gradually approaches the bulk value of 79 at about 12 A. For the field in the inner layer, the field dependence of the dielectric constant is very important. Whether the smoothness of the variation of ϵ with distance is important is not as clear (Bockris and Khan 1979, Section 1.6).

The importance of dispersion interactions (and the difficulties with theories explaining specific ionic adsorption in terms of covalent forces) follows from the increased adsorption of ions of larger ionic radius. Furthermore, ions of small radius will have a tightly held "solvation sheath" of oriented water molecules, which moves with the ion migrating through solution and prevents the ion from getting close to the electrode, thus preventing specific adsorption (Habib and Bockris 1980, Section 4.12). Calculation of adsorption energies and other thermodynamic properties is discussed by Bockris and Khan (1979, Section 1.5).

Neutral substances can also be adsorbed in the double layer. Organic molecules, being less polar than solvent molecules, are surface-active. Some of these can also interact chemically with the surface of metal electrodes, such as aromatic molecules through their π-electron clouds. If the electric field at the surface becomes large, the electrostatic attraction of polar solvent molecules will lead to the squeezing out or desorption of surfactant molecules from the electrode region. Thus the surface excess of neutral molecules as a function of electrode potential often shows a single maximum, as does the difference between the interfacial tensions in the presence and the absence of surfactant as a function of potential. If the surfactant adsorbs only by virtue of the smallness of its dipole moment, adsorption is symmetric about the ecm, while chemisorptive interactions are evidenced by a shift of the adsorption region to one side or the other. The adsorption of neutral organic species is discussed by Van Laethem-Meuree (1978, Chapter VI).

E. METALS

The typical electrochemical interface is between a metal and an ionic conductor. A crystalline metal is a regular array of ions, corresponding to the cores of the atoms from which the solid is formed, in a sea of conduction electrons, whose parentage is the valence electrons of the atoms. The electrons of the ionic cores are strongly bound or localized, whereas the conduction electrons move easily over distances large compared to interatomic spacing, which is responsible for the high electrical and thermal conductivities of metals. The spatial delocalization of the conduction electrons means they have low kinetic energies and low-energy excitations, which are responsible for the metallic appearance. The chemical and electrical properties of metals are understood largely in terms of the conduction electrons, just as the chemical behavior of molecules is explicable in terms of wave functions for the valence electrons. In a liquid metal, the ion cores are not arranged regularly, but one still has core and conduction electrons with the same general properties. One should note, however, that in some metals the distinction between core and conduction electrons is less sharp than in others.

Chemisorption on a metal surface involves the overlap of the electronic "tail" with molecular orbitals of adsorbate molecules. The overlap between a localized orbital on a molecule and a diffuse cloud on the metal is not a typical covalent bond, but rather a polarization of the orbital of the adatom. True covalent bonding can also occur with atomic orbitals on metal cores. The electrons in these orbitals would then be localized and lose some of their mobility, perhaps lowering the conductivity. It is also possible for cores of an adsorbing species (in particular, hydrogen) to enter the lattice of the metal, with the valence electrons becoming part of the delocalized electron gas.

Even if one considers a metal to consist of metal ions and conduction electrons, one must recognize that, although this is a system of two charged components, the electrons must be treated quite differently from the ions. Although it may be possible to describe the latter by classical statistical mechanics, the electrons must clearly be treated quantum mechanically, for example, by solution of a Schrödinger equation. Most quantum mechanical calculations of the conduction electrons are based on one-electron theories, such as Hartree–Fock, in which a wave function or spin orbital is assigned to each electron, with the antisymmetric many-electron wave function being a determinant of the one-electron wave functions. The one-electron wave functions are solutions to a Schrödinger eigenvalue equation in which the potential includes the interaction of the electron with the ion cores as well as with all the other electrons. The set of equations, one for each one-electron eigenstate, must be solved self-consistently, since the solutions are needed to construct the potential in the Schrödinger equation representing the interaction of one electron with the others. For a crystalline metal, the self-consistent potential is periodic in space and the eigenfunctions reflect this. Whether the ions are distributed periodically, as in a crystal, or not, as in a liquid, the electronic eigenenergies form a continuum or energy band because the electrons are delocalized.

Because of the mass disparity between electrons and ions, it is appropriate to consider the distribution of electrons for a particular arrangement of the ion cores, as if the ions were stationary. Thus one can determine the energy of the system as a function of ionic positions. This energy may then be used as the potential energy for the ions, and one could treat the system as involving only one component, with the motion of the ions treated by classical statistical mechanics. In principle, one can determine the equilibrium ionic configuration from the total energy as a function of ionic positions. However, the ionic Hamiltonian includes potential energy terms other than the isotropic direct pair interactions between ions: There is also an indirect interaction, which is really the interaction of each ion with the electronic cloud, distorted by all the other ions. As a result of such a treatment one can write the potential energy of a homogeneous metal, to a good approximation, as a structure-independent term, mostly due to the electrons, plus pair interactions, the latter depending on electron density.

At the surface of a crystalline metal the periodicity of the potential is broken in one direction (perpendicular to the surface) but not in the other two. The electron density decays from its bulk value to zero over a distance of several angstroms. Thus the energy of an ion, which involves interaction with the electronic cloud, will depend on its position. The potential seen by the ionic cores near the surface differs from that in the bulk, which may lead to some change in ionic configuration (changed interatomic spacing for a crystal, nonuniform density profile for the ion cores in a liquid metal). Of greater consequence for electrochemistry is the fact that, for a crystal, there will be an electronic atmosphere outside the ionic lattice.

For a liquid metal surface, one must consider density profiles for the ions *and* the electrons, and there is no reason to suppose they coincide. (Indeed, one can argue that an electron density that decreases monotonically from the bulk density to zero through the interface will lead to forces tending to push the ions toward the bulk and, just as would a repulsive wall outside the metal, produce an oscillatory ion density profile.)

If the positively and negatively charged species have different distributions, there will be regions of space where the net electronic charge density $\rho(z)$ is nonzero. This means that there may be a difference of electrostatic potential between the inside ($z \to -\infty$) and the outside ($z \to +\infty$) of the metal:

$$V(\infty) - V(-\infty) = -4\pi \int_{-\infty}^{\infty} dz' \int_{-\infty}^{z'} dz'' \, \rho(z'')$$

$$= 4\pi \int_{-\infty}^{\infty} dz' \, z' \, \rho(z') \qquad [45]$$

The regions where $\rho(z)$ is negative are for larger z, making $V(\infty) - V(-\infty)$ negative. Then electrons escaping from inside the metal must overcome an electrostatic potential barrier in addition to short-range forces. Estimates of the barrier (Section F) show that it is of the order of volts, representing a substantial positive contribution to the work function. The nonsuperposition of electronic and ionic profiles at a metal surface also gives important electrostatic contributions to the surface energy and surface tension. Note that there is no electrostatic energy for the bulk homogeneous fluid, where the charge densities of ions and electrons are equal and opposite; the electrostatic contribution to the surface energy then is independent of the placement of the Gibbs dividing surface.

In general, the average number of electrons occupying a one-electron state i of energy ϵ_i is given by the Fermi distribution

$$n^i = \left\{ \exp\left[\beta(\epsilon_i - \mu)\right] + 1 \right\}^{-1} = f(\epsilon_i) \qquad [46]$$

where $\beta = 1/kT$ and μ is the chemical potential (Davidson 1962, Section 6-16). The value of μ is determined so as to make $\Sigma \, n^i$ equal to the total number of electrons. The total electronic energy would be $\Sigma \, n^i \epsilon_i$, plus a correction for the double counting of the interelectronic repulsion that results from the inclusion of the electron–electron interaction in the potential of each one-electron Schrödinger equation. The existence of a continuum of energy levels requires that sums over states be replaced by integrals over energy. Thus the total number of electrons is

$$N = \int d\epsilon \, f(\epsilon) \, D(\epsilon) \qquad [47]$$

4. STRUCTURE OF SURFACES

and the sum of eigenenergies is

$$S = \int d\epsilon \, \epsilon f(\epsilon) D(\epsilon) \qquad [48]$$

where $D(\epsilon)$ is the density of states (number of electronic eigenstates per unit energy range). The low-temperature (high-β) approximation, in which $\mu = \epsilon_F$ (Fermi energy), and $n^i = 0$ for $\epsilon_i > \epsilon_F$ and $n^i = 1$ for $\epsilon_i < \epsilon_F$, suffices for our purposes:

$$f(\epsilon) = 1, \quad \epsilon < \epsilon_F$$
$$f(\epsilon) = 0, \quad \epsilon > \epsilon_F \qquad [49]$$

For our systems, N will be infinite; it is convenient to consider the electron density n and $\rho(\epsilon)$, the number of electronic eigenstates per unit energy range and per unit volume. Now

$$n = \int d\epsilon f(\epsilon) \rho(\epsilon) \qquad [50]$$

For free electrons, the one-electron states are plane waves with energies $\epsilon = \hbar^2 k^2/2m$ (Ziman 1964, Chapter 3). Including spin orientations, there are $(4\pi k^2)/(4\pi^3)$ states per unit k-range per unit volume, so that the number of electrons per unit volume is

$$n = \int d\epsilon f(\epsilon) \rho(k) \frac{dk}{d\epsilon}$$
$$= \int_0^{\epsilon_F} d\epsilon \frac{k^2}{\pi^2} \left(\frac{\hbar^2 k}{m}\right)^{-1} = \frac{(2m\epsilon_F/\hbar^2)^{3/2}}{3\pi^2} \qquad [51]$$

and the density of (kinetic) energy is

$$n_k = \int_0^{\epsilon_F} d\epsilon \frac{m}{\pi^2 \hbar^2} \left(\frac{2m\epsilon}{\hbar^2}\right)^{1/2} \epsilon = \frac{(2m)^{3/2} \pi^{-2} \hbar^{-3} \epsilon_F^{5/2}}{5} \qquad [52]$$

Then substituting for ϵ_F in terms of n using equation [51] gives

$$n_k = \frac{3}{10} \frac{\hbar^2}{m} (3\pi^2)^{2/3} n \qquad [53]$$

Because of their interaction with the ion cores and with each other, the conduction electrons in a metal are not described by plane waves, unless the interaction can be considered as a weak perturbation. Although the interaction is not weak, the energy bands calculated for certain metals resemble those calculated for free electrons (Harrison 1966; Harrison 1979, Section II.5). This is because, for a conduction electron in the region of an atomic core, much of the effect of the attractive potential is canceled by an increased kinetic energy associated with orthogonality of the conduction electron wave functions to core states. As a result, one can consider pseudo wave functions, which resemble the true wave functions outside atomic core regions and are smooth inside such regions. The Schrödinger equation for a pseudo wave function involves a pseudopotential which includes a repulsive interaction that replaces the orthogonality constraint that the true wave functions must satisfy.

Perhaps the simplest model of a metal is the Sommerfeld model, in which the average potential seen by electrons (Hartree potential) is supposed to be a constant, say V_0, inside the metal and zero outside, with V_0 negative. If the magnitude of V_0 is large, the solutions to the electronic Schrödinger equation are those for a particle in a box, that is, for a one-dimensional metal of length L, $\phi_k(x) = (2/L)^{1/2} \sin(kx)$ with $k = n\pi/L$ ($n > 0$). The corresponding energies are $W = V_0 + \hbar^2 k^2/2m$. Since L is large, the energies are closely spaced with L/π states per unit interval in k. The number of states per unit energy range $\rho(W)$ is $(L/\pi)(dk/dW) = (L/\hbar\pi)(m/2)^{1/2}(W - V_0)^{-1/2}$.

More appropriate for the interior of a metal are periodic boundary conditions on the one-electron wave functions: $\phi_k(0) = \phi_k(L)$. Then the wave functions are e^{ikx} with $k = \pm 2\pi n/L$, the density of states in k-space is $L/2\pi$, and the density of states in energy space is $2(L/2\pi)(dk/dW)$, which is the same as for the sine functions. Dividing by L to get a density of states per unit volume (length) and multiplying by 2 to take into account the two possible electron spin states, we get $\rho(\epsilon) = (\pi\hbar)^{-1}(2m)^{1/2}(W - V_0)^{-1/2}$. For a three-dimensional crystal, the wave functions are $e^{i(kx + ly + mz)}$ with periodic boundary conditions and the energies ϵ are $V_0 + \hbar^2 K^2/2m$ with $K^2 = k^2 + l^2 + m^2$. The number of states per unit crystal volume in $dk\,dl\,dm$ is $(2\pi)^{-3}$, so the number of states with K between K and $K + dK$ is $(2\pi)^{-3} 4\pi K^2\,dK$ and, including spin, the number of states per unit energy range, per unit volume, is

$$\rho(\epsilon) = 2(2\pi^2)^{-1} K^2 \frac{dK}{d\epsilon} = (2\pi^2)^{-1} \left(\frac{2m}{\hbar^2}\right)^{3/2} (\epsilon - V_0)^{1/2}$$

Except for the presence of V_0, this is identical to the free-electron gas mentioned earlier.

The Fermi energy $\epsilon_F = \hbar^2 k_F^2/2m$ is found by setting the integral of $\rho(\epsilon)$ from V_0 to $V_0 + \epsilon_F$ equal to n, the number of electrons per unit volume. (ϵ_F is the energy of the most energetic electron relative to that of the lowest energy; some authors use different definitions of the Fermi energy.) The result is

$$\epsilon_F = \frac{\hbar^2}{2m} (3\pi^2 n)^{2/3} \qquad [54]$$

identical to equation [51]. The energy of the most energetic electron is $\epsilon_F + V_0$.

At equilibrium, one expects the most energetic electron to have the same energy everywhere, that is, if there is an electrical potential ϕ that varies with position, n should vary such that $\epsilon_F = (\hbar^2/2m)(3\pi^2 n)^{2/3} - e\phi$ is constant. According to equation [54], the ratio of electron densities at locations where the electrical potentials are ϕ_1 and ϕ_2 is $n_2/n_1 = (\epsilon_F + e\phi_2)^{3/2}/(\epsilon_F + e\phi_1)^{3/2}$.

The work required to bring an electron from vacuum, infinitely far removed from the metal, to the interior of the metal, keeping T and V constant, is the electrochemical potential. Since at $T = 0$ the electron must go to the lowest unoccupied energy level, the electrochemical potential is

$$\tilde{\mu}_e = \epsilon_F + V_0$$

The work function Φ is the minimum energy required to extract an electron from inside the metal to a position just outside the metal. In the Sommerfeld model, this is equal to $-(\epsilon_F + V_0)$ for a neutral metal, since the electron comes from the highest occupied energy level.

F. METAL SURFACES

One feature missing from the Sommerfeld model is the surface potential χ, which we discussed earlier (equation 45). The electrical potential inside the metal is ψ, where ψ is the potential just outside the metal. For an uncharged metal, $\psi = 0$ so that the chemical and electrochemical potentials are identical, $\mu_e = \tilde{\mu}_e$. In more realistic models the potential inside the metal, relative to vacuum infinitely far from the metal, is

$$\phi = \psi + \chi$$

For an uncharged metal, $\tilde{\mu}_e = \epsilon_F + V_0 - e\chi$, where ϵ_F is the Fermi energy calculated in bulk metal. Thus $\epsilon_F + V_0$ may be identified with the chemical potential μ_e. Note that $\tilde{\mu}_e$ is equal to Lange's "real potential" α_e and also equal to the negative of the work function Φ. For a charged metal ($\psi \neq 0$), the work in bringing an electron from infinity into the metal is the Fermi energy plus the electrical work $-e\phi$, that is,

$$\tilde{\mu}_e^{(M)} = \mu_e^{(M)} - e\psi^{(M)} - e\chi^{(M)} = \alpha_{e^-}^{(M)} - e\psi^{(M)}$$

where $\alpha_{e^-}^{(M)}$, the real potential, includes surface and short-range contributions and is supposed to be independent of $\psi^{(M)}$. The work function is still $\Phi = -\mu_e + e\chi$

$= -\alpha_e$ since it involves bringing the electron from inside only to just outside the metal, where the potential is ψ.

To change $\psi^{(M)}$, one must charge a metal by adding or subtracting electrons. In principle, this affects the chemical potential μ_e so that α_e would change with $\psi^{(M)}$. To show that the change is negligible, consider the charge that must be added to change the potential of a sphere of radius r cm by 6 V (0.02 statvolts). The total charge is (0.02 statvolt) (r cm) = 0.02 r esu, which changes the average electron density in the sphere by $0.48 \times 10^{-2} \, r^{-2}$ esu/cm^3 = $1.5 \times 10^{-18} \, r^{-2} \, e/(a_0)^3$. Since this is small compared to the densities of conduction electrons in metals (as long as r exceeds about 10^{-7} cm), it can hardly affect chemical forces.

Sometimes one writes the work function for a charged metal in terms of V_{eff}, the effective potential that enters the electronic Schrödinger equation. Note that V_{eff} is due to interactions with the ions and the other electrons; it should really depend on the energy of the electron, but not in the Hartree approximation. Then

$$-\Phi = \langle V_{\text{eff}} \rangle + \left(\epsilon'_F - \langle V_{\text{eff}} \rangle \right)$$

where $\langle \cdots \rangle$ represents an average over position, and ϵ'_F is the eigenvalue of the highest occupied electron state (including kinetic and potential energy contributions), so that the quantity in parentheses is the height of the conduction band (equal to ϵ_F). For a neutral system, $V_{\text{eff}}(z)$ is $-e\phi(z) + \mu_{\text{xc}}[n(z)]$, where μ_{xc}, representing the effects of exchange and correlation, depends on position via the local electron density (Lang, 1981). For an uncharged metal, $\langle V_{\text{eff}} \rangle = -e\chi + \langle \mu_{\text{xc}} \rangle$, dividing the potential into surface and bulk contributions. Thus the work function is

$$-\Phi = -e\chi + \langle \mu_{\text{xc}} \rangle + \left(\epsilon'_F - \langle V_{\text{eff}} \rangle \right)$$

and $\tilde{\mu}_e = \mu_e - e\phi$, with the chemical potential $\mu_e = \langle \mu_{\text{xc}} \rangle + (\epsilon'_F - \langle V_{\text{eff}} \rangle)$. Note that $\langle V_{\text{eff}} \rangle$ replaces V_0 of the Sommerfeld model.

A simple quantum mechanical model for the electrons, which shows a surface potential, is as follows: Since we expect that the ionic profile $n_+(z)$ varies more abruptly than the electronic profile $n_-(z)$, we represent the former as a step function, say

$$n_+(z) = n_b, \quad z < 0 \qquad [55]$$

The interaction of an electron with the ion cores and the other electrons, overall, is attractive, causing the electron density to decrease to zero as z becomes positive outside the metal. This suggests consideration of the electronic profile in the presence of an infinitely repulsive wall. The wall models the effect of the ions in attracting the electrons. The position of the wall, z_w, to make the system electrically neutral overall is determined. The wave functions for electrons at an infinite wall are products of imaginary exponentials (free-particle wave functions) in the x and y directions and sine functions vanishing at z_w in the z direction.

4. STRUCTURE OF SURFACES

The electron density in the presence of a wall at z_w is obtained by adding together the densities of the electronic wave functions, starting with zero energy and including enough to give the right density for $z = \to -\infty$ (bulk). The result is, for $z < z_w$:

$$n_-(z) = \rho_b \left\{ 1 + \frac{3\cos[2k_F(z-z_w)]}{4(k_F)^2(z-z_w)^2} - \frac{3\sin[2k_F(z-z_w)]}{8(k_F)^3(z-z_w)^3} \right\} \quad [56]$$

where $\rho_b = z_+ n_b$ and $(k_F)^3 = 3\pi^2 \rho_b$. For electrical neutrality, since $\rho_+(z) = z_+ n_b = \rho_b$ for $z < 0$, we have

$$\rho_b \int_{-L}^{z_w} dz \left\{ 1 + \frac{3\cos[2k_F(z-z_w)]}{4(k_F)^2(z-z_w)^2} - \frac{3\sin[2k_F(z-z_w)]}{8(k_F)^3(z-z_w)^3} \right\} = \int_{-L}^{0} dz\, \rho_b$$

where $L \to \infty$, or

$$\int_{z_w}^{0} dz - \frac{3}{2k_F} \int_{-\infty}^{0} dx (x^{-2}\cos x - x^{-3}\sin x) = -\frac{3}{4k_F} \int_{-\infty}^{0} dx\, x^{-1}\sin x$$

Then to have electroneutrality one must have

$$z_w = \frac{3\pi}{8k_F} \quad [57]$$

The electrostatic potential difference according to equation [45] is then

$$V(\infty) - V(-\infty) = 4\pi \int_{-L}^{0} dz\, z\rho_b - 4\pi \int_{-L}^{z_w} dz\, zn_-(z)$$

$$= 4\pi\rho_b \left[-\int_{0}^{z_w} z\, dz - \frac{3}{4(k_F)^2} \int_{-\infty}^{0} dx(x + 2k_F z_w) \right.$$

$$\left. \cdot (x^{-2}\cos x - x^{-3}\sin x) \right]$$

$$= 4\pi\rho_b \left[\frac{-(z_w)^2}{2} - \frac{3}{4(k_F)^2} + \frac{3z_w k_F \pi}{8(k_F)^2} \right] \quad [58]$$

which, on substituting for ρ_b and z_w, is

$$V(\infty) = V(-\infty) = \frac{k_F}{\pi}\left(\frac{3\pi^2}{32} - 1\right)$$

F. METAL SURFACES

The surface potential of the metal,

$$\chi = V(-\infty) - V(\infty) \qquad [59]$$

is thus positive. For example, if $\rho_b = 0.01\ e/(a_0)^3$, so that $k_F = 0.67\ (a_0)^{-1}$, χ is 0.016 esu = 4.8 V according to this model.

This model may be used to estimate the capacitance per unit area of a metal surface by calculating χ for nonzero surface charge densities. The surface charge density on the metal is

$$q^M = \int dz [n_+(z) - n_-(z)] = -\rho_b(z) \int_0^{z_w} dz$$

$$-\rho_b \int_{-\infty}^0 dx \left[\frac{3 \cos 2k_F x}{4(k_F)^2 x^2} - \frac{3 \sin 2k_F x}{8(k_F)^3 x^3} \right]$$

$$= -\rho_b \left(z_w - \frac{3\pi}{8k_F} \right) \qquad [60]$$

and a compensating surface charge density of $-q^M$ is placed at $z = z_w$. One can then calculate the potential difference across the interface according to equations [45] and [58], with an additional term from $-q^M$:

$$\Delta V = V(-\infty) - V(\infty) = 4\pi q^M z_w + 4\pi\rho_b \left[\frac{z_w^2}{2} + \frac{3}{4\pi(k_F)^2} - \frac{3z_w\pi}{8k_F} \right]$$

$$= 4\pi\rho_b \left[\frac{-z_w^2}{2} + \frac{3}{4\pi(k_F)^2} \right]$$

The capacitance is

$$C = \frac{dq^M}{d\Delta V} = \frac{dq^M/dz_w}{d\Delta V/dz_w} = (4\pi z_w)^{-1} \qquad [61]$$

which corresponds to an ideal plane capacitor with plates separated by z_w: In this model the electron density profile is only shifted, with no change in shape, so that the surface potential χ is always the same and does not contribute to the capacitance. In a more sophisticated model, one must recalculate the electronic density profile when the surface charge is changed; the surface potential may then change with surface charge and the capacitance will include a contribution from the metal surface.

Much more realistic than the hard wall for the electrons at a metal surface are the self-consistent jellium models, in which the ion cores are represented by a semi-infinite background and the electronic profile is calculated taking into account electrostatic and nonelectrostatic interactions between the electrons and the ion cores, as well as interelectronic interactions. In these calculations, one considers a reference system of noninteracting electrons in a potential V_{eff} such that their electron density is the same as the electron density of the interacting electrons, n_-. The energy of the system of interacting electrons can be shown (Lang and Kohn 1970) to be

$$\int V(\mathbf{r}) n_-(\mathbf{r}) \, d\mathbf{r} + \left(\frac{e^2}{2}\right) \int n_-(\mathbf{r}) n_-(\mathbf{r}') \left|\mathbf{r} - \mathbf{r}'\right|^{-1} d\mathbf{r} \, d\mathbf{r}' + T_s + E_{\text{xc}}$$

where V is the true external potential (due to ions of a metal), T_s is the kinetic energy of the reference system, and E_{xc}, representing electronic exchange and correlation energy, is written as a functional of n_-. Then n_- can be calculated by adding the squares of the occupied one-electron functions for the reference system, which are obtained from a Schrödinger equation with potential

$$V_{\text{eff}} = V + \int n_-(\mathbf{r}') \left|\mathbf{r} - \mathbf{r}'\right|^{-1} d\mathbf{r}' + \frac{\delta E_{\text{xc}}}{\delta n_-}$$

The electron density n_- is needed to construct V_{eff}, whereas V_{eff} is required to obtain the one-electron functions which yield n_-; a self-consistent solution to the equations must be found.

The electronic profile from a self-consistent jellium calculation for a step-function profile (Figure 10) exhibits Friedel oscillations just inside the jellium surface, which are of greater importance for lower bulk electron density. For a solid metal, the actual known positions of the ion cores may be used to construct a perturbative correction to the jellium potential. For a liquid metal, one requires an ionic profile (see Croxton 1980, Chapter 5, for a review of calculated properties of the liquid metal surface). Allen and Rice (1978) have performed calculations in which electronic and ionic densities are self-consistently adjusted to minimize the surface energy. Their results yield oscillatory profiles, the electronic and ionic profiles being very similar. Computer simulations on a pseudoatom theory also yield oscillatory profiles (D'Evelyn and Rice 1983).

It is also possible to construct "pseudo-ion" models, in which one deals only with ions, interacting with a potential which incorporates the interactions of the electron gas. As mentioned, a one-particle potential, due to the nonuniformity of the electron gas at the surface, needs to be included. Croxton (1980, Section 5.4) reviews results for surface tensions and surface energies from this and other models.

In a jellium model, the surface potential χ is calculated according to equation [45]. The bottom of the conduction band is at $V_0 - e\chi$, where V_0 is the potential seen by a conduction electron in bulk metal, due to ions and other electrons. The top of the band is at $V_0 - e\chi + \epsilon_F$. Thus the electrochemical potential of the

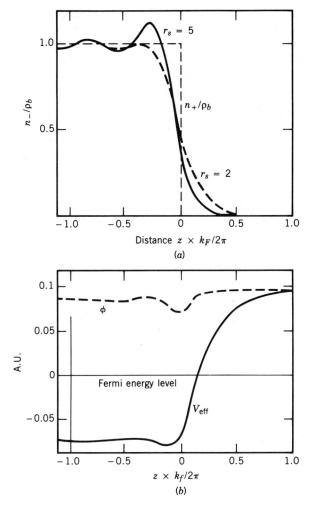

FIGURE 10. Results of jellium calculations. (a) Electronic profiles for step-function ionic profile (dashed rule); $4\pi(r_s)^3/3 = (\rho_n)^{-1}$, ρ_n = bulk density. (b) Electrostatic potential and total effective potential for electrons in atomic units.

electrons is $\tilde{\mu}_e = \mu_e = e\phi$, with $\mu_e = \epsilon_F + V_0$: $\phi = \chi$ for a neutral metal and $\phi = \chi + \psi$ for a charged metal. The work function Φ is $-\mu_e + e\chi$.

As seen in Figure 10, which gives some results of Lang and Kohn (1970), the electron density and hence the exchange-correlation potential is oscillatory, as is the electrostatic potential $\phi(z)$, but V_{eff} is not. Such calculations give (Lamy 1978, p. 225) work functions in good agreement with experimental values for solid metals: Φ is 2.3 eV for Na ($n = 2.5 \times 10^{22}$ cm^{-3}, $r_s = 4.0$ Å), 4.4 eV for Cu ($n = 8.5 \times 10^{22}$ cm^{-3}, $r_s = 2.7$ Å). From measured work functions, values for the surface potential can be derived using calculated values of μ_e, a bulk property.

Note that Φ varies much less from metal to metal than does either μ_e or χ; μ_e changes by some 5 eV between Cu and Na. The importance of χ relative to Φ increases as one goes from monovalent to many-electron metals.

The jellium values of χ are really upper bounds (Trasatti 1980); they agree with experiment (from $\Phi + \mu_e$) for low-melting metals and for crystal faces of high surface density of metal atoms. The actual metal surface is of course not smooth because of the atomic-scale roughness; one may represent this by an ionic density profile that decreases from bulk density to zero with a noninfinite slope. The more negative the slope, the larger is χ. Thus χ is largest for the smoothest surfaces, which are those of high atomic surface density. The chemical potential μ_e is a bulk property and is always the same, so one can predict that work functions ($\Phi = -\mu_e + e\chi$) increase with atomic-scale smoothness, as is usually the case. Polycrystalline surfaces can be considered as a mosaic of patches of different crystal faces, with low-index faces, which have lower atomic density, being most probable (Trasatti 1980). The observed work function would thus be a weighted mean of work functions of various faces, close to that for low-index faces.

The preceding discussion is for zero temperature. At temperatures above absolute zero, the electrochemical potential of an electron, again defined as the work required to bring an electron to the interior of the metal from vacuum infinitely far away, is related to the Fermi distribution as follows. If there are $\rho(\epsilon)\, d\epsilon$ one-electron states per unit volume with energies between ϵ and $\epsilon + d\epsilon$, the number of electrons per unit volume with energies between ϵ and $\epsilon + d\epsilon$ is (see equations [46]–[50])

$$N(\epsilon)\, d\epsilon = \rho(\epsilon)\left\{\exp\left[\beta(\epsilon - \mu)\right] + 1\right\}^{-1}$$

where $\beta = 1/kT$ and μ is chosen to make $\int N(\epsilon)\, d\epsilon$ equal to the total electron density ρ_b. The free energy per unit volume is (Davidson 1962, Section 6-16)

$$A_V = \rho_b \mu - \beta^{-1} \int \rho(\epsilon) \ln\left\{1 + \exp\left[\beta(\mu - \epsilon)\right]\right\} d\epsilon$$

The change in free energy on addition of one electron to a uniform system of volume V is V times the change in free-energy density on addition of V^{-1} electrons to unit volume of the system. Thus the electrochemical potential is

$$\tilde{\mu}_e = V\left(\frac{\partial A_V}{\partial \rho_b}\right)_{T,V,\phi V^{-1}} = \mu + \rho_b \frac{\partial \mu}{\partial \rho_b}$$

$$- \frac{\partial \mu}{\partial \rho_b} \int \rho(\epsilon)\left\{1 + \exp\left[\beta(\mu - \epsilon)\right]\right\}^{-1} \exp\left[\beta(\mu - \epsilon)\right] d\epsilon \quad [62]$$

The last integral is just $\rho_b = \int N(\epsilon)\, d\epsilon$, so that μ is exactly the electrochemical potential and we write

$$N(\epsilon)\, d\epsilon = \rho(\epsilon)\left\{\exp\left[\beta((\epsilon - \tilde{\mu}_{e-}))\right] + 1\right\}^{-1} d\epsilon \qquad [63]$$

To treat chemisorbed species on metal surfaces, one can consider electronic wave functions that are combinations of the metal states ψ_{ks} (belonging to a continuum or energy band) and the states of the adsorbate ψ_{as} (which are discrete levels). Here, k and a are energy indices, and the subscript s is a spin index ($\pm\frac{1}{2}$). The Anderson model (Anderson 1961) writes the electronic Hamiltonian in terms of matrix elements between these states. It includes a sum over occupied states,

$$\sum_{ks} \epsilon_{ks} n_{ks} + \sum_{s} \epsilon_{as} n_{as}$$

where ϵ_{ks} and ϵ_{as} are the energies of ψ_{ks} and ψ_{as}, and n_{ks} and n_{as} are the numbers of electrons in these states; only a single adsorbate level is considered. The occupation numbers n_{ks} and n_{as} are taken as operators, the Hamiltonian being written in second-quantized form. To the terms given above, one adds terms for the interactions between adsorbate states and metal states (interelectronic interactions on the metal are already taken into account in forming the ψ_{ks}) and a term $U_{as} n_{as}$ representing the Coulombic repulsion energy between electrons in the adatom state. If the atom is far from the metal surface so that atom–surface interactions are small, one has just the band of energy levels on the metal, occupied up to the Fermi level, and a single energy level on the atom. As the adatom approaches the surface, the atomic state shifts in energy and broadens into a narrow band of levels, due to its mixing with the states of the metal, that is, there begin to be states with a range of energies that have an important component of their electron density on the adatom. If the energies of these states are below the Fermi level, they will be filled, corresponding to full occupation of the atomic state. These states are surface states caused by of the adatom, which is chemisorbed, since the electronic energy of the combined system is less than that of the metal separated from the atom. Some calculations of surface states and properties of adsorbed atoms are summarized by Bockris and Khan (1979, Sections 3.5–3.9).

G. SEMICONDUCTORS

The electrochemical interface between a semiconductor and an electrolyte is of increasing interest. The properties of such an interface depend on the electronic structure of the solid to a greater extent than do the properties of the metal–electrolyte interface (Pleskov 1980, Chapter 6). As compared with a metal, a semiconductor has a low concentration of mobile charged species and a finite dielectric constant. The consequence of this is that a diffuse double layer, called a *space charge*, with an exponentially decreasing charge density and electrical potential, is found at the surface of a semiconductor. Between the semiconductor surface at $z = 0$ and the diffuse layer in the electrolyte, there is still the compact layer, through which the electrical potential varies linearly. If there is no specific adsorption of anions of the electrolyte on the semiconductor, the compact layer contains adsorbed solvent only, but it may also include adsorbed ions or ionized sites on the surface of the solid. The potential difference across the interface is thus a sum of at least three potential differences. An additional layer may have to be considered for some semiconductors, which have surface states. Charge carriers in such states will also contribute to the overall potential difference across the interface.

Within the body of a semiconductor, current is transported by movement of electrons and by the apparent movement of holes, which are vacancies in the valence band. This band is formed from the electron states below those of the conduction band in energy. If an electron on atom A jumps into the vacancy on atom B, the hole or vacancy has apparently moved from B to A, corresponding to a transport of positive charge from B to A, in direction opposite to the movement of the electron. The mobility of an electron or a hole is defined in the usual way, as the ratio of the velocity of displacement under the influence of an electric field to the strength of the field. The conductivity is a sum of contributions of the charge carriers, each contribution being the mobility times the concentration. Since the concentration of charge carriers in a semiconductor is usually several orders of magnitude lower than in a 0.01 M electrolyte solution, the diffuse layer in a semiconductor can be expected to be more extended, and the electric fields can be expected to be smaller, than in an electrolyte.

We did not consider the valence bands in our discussion of metals. All the states in these bands are filled with electrons at $T = 0$. Let ϵ_v be the highest energy of a state in the band just below the conduction band and let ϵ_c be the lowest energy of a state in the conduction band. Then at $T = 0$ there are no electrons with energy between ϵ_v and ϵ_c, and, in the free-electron model for a metal, electrons with all energies between ϵ_c and $\epsilon_c + \hbar^2(k_F)^2/2m$, the Fermi energy ϵ_F. In an insulator, the valence bands are filled and the conduction band is empty at $T = 0$, so there are no low-energy excitations (which are required for current flow) possible to states with nonzero momentum. At higher temperatures some electrons may be excited from the valence bands to the conduction band, leaving holes in the valence band. For a metal at $T = 0$ the minimum energy required for this process is $(\epsilon_c - \epsilon_v) + \hbar^2(k_F)^2/2m$, where $(\epsilon_c - \epsilon_v)$ is the band gap.

The ionization of an atom to create an electron and a hole which can move independently of each other always occurs to some extent for $T > 0$ because the populations of electronic states are governed by the Fermi distribution

$$f_i = f(\epsilon_i) = \left[1 + \exp\left(\frac{\epsilon_i - \epsilon_F}{kT}\right)\right]^{-1} \quad [64]$$

Here, f_i is the fractional occupation of the state i with energy ϵ_i, and ϵ_F is the Fermi energy. At 0 K, all electronic levels are occupied up to the Fermi energy ϵ_F, and all levels above ϵ_F are empty. At higher temperature, one gets a nonzero population of higher electronic levels and a decreased population of levels below ϵ_F, or an increased population of holes in the valence band of an insulator. Thus for silicon, with atomic concentration $\sim 10^{22}$ cm^{-3}, holes and electrons exist, each having a room-temperature equilibrium concentration of 3×10^{10} cm^{-3}. At temperatures above zero, of course, ϵ_F is no longer the dividing line between filled and unfilled levels, but, as shown above, ϵ_F is equal to the electrochemical potential $\tilde{\mu}_{e^-}$. If the excitation of an electron from a filled level to an empty one is represented as the chemical reaction $0 \rightarrow h^+ + e^-$, one can assign an electrochemical potential to the holes by noting that at equilibrium one must have $\tilde{\mu}_{e^-} + \tilde{\mu}_{h^+} = 0$. Thus $\tilde{\mu}_{h^+} = -\epsilon_F$.

Apart from intrinsic semiconductors such as silicon, with equal concentrations of electrons and holes, there are extrinsic semiconductors, which have charge carriers of mostly one kind or the other, due to substitution of heteroatoms or impurities in the lattice (doping). If the heteroatom has more valence electrons than the host (e.g., P or As in Si), there are extra electrons that can carry current, the atomic cores with residual positive charge remaining fixed; if the heteroatom has fewer valence electrons (e.g., B or Al in Si), the heteroatom can capture an electron from the conduction band and leave a mobile hole. The terms n-type (negative charge carrier) and p-type (positive charge carrier) semiconductors are used to label these situations.

If $\epsilon_i \gg \epsilon_F$, the occupation of state i is low and f_i is approximately $\exp[-(\epsilon_i - \epsilon_F)/kT]$, that is, a Boltzmann distribution. If $\epsilon_i \ll \epsilon_F$, the occupation of the state i is close to unity and, since the exponential is small, $f_i = 1 - \exp[(\epsilon_i - \epsilon_F)/kT]$, that is, the probability of finding a hole (empty state) in state i is Boltzmann-like. The full Fermi distribution must be used for states in the forbidden energy region between the bands, such as the surface states (see Section I).

Consider N sites where an electron can go and N_- electrons, $N_- \ll N$ (as is usually the case). Let ϵ_c be the energy of the states at the bottom of the (empty) conduction band, which is well above ϵ_F. Then $(\epsilon_c - \epsilon_F) \gg kT$ and the probability is Boltzmann-like:

$$f(\epsilon_c) \cong \exp\left(\frac{\epsilon_F - \epsilon_c}{kT}\right) = \frac{N_-}{N} \qquad [65]$$

or $\epsilon_F = \epsilon_c + kT \ln(N_-/N)$. Let ϵ_v be the top of the (filled) valence band. If $(\epsilon_F - \epsilon_v) \gg kT$ the fraction of holes is also given by a Boltzmann-like expression:

$$1 - f(\epsilon_v) \cong = \exp\left(\frac{\epsilon_F - \epsilon_v}{kT}\right) = \frac{N_+}{N} \qquad [66]$$

or $\epsilon_F - \epsilon_v = -kT \ln(N_+/N)$. Multiplying equations [65] and [66], we get

$$\frac{N_+ N_-}{N^2} = \exp\left(\frac{\epsilon_c - \epsilon_v}{kT}\right) \qquad [67]$$

For an intrinsic semiconductor, $N_+ = N_-$, which means $\epsilon_F - \epsilon_c = \epsilon_v - \epsilon_F$ or $\epsilon_F = \frac{1}{2}(\epsilon_c + \epsilon_v)$. In the presence of an electrical potential ϕ, $\epsilon_c = \epsilon_c^0 - e\phi$ and $\epsilon_v = \epsilon_v^0 - e\phi$, where the superscript 0 refers to band positions in the absence of ϕ. In the region of the surface, where ϕ varies as a function of position, the bands are said to be curved or bent. The Fermi energy in the presence of ϕ is $\epsilon_F = \epsilon_c^0 - e\phi + kT \ln(N_-/N)$ from equation [65] or $\epsilon_F = \epsilon_v^0 - e\phi - kT \ln N_+/N$ from equation [66] and $\epsilon_F = \epsilon_F^0 - e\phi$. Since $N_+ = N_-$ for an intrinsic semiconductor, $\epsilon_F = \frac{1}{2}(\epsilon_c^0 + \epsilon_v^0)$, independent of ϕ.

For an extrinsic semiconductor, the concentration of ionized donors or acceptors depends on the local potential. Suppose the donors have an energy level ϵ_D with degeneracy g_D. Using equation [64] for the number of electrons in one of the states of such a level, the charge on a donor is

$$g_D - g_D\left[1 + \exp\left(\frac{\epsilon_D - e\phi - \epsilon_F}{kT}\right)\right]^{-1} = g_D\left[1 + \exp\left(\frac{\epsilon_F - \epsilon_D + e\phi}{kT}\right)\right]^{-1}$$

Multiplied by the donor concentration, this gives the number of charges per unit volume due to ionized donors. Similarly, the charge on an acceptor with energy level ϵ_A and degeneracy g_A is

$$-g_A\left\{1 + \exp\left[\frac{(\epsilon_A - e\phi - \epsilon_F)}{kT}\right]\right\}^{-1}$$

For a donor, ϵ_D must be above ϵ_F and, if ϵ_D is well above ϵ_F or close to the conduction band edge, the donor will be fully ionized. Similarly, ϵ_A is to be below ϵ_F, and an acceptor will be fully ionized (with a negative charge) if ϵ_A is close to the (upper) edge of the valence band.

H. SEMICONDUCTOR SURFACES

If there is a surface charge on a semiconductor, there may be an excess of electrons or holes in the region near the surface, giving the *space charge* which compensates the surface charge. If all the donors and acceptors are fully ionized, the charge density is $-en_- + en_+ + en_D - en_A$, where n_- is the volume number density of electrons, n_+ is the density of holes, n_D is the density of donor atoms, and n_A is the density of acceptor atoms. The electron and hole densities may depend on position. For an *n*-type semiconductor, $n_D \sim n_-$ and $n_- \gg n_+$ or n_A. For a *p*-type semiconductor, $n_A \sim n_+$ and both are $\gg n_-$ or n_D. For an intrinsic semiconductor, $n_D = n_A = 0$. In the case of an intrinsic semiconductor with mobile + and − charges, the problem of finding the potential and charge density near a surface is the same as in the Gouy–Chapman diffuse-layer theory.

To treat all the cases together, it is convenient to define a parameter p by the relations

$$n_- = n_0 e^{p/kT}, \quad n_+ = n_0 e^{-p/kT} \quad [68]$$

where

$$n_0 = n \exp\left(\frac{-(\epsilon_c - \epsilon_v)}{2kT}\right) \qquad [69]$$

This is suggested by equation [67]: $n = N/V$ is the concentration of energy states and $(\epsilon_c - \epsilon_v)$ the band gap. The sign of p determines the sign of the majority carrier: $p > 0$ means there is an excess of electrons and $p < 0$ means an excess of holes. In a region of nonzero electrical potential ϕ, such as near a surface,

$$n_- = n_0 e^{(p+e\phi)/kT}, \qquad n_+ = n_0 e^{-(p+e\phi)/kT}$$

If p and ϕ have the same sign the charge carriers present in excess are more concentrated in the space-charge region near the surface than in the bulk region of the semiconductor. One speaks of an *accumulation layer*. If p and ϕ have opposite signs, the majority charge carrier has a lower concentration in the space-charge region than in bulk. Then one has a *depletion layer*, as long as $e\phi$ is smaller in magnitude than p. If p and ϕ have opposite signs and $|e\phi| > |p|$ the difference between electron and hole concentrations will have different signs in the bulk and surface regions; this is called an *inversion layer*. We also write

$$n_D = n_0 e^{q/kT} \quad \text{and} \quad n_A = n_0 e^{-q/kT} \qquad [70]$$

In terms of the quantities defined in equations [68]–[70], the charge density at any point is

$$\rho = -en_0[e^{(p+e\phi)/kT} - e^{-(p+e\phi)/kT} - e^{q/kT} + e^{-q/kT}] \qquad [71]$$

Inserting this into the Poisson equation, we have

$$\frac{d(\epsilon\, d\phi/dz)}{dz} = -4\pi\rho = 8\pi e n_0 \left[\sinh\frac{p+e\phi}{kT} - \sinh\frac{q}{kT}\right]$$

Taking ϵ independent of z and multiplying by $d\phi/dz$, the left-hand side is $\frac{1}{2}\epsilon\, d(d\phi/dz)^2/dz$. Then we may integrate to give

$$-E = \frac{d\phi}{dz} = \left(\frac{16\pi e n_0}{\epsilon}\right)^{1/2}\left(\frac{kT}{e}\cosh\frac{p+e\phi}{kT} - \frac{kT}{e}\cosh\frac{p}{kT} - \phi\sinh\frac{q}{kT}\right)^{1/2}$$

[72]

The integration constant has been chosen as $-(kT/e)\cosh(p/kT)$ so that $d\phi/dz = 0$ where $\phi = 0$, at $z \to -\infty$. The surface charge density of the space charge is

$$q^{sc} = -\int_0^{-\infty} \rho\, dz = -\frac{\epsilon E_0}{4\pi} \qquad [73]$$

where E_0 is the field at $z = 0$. Then from equation [72]

$$q^{sc} = \frac{2^{1/2}\epsilon kT}{4\pi l_D e}\left(\cosh\frac{p+e\phi_0}{kT} - \cosh\frac{p}{kT} - \frac{e\phi_0}{kT}\sinh\frac{q}{kT}\right)^{1/2} \quad [74]$$

where we have introduced the Debye length l_D, defined by

$$(l_D)^2 = \frac{\epsilon kT}{8\pi n_0 e^2}$$

and ϕ_0 is the potential at the surface ($z = 0$): The sign of the square root in equations [72] and [74] depends on the sign of ϕ_0. The size of $l_D = (\epsilon kT/8\pi n_0 e^2)^{1/2}$ with $n_0 = 10^{13}$ cm^{-3}, $T = 300$ K, and $\epsilon = 10$, is 8.5×10^{-5} cm. Thus the thickness of the diffuse layer in semiconductors is typically several orders of magnitude greater than in all but very dilute solutions; since the electron density in a metal is of the size 10^{22} cm^{-3}, the Debye length for a metal would be below atomic size. To estimate the size of q^{sc}, we note that the coefficient of the term in parentheses in equation [74] is $(\epsilon kTn_0/\pi)^{1/2} = 1.1$ esu/cm^2 or 2.4×10^9 electronic charges/cm^2.

From equation [74] one can calculate the space charge as a function of the surface potential for any choice of p. If $p = 0$ and $q = 0$ one has an intrinsic semiconductor (n_A and n_D negligible) and we have the Gouy–Chapman result

$$q^{sc} = \pm\frac{2^{1/2}\epsilon kT}{4\pi l_D e}\left(\cosh\frac{e\phi_0}{kT} - 1\right)^{1/2} \quad [75]$$

For small values of x, $\cosh x \sim 1 + x^2/2$ so that

$$q^{sc} = \frac{\epsilon kT}{4\pi l_D e}\frac{e\phi_0}{kT}$$

and the space charge is proportional to the potential, the capacitance being

$$C_{sc} = \frac{dq^{sc}}{d\phi_0} = \frac{\epsilon}{4\pi l_D}$$

In general, differentiating equation [75] with respect to ϕ_0 gives for the capacitance

$$C_{sc} = \frac{2^{1/2}\epsilon}{8\pi l_D}\sinh\frac{e\phi_0}{kT}\left(\cosh\frac{e\phi_0}{kT} - 1\right)^{-1/2}$$

$$= \frac{\epsilon}{4\pi l_D}\cosh\frac{e\phi_0}{2kT}$$

H. SEMICONDUCTOR SURFACES

A curve of $\ln C_{sc}$ vs. ϕ_0 is therefore V-shaped and symmetrical about $\phi_0 = 0$. In fact, for large $|\phi_0|$ the capacitance for a real semiconductor surface levels off because the space-charge density is limited by the density of states.

If there are ionized donors and acceptors, we have for the capacitance, from equation [74],

$$C = \frac{2^{-1/2}\epsilon}{4\pi l_D} \frac{\pm\{\sinh[(p+e\phi_0)/kT] - \sinh[p/kT]\}}{\{\cosh[(p+e\phi_0)/kT] - \cosh[p/kT] - (e\phi_0/kT)\sinh[p/kT]\}^{1/2}} \quad [76]$$

(Myamlin and Pleskov 1967, Section 12). We have used the electroneutrality condition for bulk semiconductor (far from the surface), $n_D - n_A = n_0 e^{p/kT} - n_0 e^{-p/kT}$ or

$$n_0 \sinh \frac{q}{kT} = n_0 \sinh \frac{p}{kT}$$

to replace $\sinh(q/kT)$ by $\sinh(p/kT)$ in equation [76]. The sign of the square root is to be taken so that C is positive.

At very large positive potentials, exponentials of $e\phi_0/kT$ will dominate and the second factor in equation [76] will approach $\exp[(p+e\phi_0)/2kT]$. At very large negative potentials, one need keep only exponentials in $-e\phi_0/kT$, so that this factor will approach $\exp[-(p+e\phi_0)/2kT]$. Thus $\ln C_{sc}$ becomes linear in $|\phi_0|$ for large $|\phi_0|$, as in the intrinsic case just discussed. Again, the limited density of states limits the charge density and eventually makes $\ln C_{sc}$ drop below the lines $\pm(p+e\phi_0)/2kT$. It is also clear that, unless $p = 0$ (intrinsic semiconductor), the capacitance is not a symmetrical function of the potential. This can also be seen by using $\sinh(x+y) \sim \sinh x + y \cosh x + (y^2/2) \sinh x$ and $\cosh(x+y) \sim \cosh x + y \sinh x + (y^2/2) \cosh x$, valid for small y, in equation [76], which gives

$$C_{sc} = (\epsilon/4\pi l_D) \frac{\cosh(p/kT) + (e\phi_0/2kT)\sinh(p/kT)}{[\cosh(p/kT)]^{1/2}}$$

There is thus a minimum in C_{sc} for $e\phi_0/2kT = -\tanh(p/kT)$. Aside from the shift of the curve along the ϕ_0 axis, C_{sc} as a function of ϕ_0 resembles that for an intrinsic semiconductor.

Now suppose p/kT and $(p+e\phi_0)/kT$ are both large compared to 1 (a heavily doped n-type semiconductor). The capacitance [76] is

$$C = \frac{(\epsilon/8\pi l_D) e^{p/2kT} [\exp(e\phi_0/kT) - 1]}{[\exp(e\phi_0/kT) - 1 - (e\phi_0/kT)]^{1/2}}$$

For potentials small enough to expand the exponentials in ϕ_0 as power series, the capacity will be a linear function of the potential, with the value

$$C = \frac{2^{1/2}\epsilon}{8\pi l_D} e^{p/2kT}$$

for $\phi_0 = 0$. On the other hand, consider the same semiconductor for potentials negative enough for $\exp[(p + e\phi_0)/kT]$ to be small compared to $e^{p/kT}$, but not so negative that $p + e\phi_0 < 0$. This means that near the surface both n_+ and n_- are less than n_D, that is, this is a depletion layer. Then the capacitance [76] is

$$C = \frac{(\epsilon/8\pi l_D) e^{p/kT}}{\left[-e^{p/kT}(1 + e\phi_0/kT)\right]^{1/2}}$$

In this potential range, $1/C^2$ is linear in ϕ_0. If the reciprocal of the squared capacitance of a semiconductor electrode varies linearly with the electrode potential in some range of potential, one can extrapolate to $1/C^2 = 0$ and determine the electrode potential such that $1 + e\phi_0/kT = 0$. This electrode potential differs by kT/e from the potential that makes $\phi_0 = 0$, referred to as the *flat band potential*, because when $\phi_0 = 0$ the boundaries of the energy bands are not deformed by electric fields in the space-charge region. The flat band potential corresponds to $q_{sc} = 0$, so that it is equivalent to the pzc of a metal electrode.

The linearity of C^{-2} with potential also results from a Thomas–Fermi model for electrons in a solid. Consider a density n_1 of fixed positive charges (donors) and a spatially varying density of electrons given by $(\hbar^2/2m)(3\pi^2 n)^{2/3} = \epsilon_F + e\phi$ (see equation [54] and thereafter). If the potential in the region (bulk) where $n = n_1$ is $\phi = \phi_1$,

$$\left(\frac{n}{n_1}\right) = \left(\frac{\epsilon_F + e\phi}{\epsilon_F + e\phi_1}\right)^{3/2}$$

Then the Poisson equation is

$$\frac{d^2\phi}{dz^2} = 4\pi e(\epsilon_1)^{-1} n_1 \left\{\left[1 + \frac{e(\phi - \phi_1)}{\epsilon_F}\right]^{3/2} - 1\right\}$$

$$\sim 4\pi e(\epsilon_1)^{-1} n_1 \left[\frac{3e(\phi - \phi_1)}{2\epsilon_F} + \cdots\right] \quad [77]$$

assuming $e(\phi - \phi_1)$ is small enough to linearize. The solution to equation [77] is

$$\phi - \phi_1 = A \exp\left(\frac{z}{l_{TF}}\right) \quad [78]$$

where the Thomas–Fermi screening length is given by

$$(l_{TF})^{-2} = \frac{6\pi e^2 n_1}{\epsilon_F \epsilon_1}$$

A typical size of l_{TF}, with $e = 4.8 \times 10^{-10}$ esu, $n_1 = 0.01(a_0)^{-3} = 6.7 \times 10^{22}$ cm^{-3}, $\epsilon_F = 9.7 \times 10^{-12}$ erg, and $\epsilon_1 = 1$, is 0.57 Å. The value of A in equation [78] is determined so that $(d\phi/dz)_0 = A/l_{TF}$ is equal to $-E(0)$. The potential drop across the space-charge region is $\phi_1 - \phi(0) = -A = l_{TF} q^{(1)}/\epsilon_1$, so that the capacitance for small potential drops is

$$C = \frac{dq^{(1)}}{d[\phi_1 - \phi(0)]} = \frac{\epsilon_1}{l_{TF}}$$

Now consider potential drops $\phi - \phi_1$ large enough for there to be a Schottky barrier region. Suppose $n(z) = 0$ for $-x \le z \le 0$ so that

$$\frac{d^2\phi}{dz^2} = \frac{-4\pi(en_1)}{\epsilon_1}$$

The solution is a potential quadratic in z, namely, $\phi(0) + bz - 2\pi e n_1 z^2/\epsilon_1$, where $b = -E(0) = -q^{(1)}/\epsilon_1$. Since $n(z)$ vanishes at $z = -x$, $\phi(-x) = -\epsilon_F/e$. Thus

$$-\frac{\epsilon_F}{e} = \phi(0) + \frac{q^{(1)}x}{\epsilon_1} - \frac{2\pi e n_1 x^2}{\epsilon_1} \qquad [79]$$

In the region $z < -x$ we have to solve

$$\frac{d^2\phi}{dz^2} = 4\pi e\epsilon^{-1} n_1 \left\{ \left[1 + \frac{e(\phi - \phi_1)}{\epsilon_F}\right]^{3/2} - 1 \right\}$$

with $\phi \to \phi_1$ for $z \to -\infty$, $\phi \to -\epsilon_F/e$ for $z \to -x$, and $(d\phi/dz)_{-x} = -(q^{(1)} - 4\pi e n_1 x)/\epsilon_1$. Since an exact solution is not possible, let us approximate the solution by $\phi_1(1 - e^{az})$, which behaves properly at $z \to -\infty$. To satisfy the conditions at $z = -x$, we require $\phi_1(e^{-ax} - 1) = \epsilon_F/e$ and $a\phi_1 e^{-ax} \epsilon_1 = q^{(1)} - 4\pi e n_1 x$. Eliminating a between these two equations, we get

$$-x^{-1} \epsilon_1 \left(\frac{\epsilon_F}{e} + \phi_1\right) \ln\left(\frac{\epsilon_F}{e\phi_1} + 1\right) = q^{(1)} - 4\pi e n_1 x$$

4. STRUCTURE OF SURFACES

which gives

$$x = (8\pi e n_1)^{-1} [q^{(1)} + (q^{(1)2} + S)^{1/2}]$$

$$S = 16\pi n_1 \epsilon_1 (\epsilon_F + e\phi_1) \ln(1 + \epsilon_F/e\phi_1)$$

According to equation [79] the capacitance is given by

$$C^{-1} = \frac{d[\phi_1 - \phi(0)]}{dq^{(1)}}$$

$$= (\epsilon_1)^{-1} \left(x + \frac{q^{(1)} \, dx}{dq^{(1)}} - \frac{4\pi e n_1 x \, dx}{dq^{(1)}} \right)$$

$$= (16\pi e n_1 \epsilon_1)^{-1} \left\{ 2q^{(1)} + (q^{(1)})^2 [(q^{(1)})^2 + S]^{-1/2} + [(q^{(1)})^2 + S]^{1/2} \right\}$$

For the values of the parameters used above and $\phi_1 = 3$ V $= 0.01$ statvolt, $S = 5.4 \times 10^{13}$ esu^2/cm$^4 = 6.0 \times 10^6$ (μC/cm^2)^2, which is much larger than accessible values of $(q^{(1)})^2$. Therefore $[(q^{(1)})^2 + S]^{1/2} \sim S^{1/2} [1 + (q^{(1)})^2/2S]$ and

$$C^{-1} = \frac{d[\phi_1 - \phi(0)]}{dq^{(1)}} = (16\pi e n_1 \epsilon_1)^{-1} (S^{1/2} + 2q^{(1)})$$

This shows that $\phi_1 - \phi(0)$ is a quadratic function of $q^{(1)}$, so that C^{-2} is a linear function of the potential drop $\phi_1 - \phi(0)$.

Since an electric field in the semiconductor makes the electron and hole concentrations near the surface differ from the concentrations in bulk, the conductivity of the surface region will correspondingly differ from that in bulk. The extra conductivity, referred to as *surface conductivity*, is written in terms of the mobilities $\lambda_\pm$ and surface excesses $\Gamma_\pm$ of electrons and holes:

$$K_{\text{surf}} = e(\lambda_- \Gamma_- + \lambda_+ \Gamma_+)$$

For an intrinsic semiconductor, $n_- = n_+ = n_0$ in bulk and $n_\mp = n_0 e^{\pm e\phi/kT}$ near the surface. The surface excess of electrons or holes is $\int (n_\mp - n_0) \, dz$, that is, the surface concentration minus what the surface concentration would be if the electron or hole density were uniform up to the semiconductor surface. If $\phi > 0$ there is enrichment in electrons and depletion in holes, while if $\phi < 0$ the reverse is true; $n_+ n_- = (n_0)^2$ in any case. The situation for $\phi > 0$ corresponds to a downward bending of the bands (see Figure 11). For $\phi < 0$ the bands are bent upward. In either case, the concentration of one charge carrier rises rapidly with $|\phi|$ so that K_{surf} exhibits a minimum near $\phi = 0$ (the flat-band potential) unless λ_+ and λ_- are very different. Something similar must occur in the diffuse layer of an electrolyte, but the charge carriers are the ions, with mobilities much lower than

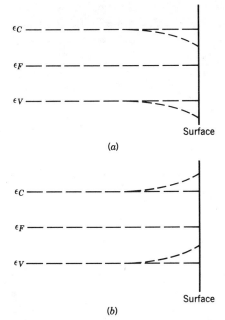

FIGURE 11. Band bending. (a) Surface potential $\phi > 0$. (b) Surface potential $\phi < 0$. ϵ_C = bottom of conduction band, ϵ_V = top of valence band.

those of the electrons and holes of a semiconductor, and the thickness of the surface region is much smaller.

For an extrinsic semiconductor, the situation with respect to surface conductivity is a bit more complicated, as the concentration of one current carrier (electrons or holes) may be much larger than the concentration of the other. For an n-type semiconductor ($p > 0$ in equation [68]) the concentration of electrons (majority carrier) increases as ϕ becomes more positive and the concentration of holes (minority carrier), already smaller than that of electrons, decreases. The surface conductivity rises the same as for an intrinsic semiconductor, since in this range of ϕ the surface conductivity is due to the electrons. For *very* negative values of ϕ, the concentration of holes near the surface will exceed that of electrons. When the local concentration of minority carriers exceeds that of majority carriers, one has an inversion layer. With ϕ becoming more negative, the hole concentration grows exponentially and leads to a surface conductivity that increases as ϕ becomes more negative.

At moderately negative values of ϕ, the concentration of electrons decreases from its bulk value whereas the concentration of holes, though increased from its bulk value, is still much less than that of the electrons, so the total concentration of mobile charged species near the surface is less than that in the bulk region. One forms a Mott–Schottky layer, in which the space charge consists mainly of ionized impurity atoms (donors in the present example of an n-type superconductor) that are not mobile. The conductivity is diminished and K_{surf}, being defined as a difference, is negative. The surface conductivity as a function of ϕ may show a rather broad minimum with negative values, since there may be a range of ϕ for which the concentration of mobile charge carriers is essentially zero, with the majority carriers being depleted and the concentration of minority carriers still being small.

I. SURFACE STATES

The discussion so far has ignored the existence of surface states associated with the changed one-electron potential near the surface. These states are particularly important for understanding the structure of semiconductors. Surface states appear in the Kronig–Penney model, which considers solutions of the one-dimensional Schrödinger equation for a single electron,

$$-\frac{\hbar^2}{2m}\frac{d^2\psi}{dz^2} + V\psi = \epsilon\psi \qquad [80]$$

with the potential (see Figure 12),

$$V = V_0, \ z \geq 0; \quad V = 0, \ -(n+1)a + t < z < -na, \ n \geq 0;$$
$$V = S, \ -(n+1)a \leq z \leq -(n+1)a + t \qquad [81]$$

The interatomic spacing is a and the thickness of an atomic core is t, so that the region from $-(n+1)a$ to $-(n+1)a + t$ is the core of the $(n+1)$st atom. Inside the solid ($z < 0$), V is a series of square barriers. In a region between cores, the solutions to the Schrödinger equation (equation [80]) are linear combinations of $\sin kx$ and $\cos kx$, where $k^2 = 2m\epsilon/\hbar^2$. If t is small, ψ may be taken as constant

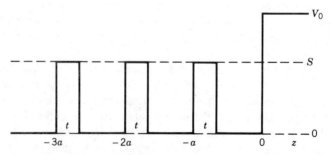

FIGURE 12. Potential for Kronig–Penney model. Square barriers are equally spaced up to the surface.

I. SURFACE STATES

over a core region, and integration of equation [80] through a core region then gives

$$\left(\frac{d\psi}{dz}\right)_{-(n+1)a+t} - \left(\frac{d\psi}{dz}\right)_{-(n+1)a} \simeq \frac{-2m(\epsilon - S)\, t\psi_{-(n+1)a}}{\hbar^2} \quad [82]$$

Since one is particularly interested in energies much below S, $\epsilon - S$ may be replaced by $-S$ in equation [82].

One also has the Bloch theorem, which states that the solutions to the Schrödinger equation for a periodic potential must be of the form $e^{iKz} u(z)$ where u is periodic, $u(z - a) = u(z)$, so that the electron density $|\psi|^2$ be periodic with period a, that is, $\psi(z - a) = e^{-iKa} \psi(z)$. Suppose that, in the region $-(n + 1)a + t \leq z \leq -na$, we write the wave function as

$$\psi(z) = A_n \sin kz + B_n \cos kz \quad [83]$$

and similarly $A_{n+1} \sin kz + B_{n+1} \cos kz$ for the region $-(n + 2)a + t \leq z \leq -(n + 1)a$.

We use the above conditions to relate the coefficients A_i and B_i. To satisfy the Bloch theorem we must have

$$e^{-iKz}(A_n \sin kz + B_n \cos kz)$$
$$= e^{-iK(z-a)} \left\{ A_{n+1} \sin [k(z - a)] + B_{n+1} \cos [k(z - a)] \right\}$$

or, on expanding $\sin [k(z - a)]$ and $\cos [k(z - a)]$ and equating coefficients of $\sin kz$ and $\cos kz$,

$$A_n e^{-iKa} = A_{n+1} \cos ka + B_{n+1} \sin ka$$
$$B_n e^{-iKa} = -A_{n+1} \sin ka + B_{n+1} \cos ka \quad [84]$$

To satisfy condition [82] with $t \to 0$ and $\epsilon \ll S$,

$$k\left\{ A_n \cos [-(n + 1)ka] - B_n \sin [-(n + 1)ka] \right\}$$
$$- k\left\{ A_{n+1} \cos [-(n + 1)ka] - B_{n+1} \sin [-(n + 1)ka] \right\}$$
$$= q \left\{ A_n \sin [-(n + 1)ka] + B_n \cos [-(n + 1)ka] \right\} \quad [85]$$

where $q = 2mSt/\hbar^2$. Finally, the equality of the wave function at $-(n + 1)a + t$ and the wave function at $-(n + 1)a$ (since $t \to 0$) is expressed by

$$A_n \sin [-(n + 1)ka] + B_n \cos [-(n + 1)ka]$$
$$= A_{n+1} \sin [-(n + 1)ka] + B_{n+1} \cos [-(n + 1)ka] \quad [86]$$

4. STRUCTURE OF SURFACES

To have a solution for the four homogeneous linear equations (equations [84]–[86]) for A_n, B_n, A_{n+1}, and B_{n+1} the determinant of the coefficients must vanish. Expanding the determinant, one gets

$$ke^{-2iKa} - qe^{-iKa}\sin ka - 2ke^{-iKa}\cos ka + k = 0$$

which may be rewritten as

$$\cos Ka = \cos ka + \frac{q}{2k}\sin ka \qquad [87]$$

For small values of ka, the right-hand side of equation [87] becomes $1 + \frac{1}{2}qa$ and, since $\cos Ka \leq 1$ for real K, there are no solutions to equation [87] for real values of K. The only solutions in this "forbidden zone" of energy are those with K imaginary. These are surface states, with wave functions decreasing in size for increasingly negative z (into the solid). To see this, we solve equations [84] for A_{n+1} and B_{n+1}, obtaining

$$A_{n+1} = e^{-iKa}(A_n \sin ka + B_n \cos ka)$$
$$B_{n+1} = e^{-iKa}(A_n \cos ka - B_n \sin ka) \qquad [88]$$

For real K, the magnitudes of A_{n+1} and B_{n+1} are the same as those of A_n and B_n, but for imaginary K the coefficients increase or decrease in magnitude by $\exp(\pm|K|a)$ for each atomic spacing further into the solid. The increasing solution leads to a divergent wave function, but the decreasing solution represents a wave function that approaches zero as one gets further from the interface, that is, a surface state.

To complete specification of such a state, we note that, in the region $z \geq 0$, where $V = V_0$, all wave functions must be of the form $\exp[-(2m\epsilon'/\hbar^2)^{1/2}z]$, where $\epsilon' = V_0 - \hbar^2 k^2/2m$. For the wave function to be continuous at $z = 0$ with $\psi(z)$ of equation [83], $A_0 \sin kz + B_0 \cos kz$, the exponential must be multiplied by B_0; for continuity of the slope at $z = 0$, we require $kA_0 = -B_0(2m\epsilon'/\hbar^2)^{1/2}$. Therefore, with $r^2 = 2m\epsilon'/\hbar^2$, our wave function in the region from $z = 0$ to $z = -a$ is

$$B_0(-rk^{-1}\sin kz + \cos kz)$$

Then, according to equation [88], we have

$$A_1 = e^{-iKa}B_0(-rk^{-1}\sin ka + \cos ka)$$
$$B_1 = e^{-iKa}B_0(-rk^{-1}\cos ka - \sin ka)$$

and subsequent A_n and B_n may be obtained by repeated use of equation [88].

The surface states may also be discussed from the point of view of atomic orbitals, if the atomic positions are far enough apart so that overlaps between orbitals on adjacent atoms are small (tight binding approximation). From a chemical point of view, one can consider the surface states as the consequence of free valence at the surface, corresponding to the bonds that must be broken to form a surface by cleaving a solid. An unpaired electron on a surface atom subsequently may be lost, in which case the atom has provided a donor surface state, or an electron may be added to form a pair, in which case the atom has provided an acceptor surface state. In a doped semiconductor, there will be acceptor atoms and donor atoms on the surface, capable of providing surface states. Surface states formed by the adsorption of foreign atoms on a semiconductor surface are called *Shockley levels*. *Tamm levels* refers to the surface states arising from the breakdown in the periodicity of the potential of the solid, as shown in equations [80]–[88].

Although there may be surface states with energies within the solid-state (valence and conduction) bands, those with energies in the forbidden band are of most interest. The population of the states is governed by the Fermi distribution (equation [64]), so that, if there are N_t states per unit area with energy ϵ_t, the number of electrons per unit area in these surface states, n_t, is given by

$$\frac{n_t}{N_t} = \left[1 + \exp\left(\frac{\epsilon_t - \epsilon_f^s}{kT}\right)\right]^{-1} \qquad [89]$$

The surface Fermi level ϵ_F^s is identical to ϵ_F, for the bulk, when electrons in the surface layer are in equilibrium with electrons in the bulk. We shall assume this is the case.

Surface states can be either donors or acceptors. As for the corresponding bulk states, the number of negative charges on N_A acceptor surface levels, each with energy ϵ_{As} and degeneracy g_{As}, is

$$N_{A^-} = N_A g_{As} \left[1 + \exp\left(\frac{\epsilon_{As} - e\phi_1 - \epsilon_F}{kT}\right)\right]^{-1}$$

and the number of positive charges on N_D donor surface levels with energy ϵ_{Ds} and degeneracy g_{Ds} is

$$N_{D^+} = N_D g_{Ds} \left[1 + \exp\left(\frac{\epsilon_F - \epsilon_{Ds} + e\phi_1}{kT}\right)\right]^{-1}$$

The energy of an electron in a level t is $\epsilon_t^0 - e\phi_1$, where ϕ_1 is the potential at the surface relative to the interior ($\phi_1 = 0$ when there is no space charge). The contribution of surface states to the surface charge density is

$$q^t = -\frac{n_{As} N_{A^-}}{N_A} + \frac{n_{Ds} N_{D^+}}{N_D} \qquad [90]$$

where n_{As} and n_{Ds} are the number of donor and acceptor levels per unit surface area. The electric field at the semiconductor surface, E_1, is determined by

$$\epsilon_1 E_1 = \epsilon E_s + 4\pi q^t$$

where ϵ_1 is the dielectric constant outside the semiconductor, ϵ is the dielectric constant inside, and E_s is due to the space charge and given by substituting ϕ_1 for ϕ in equation [72]. Thus the total surface charge density is

$$q^s = \left(\frac{\epsilon n_0}{\pi}\right)^{1/2} \left(kT \cosh \frac{p + e\phi_1}{kT} - kT \cosh \frac{p}{kT} - e\phi_1 \sinh \frac{p}{kT}\right)^{1/2} + q^t$$

The capacitance $dq^s/d\phi_1$ is then, after adding in the potential drop across surface states,

$$C = \left(\frac{e^2 \epsilon n_0}{4\pi}\right)^{1/2} \left(\sinh \frac{p + e\phi_1}{kT} - \sinh \frac{p}{kT}\right)$$

$$\cdot \left(kT \cosh \frac{p + e\phi_1}{kT} - kT \cosh \frac{p}{kT} - e\phi_1 \sinh \frac{p}{kT}\right)^{1/2}$$

$$- n_{As} g_{As} \left[1 + \exp\left(\frac{\epsilon_{As} - e\phi_1 - \epsilon_F}{kT}\right)\right]^{-2} \exp\left(\frac{\epsilon_{As} - e\phi_1 - \epsilon_F}{kT}\right) \frac{e}{kT}$$

$$- n_{As} g_{Ds} \left[1 + \exp\left(\frac{\epsilon_F - \epsilon_{Ds} + e\phi_1}{kT}\right)\right]^{-2} \exp\left(\frac{\epsilon_F - \epsilon_D + e\phi_1}{kT}\right) \frac{e}{kT} \quad [91]$$

Capacitances for a number of related models are given by Myamlin and Pleskov (1977, Section 12).

If the number of surface levels is large, the first (space-charge) term may be neglected. The magnitude of each of the remaining terms goes through one maximum as a function of ϕ_1, suggesting the capacitance of the surface states should show two maxima. The actual capacitance is $dq^s/d\phi$, where ϕ is ϕ_1 plus a contribution of the charge density of the surface states.

Semiconductor electrochemistry is reviewed by Gerischer (1970). The interface between a semiconductor and an electrolyte involves two diffuse layers of charge, namely, the space charge in the semiconductor and the diffuse layer of the electrolyte, as well as localized charged layers associated with surface states on the semiconductor and the Helmholtz (compact) layer of the electrolyte at the semiconductor surface. A further complication is the fact that there is often a layer of oxide on the semiconductor surface. The change in potential going from the bulk semiconductor to the solution may be written as a sum:

$$\Delta \phi = \phi_{sc} - \chi_{s-ox} - \phi_{ox} - \chi_{cl} - \phi_{dl}$$

I. SURFACE STATES

where ϕ_{sc} is the space-charge potential (with potential equal to zero in the bulk semiconductor), χ_{s-ox} is the dipolar potential difference across the semiconductor–oxide interface (semiconductor potential minus oxide potential), ϕ_{ox} is the potential drop across the oxide layer, χ_{cl} is the potential drop (oxide minus solution) across the compact layer of the oxide–solution interface, and ϕ_{dl} is the diffuse-layer potential drop in the electrolyte. The charge on the electrolyte side of the interface is $q^{el} = q^{cl} + q^{dl}$ (compact- plus diffuse-layer contributions) and is equal and opposite to the charge on the semiconductor side, q^{so}, which is the sum of three contributions: q^{sc} is from the space charge, q^{ss} is from surface states, and q^{ox} is from the oxide layer. If one assumes the compact layer (dipolar) potential drops are not changed by charging, the capacity is given by

$$C^{-1} = \frac{d(\Delta\phi)}{dq^{el}} = \frac{d\phi_{sc}}{dq^{el}} - \frac{d\phi_{ox}}{dq^{el}} - \frac{d\phi_{dl}}{dq^{el}}$$

This may be written as

$$C^{-1} = (C_{sc})^{-1} + (C_{ox})^{-1} + (C_{el})^{-1} \qquad [92]$$

as appropriate for three capacitances in series. Since, for example, the charge q^{so} is a sum of three contributions, the capacitance C_{sc} is a sum of contributions, corresponding to capacitances in parallel (Spaarnay 1978).

Gerischer (1970, Section 5.II.A.3) discusses capacitances for several simpler situations in which there is no oxide layer. If there are no surface states or adsorbed species in the Helmholtz layer of the electrolyte, the potential difference between the outer Helmholtz plane and the interior of the semiconductor is

$$\Delta\phi = \phi_{sc} + \chi_{dip} + \frac{\epsilon}{\epsilon_1}\left(\frac{d\phi}{dz}\right)_0 d_1 \qquad [93]$$

where χ_{dip} is the contribution of the oriented dipoles in the Helmholtz layer, ϵ_1 is the dielectric constant, and d_1 is the thickness of the Helmholtz layer. The magnitude of the field at a semiconductor surface, $|(d\phi/dz)_0|$, is usually much less than that at a metal surface. If χ_{dip} does not change appreciably with charging, one has for the capacitance of the compact layer of the electrolyte (which is in series with that of the semiconductor)

$$C_c = \frac{\partial q^{sc}}{\partial(\Delta\phi - \phi_{sc})} = \frac{-\epsilon}{4\pi} \frac{-\epsilon_1}{\epsilon d_1}$$

If there are electronic surface states, their charge density (q^t, equation [90]) must be added to q^{sc}.

Gerischer distinguishes between the charges of surface states, formed by population or depopulation of electronic levels on the semiconductor surface, and charges formed on the surface by chemical reaction with the electrolyte. The latter charges, which belong to ionic groups bonded to the surface, may be produced by dissociation equilibria such as

$$RY \leftrightarrow R^-(s) + Y^+(solv)$$

They are grouped with charges associated with adsorbed ions of the Helmholtz layer, formed according to, for example,

$$X^-(solv) \leftrightarrow X^-(ads)$$

The surface charge density in either case is calculated from an equilibrium between adsorbed and solvated species.

The net charge of the electrolyte is the sum of the adsorbed charge and the ionic charge in the solution double layer, whereas the net charge of the semiconductor is the sum of the space charge and the charge in surface states, that is, $q^{sc} + q^{ss} = -(q^{ad} + q^{dl})$, where ss stands for surface states and dl stands for electrolyte double layer. Thus the interface may be modeled as two capacitors in series (electrolyte and semiconductor), each of which consists of two capacitors in parallel. The capacity of the whole interface then is given by

$$C^{-1} = (C_{sc} + C_{ss})^{-1} + (C_{ad} + C_{dl})^{-1}$$

The ratio of the potential difference between the interior of the semiconductor and its surface, to the potential difference between the semiconductor surface and the interior of the electrolyte, is $(C_{dl} + C_{ad})/(C_{sc} + C_{ss})$.

5
INTERFACES

A. METAL–METAL INTERFACE

If two metals are placed in contact in such a way that electrons can pass from one to the other, equilibrium will be attained at $T = 0$ when the energy of the most energetic electrons is the same in both phases, that is, when the electrochemical potentials are equal. At temperatures above absolute zero there is a nonzero probability of occupation of all electronic energy levels but equilibrium still corresponds to equal electrochemical potentials because the free energy per electron must be the same in each phase (see Chapter 1, Section C). The electrochemical potential of electrons in metal M is given by

$$\tilde{\mu}_e^M = \mu_e^M - e\chi^M - e\psi^M = \mu_e^M - e\phi^M = \alpha_e^M - e\psi^M \qquad [1]$$

where ϕ^M is the inner potential and α_e^M is the real potential. The work function Φ^M is $e\chi^M - \mu_e^M = -\alpha_e^M$. If the electrochemical potentials of the electrons are different for the two metals, the equilibrium is established by transfer of electrons from one phase to another. The electron transfer leads to charging of both phases and the establishment of a difference in outer potentials between them. As noted in Chapter 4, Section F, an amount of electrons large enough to change the outer potential ψ substantially produces negligible changes in the chemical potential μ_e^M and the surface potential χ^M. The outer potential difference, which is measurable since it is between two points in the same phase, is just the contact potential.

The contact potential between two metals is equal to the difference of their work functions, according to the following thermodynamic argument (Frumkin, Petrii, and Damaskin 1980, Section 5.6). The work required to bring an electron from inside M_1 to inside M_2 is zero because of the contact equilibrium; this is also the work along a path from the inside of M_1 to vacuum just outside M_1, to infinity, to vacuum just outside M_2 and back to the inside of M_2, that is,

$$\Phi_1 + e\psi_1 - e\psi_2 - \Phi_2 = 0$$

Here, ψ_i is the Volta (outer) potential, Φ_i is the work function of metal i, and the contact potential difference is $C = \psi_2 - \psi_1 = -\Delta_2^1\psi = e^{-1}(\Phi_1 - \Phi_2)$. For a path from just outside M_1 to just outside M_2, but passing through the metal–metal boundary, one has $\psi_2 - \psi_1 = -\chi_2 + \Delta_1^2\phi + \chi_1$ where the χ_i are surface potentials ($\chi_i > 0$ if the inside of the metal is at higher potential than the outside).

Now consider the cell-like arrangement $M_2^{(A)}|\text{vacuum}|M_1|M_2^{(B)}$. The cell potential is

$$\phi_B - \phi_A = (\Delta_1^2\phi + \phi_1) - \phi_2^{(A)}$$

where $\Delta_1^2\phi$, the potential difference across the metal-metal interface, is $C + \chi_2 - \chi_1$. Supposing both metals are uncharged, the cell potential $\phi_B - \phi_A$ is the difference of inner potentials at zero surface charge, $\phi_1 - \phi_2^{(A)}$, plus the contact potential difference, C, plus the difference of surface potentials, $\chi_2 - \chi_1$.

If the same metals are now immersed in an electrolyte to create the cell $M_2^{(A)}|S|M_1|M_2^{(B)}$, the two metals still supposed to be uncharged, the cell potential will have the above contributions plus a potential difference related to the presence of solution on the metal surfaces. Let $\delta\chi_i$ be the change in metal surface potential due to the solution and let $\chi^{S(i)}$ be the surface potential of solution on metal i, which is positive if the positive end of the surface dipole is toward the solution. The difference of Volta potentials between metal 1 and solution at zero surface charge is $\psi_1 - \psi_S = -\chi_1 + (\chi_1 + \delta\chi_1) - \chi^{S(1)} + \chi_S$, where χ_S is the (free) surface potential of the solution, and is expressed similarly for metal 2.

Since the terminals are of the same metal and have the same surface potential, the cell potential at zero surface charge is

$$E_0 = \phi_2^{(B)} - \phi_2^{(A)} = \psi_2^{(B)} - \psi_2^{(A)} = C + (\psi_1 - \psi_S)_0 - (\psi_2 - \psi_S)_0 \quad [2]$$

As discussed by Frumkin et al. (Frumkin, Petrii, and Damaskin 1980, pp. 269–270), $(\psi_{Me} - \psi_S)_0$ is measurable, being -0.26 V for Hg in water, -0.51 V for Hg in methanol, and so on. These values are substantially different from zero, and there is no reason to suppose they are independent of metal. The hope of E_0 being numerically equivalent to $C = \Phi_1 - \Phi_2$, which was the subject of many investigations, is thus laid to rest.

However, a variety of linear relationships between E_0 and C have been proposed. Since the cell potential at zero surface charge, E_0 is $E_0^{(1)} - E_0^{(2)}$, where $E_0^{(1)}$ and $E_0^{(2)}$ are potentials of zero charge (pzc's) against some common reference electrode, this implies linear relationships between the pzc and the work function of a metal. As shown by Trasatti (1980), several such relationships are necessary to cover all the metals. One can expect different degrees of orientation of water on different metal surfaces, and hence different potentials $\chi^{S(i)}$. Since $(\psi_1 - \psi_S)_0 = \delta\chi_1 - \chi^{S(1)} + \chi_S$, equation [2] gives

$$E_0^{(1)} - E_0^{(2)} = \chi^{S(2)} - \chi^{S(1)} - \delta\chi_2 + \delta\chi_1 + \Phi_1 - \Phi_2$$

The pzc for a metal m (relative to some standard electrode) can thus be written as $\delta\chi_m - \chi^{S(m)} + \Phi_m + k$, with the value of k arbitrary. If $\delta\chi_m$ is about the same for all metals (it is probably small in any case), one has

$$E_0^{(m)} = \Phi_m + k - \chi^{S(m)}$$

and $\chi^{S(m)}$, which depends on the interaction of the solution with the metal, should be the same for chemically similar metals. The relation

$$E_0^{(m)} = \Phi_m - 4.61 - 0.40\,\alpha$$

was proposed for aqueous solutions (all quantities in volts), where α, the degree of orientation of water molecules, is supposed to increase in the sequence Au, Cu < Hg, Ag, Sb, Bi < Pb < Cd < Ga (see Section D). Some critique of the theory is given by Frumkin et al. (Frumkin, Petrii, and Damaskin 1980, pp. 271–272). We consider it further in Section B.

Returning to the metal–metal interface, we expect that the electron density varies through the metal–metal interface, from the bulk value corresponding to one metal to the bulk value of the other. The actual profile may be complicated, since it depends on the arrangements of heavy particles in the two metals and how each metal's heavy-particle arrangement is changed in the interfacial region by the presence of the other metal. The difference in inner potential from phase M to phase N, $\Delta_M^N \phi$, is determined by the difference of chemical potentials, since at equilibrium $\tilde{\mu}_e^M = \tilde{\mu}_e^N$. Using equation [1], we get

$$\mu_e^M - \mu_e^N = e(\phi^M - \phi^N) = e\Delta_N^M\phi = -e\Delta_M^N\phi$$

Alternatively, since $\phi^M = \psi^M + \chi^M$ and $\alpha_e^M = \mu_e^M - e\chi^M$, we have

$$\alpha_e^M - \alpha_e^N = e(\psi^M - \psi^N)$$

that is, the transfer of charge required to produce equality of electrochemical potentials is such that the difference in outer potentials after transfer is equal to the difference in real potentials before transfer. Since the real potential is the negative of the work function, we have

$$\psi^M - \psi^N = \frac{\Phi^N - \Phi^M}{\mathcal{F}} \qquad [3]$$

If $\Phi^N > \Phi^M$, contact between metals produces a transfer of electrons from M to N, charging M positively and N negatively.

One can also write $\Delta_M^N \phi$ as $\Delta_M^N \chi + \Delta_M^N \psi$, where $\Delta_M^N \chi$ is the difference of surface potentials $\chi^N - \chi^M$: these dipolar potentials exist at the free surfaces of the metal phases, not at the interface between the metals. At the interface, the electronic profile is certainly something other than the two individual metals' electronic profiles end to end. The potential difference across the interface is $g_M^N = \Delta_M^N \phi$. Trasatti (1980) denotes this by g_M^N (ion) to indicate that this is due to free charges and not to a dipole layer. Note that $g_M^N = \Delta_M^N \phi \neq \Delta_M^N \psi$; that is, the potential difference across the intermetallic surface region is not the same as the difference of outer potentials.

For an idealized model of the interface, corresponding to the Sommerfeld model (Chapter 4, Section E) for a single metal, we may calculate g_M^N. Suppose metal M has bulk electron density n_M and metal N has bulk electron density n_N, and the potentials inside the uncharged metals before contact are V_0^M and V_0^N, so that the work functions are $-(\epsilon_F^M + V_0^M)$ and $-(\epsilon_F^N + V_0^N)$, where $\epsilon_F^M = (\hbar^2/8\pi^2 m)(3\pi^2 n)^{2/3}$ (see Figure 13). After equilibrium is established, $e\Delta_M^N \psi = \Delta_N^M \Phi$. Since there is no surface dipole potential in this model, $\Delta_M^N \phi = \Delta_M^N \psi$. The inner potential varies continuously across the metal–metal interface. The Poisson equation in each metal is

$$\frac{d^2 \phi}{dz^2} = 4\pi [n(z) - n_b]$$

where n_b is the value of n far from the interface; n_b represents the positive charge density which, in this model, is constant within the metal.

The electron density varies continuously within each metal, such that $(\hbar^2/2m)(3\pi^2 n)^{2/3} - e\phi$ is constant and equal to its value in bulk. When ϕ_b is the bulk potential, far from the interface, we have

$$\frac{d^2 \phi}{dz^2} = 4\pi \left\{ \left[n_b^{2/3} - \frac{2me}{\hbar^2} (3\pi^2)^{-2/3} (\phi_b - \phi) \right]^{3/2} - n_b \right\}$$

Similarly to equations [76]–[82] of Chapter 2, Section I, we assume $\phi_b - \phi$ is small enough so that the equation may be linearized by taking the first two terms in the expansion of the square-bracketed expression:

$$\frac{d^2 \phi}{dz^2} = 4\pi \left\{ n_b [1 - a(\phi_b - \phi)]^{3/2} - n_b \right\}$$

$$= 4\pi n_b \frac{3a}{2} (\phi_b - \phi) \qquad [4]$$

where $a = (2me/\hbar^2)(3\pi^2 n_b)^{-2/3}$.

A. METAL-METAL INTERFACE

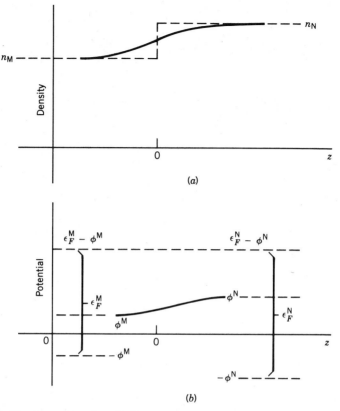

FIGURE 13. Metal–metal junction. (a) Electron density across interface. (b) Variation of potential across interface.

The general solution to equation [4] makes $\phi - \phi_b$ a linear combination of e^{cz} and e^{-cz}, where $c^2 = 6\pi n_b a$. For metal M ($z < 0$), ϕ is to approach ϕ_b (ϕ^M) for $z \to -\infty$, so

$$\phi = \phi^M + k_M \exp\left[\left(\frac{12me}{\hbar^2}\right)^{1/2}\left(\frac{n_M}{9\pi}\right)^{1/6} z\right], \quad z < 0$$

where k_M is a constant to be determined. For metal N ($z > 0$), ϕ is to approach ϕ_b (ϕ^N) for $z \to +\infty$, so

$$\phi = \phi^N + k_N \exp\left[-\left(\frac{12me}{\hbar^2}\right)^{1/2}\left(\frac{n_N}{9\pi}\right)^{1/6} z\right], \quad z > 0$$

Since $d\phi/dz$ is to be continuous at $z = 0$,

$$k_M \left(\frac{12me}{\hbar^2}\right)^{1/2} \left(\frac{n_M}{9\pi}\right)^{1/6} = -k_N \left(\frac{12me}{\hbar^2}\right)^{1/2} \left(\frac{n_N}{9\pi}\right)^{1/6} \qquad [5]$$

Since ϕ is to be continuous at $z = 0$, $\phi^M + k_M = \phi^N + k_N$. The equality of electrochemical potentials of the electron in the two metals requires

$$\frac{\hbar^2}{2m}\left[(3\pi^2 n_M)^{2/3} - (3\pi^2 n_N)^{2/3}\right] = e(\phi^M - \phi^N) = e(-k_M + k_N)$$

Combining this with equation [5], we get

$$\frac{\hbar^2}{2me}(3\pi^2)^{2/3}(n_M^{2/3} - n_N^{2/3}) = k_N\left[\left(\frac{n_N}{n_M}\right)^{1/6} + 1\right]$$

which is solved for k_N. Then one gets k_M and

$$\phi^M - \phi^N = \frac{\hbar^2}{2me}(3\pi^2)^{2/3}(n_M^{2/3} - n_N^{2/3}) \qquad [6]$$

According to this model, the potential difference across the interface is determined by the difference of electron densities of the separate metals. If $n_M = 0.10$ Å^{-3} and $n_N = 0.11$ Å^{-3}, then $\phi^M - \phi^N = 0.00179$ statvolt $= 0.54$ V.

The work required to eject an electron from either metal will be the same, equal to $-(\epsilon_F^M - e\phi^M)$ or $-(\epsilon_F^N - e\phi^N)$. This work is expected to be positive, so that ϕ^M and ϕ^N should actually be negative, unlike what is shown in Figure 13. Self-consistent calculations of electron density for the interface between two metals, using a jellium background, have also been performed (Lang 1981), giving realistic values for contact potentials.

B. METAL–POLAR-LIQUID INTERFACE

The equilibrium between a metal phase and a polar liquid usually does not involve equalization of the electrochemical potentials of the electron in the two phases, unless there are free electrons in the equilibrium liquid. Before metal–liquid contact, the electrochemical potential of the electron in the liquid is not defined. If electrons are stable in the liquid, electrons enter the liquid from the metal only simultaneously with positive ions, since there must be electrical neutrality on the thermodynamic scale. The relevant equilibrium is

$$M^+(M) + e^-(M) \leftrightarrow M^+(S) + e^-(S)$$

which is like the dissolution of a sparingly soluble salt. There is an equilibrium condition for each ion: $\tilde{\mu}_{M^+}^M = \tilde{\mu}_{M^+}^S$ and $\tilde{\mu}_{e^-}^M = \tilde{\mu}_{e^-}^S$ (although individual electrochemical potentials are not measurable). Writing each electrochemical potential as a chemical potential plus an electrical part, one has

$$\Delta_S^M \phi = - \frac{\Delta_S^M \mu_{M^+}}{e} \qquad [7]$$

$$\Delta_S^M \phi = \frac{\Delta_S^M \mu_{e^-}}{e} \qquad [8]$$

The electrical potential drop is determined by the chemical potentials in the two phases and is characteristic of the interface.

The addition of the positive ion M^+ to the solution in contact with the metal M makes the potential of the metal relative to the solution more positive if the interface is polarizable. This may be due to a change in either the free-charge or the dipolar contribution (see below), although the latter, $\delta \chi_S^M$, is usually assumed not to be altered. For a polarizable interface, M^+ does not cross into the metal, but the same change in electrode potential is produced using an external voltage source to feed the free charges onto the metal. If this is done for a nonpolarizable electrode, the charges pass through the interface into the solution, so only a composition change can alter the electrode potential.

The electrode is part of an electrochemical cell, and the cell potential for which there is no free-charge component in the electrode–solution potential difference, the pzc, is of central importance. Corresponding to this value of the cell potential, which is measurable, is a particular value of the electrode–solution potential difference, which is not a measurable quantity. Methods of determining pzc are reviewed by Frumkin, Petrii, and Damaskin (1980, Section 5.4). A table of pzc's is given: For metals adsorbing hydrogen, the table gives potentials of zero free charge and zero total charge; pzc's can generally be determined to ~ 0.1 V. For many ions in sufficiently dilute solution the pzc does not change by more than this when the ions are changed, so that specific adsorption is not important, and one can study the influence of the metal on the pzc.

Specific anion adsorption shifts the pzc in the negative direction; specific cation adsorption shifts the pzc in the positive direction. Esin and Markov found that the pzc shifts by 100–200 mV when there is a change of one order of magnitude in the activity of surface-active anions in an electrolyte. Double-layer theories that assume a "uniformly diffuse" charge distribution predict half of this shift. Grahame called the large shift the *Esin–Markov effect*. It must be explained by models invoking the discrete nature of adsorbed ions. Study of pzc's also gives important information about the adsorption of organic compounds on electrodes (Frumkin, Petrii, and Damaskin 1980, Section 5.5.1).

Whether or not electrons are stable in the liquid, at equilibrium a potential difference $\Delta_S^M \phi$ will exist between metal and solution. It may be thermodynamically defined by equations [7] and [8] or by some other interphase equilibrium involving charged particles of metal and solution. Structurally one can distinguish between dipolar and free-charge contributions, the latter being due to charge densities of opposite sign in the metal and the polar liquid. Since there are usually no components common to metal and solution, one can divide the dipolar part into contributions of components of the metal and components of the solution.

In the interface, the surface potential of the metal differs from its value for a free metal surface because the electrons encounter a dielectric medium and the electronic cloud of the solvent molecules, while the solvent molecules encounter the electrons and ions of the metal surface, changing the surface potential of the solvent from that of a free solvent surface. Thus χ^M is changed to $\chi^M + \delta\chi_S^M$ and χ^S is changed to $\chi^S + \delta\chi_M^S$, the overall dipolar contribution to $\Delta_S^M \phi$ being

$$g_S^M(\text{dip}) = g^M(\text{dip}) - g^S(\text{dip}) = (\chi^M + \delta\chi_S^M) - (\chi^S + \delta\chi_M^S) \quad [9]$$

The remainder of $\Delta_S^M \phi$, $g_S^M(\text{ion})$, is referred to as a free-charge contribution, but includes (Trasatti 1980, Section 3.2.2) the electronic polarization of the solvent molecules (the changed orientational polarization is included in $\delta\chi_M^S$). Now

$$\Delta_S^M \phi = g_S^M(\text{dip}) + g_S^M(\text{ion}) \quad [10]$$

and also

$$\Delta_S^M \phi = \Delta_S^M \chi + \Delta_S^M \psi \quad [11]$$

where $\Delta_S^M \chi = \chi^M - \chi^S$ (free-surface dipole potentials) and $\Delta_S^M \psi$ is the measurable Volta potential difference. Note that $\Delta_S^M \chi \neq g_S^M(\text{dip})$, since the latter term refers to the metal–solution interface and the former term refers to the free surfaces of the two phases, and $g_S^M(\text{ion}) \neq \Delta_S^M \psi$. The potential $g_S^M(\text{ion})$ vanishes when the surface charge at the metal–liquid surface vanishes, so that equations [9]–[11] then give

$$\Delta_S^M \psi = g_S^M(\text{dip}) - \Delta_S^M \chi = \delta\chi_S^M - \delta\chi_M^S \quad [12]$$

Thus at the pzc for the interface, the free surfaces of the two phases may be charged, so that $\Delta_S^M \psi$ is nonzero.

A difference of outer or Volta potentials is measurable, and a value of -0.26 V has been determined for $\Delta_S^M \psi$ for a mercury–aqueous-solution interface at the pzc (Trasatti 1979, pp. 115–116). Referring to equation [12] and to the value of -0.06 V for $\delta\chi_{Hg}^{H_2O}$ (estimated from various experimental and theoretical evidence—see below), the change in the surface dipole potential of mercury due to adsorbed water becomes $\delta\chi_{H_2O}^{Hg} = -0.32$ V.

This is a substantial change, which probably should be interpreted in terms of the tail of the conduction electron density, extending into the region of water molecules. The interaction of the conduction electrons with the inner-shell electrons of the water molecules is a closed-shell exchange repulsion, which, pushing the tail back toward the metal, lowers $g^M(\text{dip})$. Alternatively, the decrease in $g^M(\text{dip})$ could be explained in terms of the existence of the electronic dielectric constant in the region of water molecules: According to equation [77] of Chapter 2, the contribution of the electronic charge density in the region of dielectric constant ϵ, for $z > a$, where a is beyond the region of ions of the metal, is

$$\phi(a) - \phi(\infty) = \lim_{z \to \infty} \left[(z-a) E(a) + 4\pi\epsilon^{-1} \int_a^z dz' \int_a^z dz'' \rho(z'') \right]$$

$$= \lim_{z \to \infty} \left[(z-a) E(a) + 4\pi\epsilon^{-1} z \int_a^z dz'' \rho(z'') - 4\pi\epsilon^{-1} \int_a^z dz' \, z' \rho(z') \right]$$

on integrating by parts. Since $4\pi\epsilon^{-1} \int_a^\infty dz'' \rho(z'') = -E(a)$, one can pass to the limit $z \to \infty$. The charge density $\rho(z)$ is negative (electrons) for $z > a$, so that an increase in ϵ will decrease $\phi(a) - \phi(\infty)$, lowering $g^M(\text{dip})$. The increase in ϵ would also constitute an attraction for the metal's conduction electrons, elongating the tail, but this effect is small in model calculations (Badiali, Rosinberg, and Goodisman 1981). A value of several tenths of a volt for $\delta\chi_S^M$ arises, in these calculations, from the closed-shell repulsions or from the electronic dielectric constant.

The value of $\delta\chi_S^M$ should depend strongly on the metal, through the electron density, but less strongly on which species is adsorbed on the metal. The work function of Hg decreases by about 0.25 V on adsorption of rare gases, which could be interpreted as due to $\delta\chi_G^M$, if it is assumed that there is no potential drop across the layer of adsorbate. This is comparable to -0.32 V for water.

The electrochemical cell $M^{(1)}|\text{Ref}|S|M^{(2)}$, with leads of the same metal M, was analyzed by Bockris and co-workers to show how an estimate of an actual electrode–solution potential difference could be obtained (Habib 1977, pp. 166–169). The cell potential, assuming that the reference electrode is nonpolarizable and that the electrode $M^{(2)}$ is at its pzc, is

$$E_{\text{pzc}} = \Delta_M^{\text{Ref}} \phi + \Delta_{\text{Ref}}^S \phi + \Delta_S^M \phi$$

$$= \mathcal{F}^{-1}(\mu_e^{\text{Ref}} - \mu_e^M) + \Delta_{\text{Ref}}^S \phi + g^M(\text{dip}) - g^S(\text{dip}) \qquad [13]$$

The equality of the electrochemical potentials of the electron in $M^{(1)}$ and the reference electrode has been used in evaluating $\Delta_M^{\text{Ref}} \phi$. The quantity $\Delta_{\text{Ref}}^S \phi + \mathcal{F}^{-1} \mu_e^{\text{Ref}}$ is measurable; a value derived by applying a thermodynamic cycle to the normal hydrogen electrode (NHE) is -4.31 V. The chemical potential of an electron in a simple metal can be calculated reliably, so that a measurement of E_{pzc} gives $g^M(\text{dip}) - g^S(\text{dip})$, the metal–solution potential difference.

Since the work function Φ^M is $\chi^M - \mu_e^M/\mathcal{F}$, with $g^M(\text{dip}) = \chi^M + \delta\chi^M$, equation [13] may also be written

$$E_{\text{pzc}} = \Delta_{\text{Ref}}^S \phi - g^S(\text{dip}) + \delta\chi^M + \frac{\mu_{e^-}^{(\text{Ref})}}{\mathcal{F}} + \Phi^M \qquad [14]$$

This shows that, to have a linear relationship between the work functions of metals M and their pzc's, the change in the surface potential on immersion of M in S, $-g^S(\text{dip}) + \delta\chi^M$, must be the same for all metals. Furthermore, we see that the value of $\Delta_{\text{Ref}}^S \phi$, an unmeasurable difference of Galvani potentials, can be found if one can calculate $\mu_{e^-}^{(\text{Ref})}$ and estimate $-g^S(\text{dip}) + \delta\chi^M$. A problem is that calculations of chemical potentials are reliable only for simple metals, for which the free-electron model is good, but not for the metals (e.g., Pt) against which pzc's are measured. As discussed by Bockris and Khan (1979, Section 1.3.6), one can refer the measured pzc to a standard Na$^+$|Na electrode: $\mu_{e^-}^{(\text{Na})}$ is reliably calculated as -3.2 eV. The pzc of mercury that is referred to Na$^+$|Na is 2.51 V, and Φ^{Hg} is 4.51 V. Then

$$-\Delta_{\text{Ref}}^S \phi = -2.51 \text{ V} - g^S(\text{dip}) + \delta\chi^{\text{Hg}} - 3.20 \text{ V} + 4.51 \text{ V}$$

With estimates of $g^S(\text{dip}) \sim 0.20$ V and $\delta\chi^{\text{Hg}} \sim 0.26$ V, $\Delta_{\text{Na}}^S \phi$ becomes 1.14 V (Bockris and Argade 1968). A similar calculation for the absolute Galvani potential difference across the standard hydrogen–nickel electrode surface, with $\mu_{e^-}^{(\text{Ni})} = 5.5$ eV, gave $\Delta_{\text{Ni}}^S \phi$ as 0.74 V.

The metal–solution potential difference reflects the equilibrium between metal ion in solution and metal ion in the metal, for example, $\tilde{\mu}_{\text{Na}^+}^{(\text{Na})} = \tilde{\mu}_{\text{Na}^+}^{(S)}$. Separating chemical and electrical parts, we see that $\mu_{\text{Na}^+}^{(\text{Na})} - \mu_{\text{Na}^+}^{(S)} = \mathcal{F}\Delta_{\text{Na}}^S \phi$. The tendency of Na$^+$ to leave the metal and become solvated means its chemical potential is lower in the solution, so $\Delta_{\text{Na}}^S \phi$ is positive. The positive electrical potential difference compensates the tendency of ions to leave the metal; the situation is quite analogous to the electronic equilibrium that leads to the contact potential.

If the solvent–dipole surface potential can be calculated, one obtains from the metal–solution potential difference a value for the surface potential of the metal (due to the conduction-electron tail). Thus, assuming that $g^S(\text{dip})$ at the pzc was due to different dispersion interactions of the oxygen and hydrogen ends of a water molecule with the metal, Bockris and Habib (1977) calculated a value of 0.03 V for the surface potential of water on a mercury surface. Other workers have estimated this to be about 0.1 V (Trasatti 1979). Values of $g^M(\text{dip})$ are of the order of volts so that in fact the electrode–solution potential differences found from equation [13] are essentially equal to the values of $g^M(\text{dip})$.

Returning to equation [14], but with an NHE as reference electrode, we write

$$E_{\text{pzc}} = \Delta_{\text{Ref}}^S \phi - \chi^S + \frac{\mu_{e^-}^{(\text{Ref})}}{\mathcal{F}} + \delta\chi_S^M - \delta\chi_M^S + \Phi^M$$

$$= -E_k(\text{H}_2/\text{H}^+) + \delta\chi_S^M - \delta\chi_M^S + \Phi^M$$

Here, E_k is the "absolute potential" of the hydrogen electrode, equal to the energy required to extract an electron from the electrode to vacuum, divided by $-e$; its value is 4.44 V (Trasatti 1974). To get $\delta\chi_M^S$, one can compare potential shifts on adsorption of aliphatic alcohols at the air–solution and mercury–solution interfaces. The shifts vary linearly with chain length. The difference between the parallel lines formed by plots of potential shift vs. chain length is taken to be $\delta\chi_M^S$, and its value turns out to be -0.05 V. If E_{pzc} is -0.19 V (NHE) and $\Phi = 4.50$ eV for Hg, then $\delta\chi_S^M = -0.19 + 4.44 - 0.05 - 4.50 = -0.30$ V (Trasatti 1983), which is interpreted to mean that the tail of the electronic distribution is pushed back by the solvent, as also appears from theoretical calculations (Badiali, Rosinberg, and Goodisman 1981). Without information on $\delta\chi_M^S$ for Hg, one cannot obtain a value for $\delta\chi_S^M$ for other metals. However, one can get from experiment the variation of $\delta\chi_S^M - g^S(\text{dip})$ at the pzc from metal to metal. Since $g^S(\text{dip})$ is expected to be positive, the values calculated from various models for $\delta\chi_S^M$ are shown to be too large (Trasatti 1983).

The work function $\Phi^{M/S}$ is the minimum work necessary to extract an electron from metal into liquid:

$$\Phi^{M/S} = \tilde{\mu}_{e^-}^S - \tilde{\mu}_{e^-}^M \qquad [15]$$

The value of this quantity clearly depends on the potential difference between metal and liquid phases and hence on the surface charge, unlike the work function Φ^M for a metal (Chapter 4, Section F). In terms of the potentials defined above, we have

$$\begin{aligned}\Phi^{M/S} &= \mu_e^S - \mu_e^M - e(\phi^S - \phi^M)\\ &= \mu_e^S - (-\Phi^M + e\chi^M) + e[g_S^M(\text{dip}) + g_S^M(\text{ion})]\\ &= \mu_e^S + \Phi^M + e(\delta\chi_S^M - \chi^S - \delta\chi_M^S) + eg_S^M(\text{ion})\end{aligned}$$

If there is thermodynamic equilibrium between metal and solution with respect to electron transfer, $\tilde{\mu}_e^S = \tilde{\mu}_e^M$ and $\Phi^{M/S}$ vanishes. Otherwise, the value of $\Phi^{M/S}$ must depend on the final state of the emitted electrons. They may be fully solvated (see Chapter 6, Section C), or delocalized. In the latter case, a nonequilibrium work function is appropriate:

$$_{ne}\Phi^{M/S} = \mu_{em}^{*S} + \Phi^M + e(\delta\chi_S^M - \chi^S - \delta\chi_M^S) + eg_S^M(\text{ion})$$

where μ_{em}^{*S} is the standard chemical potential for a delocalized electron in liquid (see Chapter 6, Section C).

If the surface charge on the metal is zero, $g_S^M(\text{ion})$ vanishes and, using equation [12],

$$_{ne}\phi^{M/S} = \Phi^M - \Phi^S + e\Delta_S^M\psi \qquad [16]$$

where the work function of the solution is $\Phi^S = e\chi^S - \tilde{\mu}_{em}^{*S}$ (Trasatti 1980, Section 3.3.2). These equations apply if the theoretical equilibrium concentration of electrons is less than the concentration produced by a quantum of light via the photoelectric effect, so that equilibrium is not attained. The electrode is then polarizable with respect to electrons, although not necessarily with respect to other ions. Trasatti (1980) gives the example of a silver electrode in contact with water, for which $a_{Ag^+}a_{e^-} \sim 10^{-62}$ mol/dm^3, so that, in the absence of added Ag$^+$, there would be 1 electron in 10^7 dm^3 of solution. If Ag$^+$ is present in 1 m solution, $a_{e^-} \sim 10^{-62}$ mol/dm^3, so $\tilde{\mu}_e^M$ and $\tilde{\mu}_e^S$ are certainly not equal and the electrode is polarizable with respect to electrons although nonpolarizable with respect to Ag$^+$. Electrodes nonpolarizable with respect to electrons include alkali metals in contact with ammonia and other nonaqueous solvents.

The work function (for emission to vacuum) can be measured for a metal with adsorbed water (Habib 1977, pp. 174–176) and is found to decrease as the amount of water increases. This difference in the value of the work function compared to that of clean metal is due to the change in the potential difference between the interior of the metal and the exterior, since the chemical potential of the electron does not change. For water adsorbed on Ni, Fe, Hg, and other metals, the decrease in Φ with adsorbed water is at first proportional to the amount of water but then decreases less rapidly, and Φ reaches its limiting value when the water coverage corresponds to about two monolayers. In fact, more than three-fourths of the total change in work function is achieved for monolayer coverage. (Exceptions occur when there is oxide formation or other indication of a strong interaction of water with the metallic phase.) Thus the surface dipole potential of the water comes essentially from the first layer, which justifies consideration of only the compact layer in studying the electrode–solution interface. The implication that no dipole potential is contributed by adsorbed water between the layer closest to the metal and the outermost layer has led to a three-region model, in which, between the structured adsorbed water at the metal surface and the normal bulk water structured by hydrogen bonding, there is a transition region of disordered water molecules.

It should be noted that there is some ambiguity in interpreting the change in work function of a metal on water adsorption. The fact that Φ decreases does imply that adsorption occurs with the oxygen end toward the metal, decreasing the surface potential. But for many metals the decrease in Φ is of the order of a volt, whereas $\delta\chi_S^M$ is thought to be about -0.3 V for most metals (cf. above, after equation [12]) and $g_M^{H_2O}$ (dip) is only 0.1–0.4 V at the pzc; $-\Delta\Phi$ would then be between 0.4 and 0.7 V. The problem may be that the surface charge q^M is not maintained at zero during adsorption, so that a free-charge potential must be included as well as the dipole potentials. Indeed $\Delta\Phi$ is positive for adsorption of water on metals when oxidation and hydrogen evolution occurs.

The work function of an *ion* in a metal, which is the negative of its real potential, is measurable, like the electronic work function. One has to consider the thermodynamic cycle formed from the following reactions:

$$M^{(M)} \to M^{(vac)} \qquad \Delta H = S = \text{energy of sublimation}$$

$$M^{(vac)} \to M^{z+(vac)} + ze^{-(vac)} \qquad \Delta H = \sum I = \text{sum of ionization potentials}$$

$$M^{z+(vac)} \to M^{z+(M)} \qquad \Delta H = -w(M^{z+}) = \text{minus work function of ion in metal}$$

$$ze^{-(vac)} \to ze^{-(M)} \qquad \Delta H = -z\Phi^M = \text{minus work function of electrons}$$

Since the ΔH's must sum to zero,

$$w(M^{z+}) = S + \sum I - z\Phi^M$$

gives the work function of M^{z+} in terms of measurable quantities.

For the $R|H^+|H_2$ electrode, with the electrode metal R usually Pt, the work function of the ion in the metal is replaced by $w(H^+)$, the work required to eject an H^+ ion from R to vacuum at infinity (Antropov 1972, p. 230). To evaluate $w(H^+)$, one can consider the following reactions:

$$\tfrac{1}{2}H_2^{(g)} \to H^{(g)} \qquad \Delta H = \tfrac{1}{2}D_{H_2} \text{ (dissociation energy)}$$

$$H^{(g)} \to H^{+(vac)} + e^{-(vac)} \qquad \Delta H = I_H \text{ (ionization potential)}$$

$$e^{-(vac)} \to e^{-(Pt)} \qquad \Delta H = -\Phi^{Pt} \text{ (electronic work function)}$$

Thus the work required to eject H^+, leaving an electron on the Pt, is

$$w(H^+) = \tfrac{1}{2}D_{H_2} + I_H - \Phi^{Pt}$$

This is for the reaction $Pt + \tfrac{1}{2}H_2^{(g)} \to H^{+(vac)} + Pt + e^{-(Pt)}$.

The emf of a cell in which M is an electrode may be written in terms of $w(M^{z+})$ (Antropov 1972, pp. 238ff). The result shows the effects of the nature of the solvent and the nature of the electrode on the emf. Consider the cell

$$M'|R|L|M$$

where M and M' are the same metal, and the reference electrode $R|L$ could also be a hydrogen electrode. The emf of the cell is $\phi^{(M)} - \phi^{(M')} = \psi^{(M)} + \chi^{(M)} - \psi^{(M')} - \chi^{(M')}$. Since the surface potentials depend only on the nature of the metal, they cancel. The emf of the cell is thus a difference of Volta potentials:

$$E = \Delta_{M'}^M \psi = \Delta_{M'}^R \psi + \Delta_R^L \psi + \Delta_L^M \psi$$

For a metal-metal contact with electronic equilibrium, $\Delta_{M'}^R \psi = \mathcal{F}^{-1}(\Phi^M - \Phi^R)$

(equation [3]). For a metal–electrolyte contact with exchange of ions, $\tilde{\mu}^M_{M^{z+}} = \tilde{\mu}^L_{M^{z+}}$ leads in a similar manner to

$$\psi^M - \psi^L = (z^+\mathfrak{F})^{-1}\left[w(M^{z+}) - (-A^{M^{z+}})\right]$$

where A is the free-energy change for introduction of ions M^{z+} into the liquid from vacuum. Therefore the emf is

$$\mathfrak{F}E = (\Phi^M - \Phi^R) + \frac{w(R^{z+}) + A^{R^{z+}}}{z_+} - \frac{w(M^{z+}) + A^{M^{z+}}}{z_+}$$

where, for simplicity, the charges of the ions of R and M have both been taken as z_+. The difference $A^{R^{z+}} - A^{M^{z+}}$ may be replaced by a difference of energies of hydration, since the same surface potential of the liquid is involved in inserting both ions.

Using our results for $w(M^{z+})$ and $w(H^+)$, the emf of the cell $M'|Pt(H_2)|L(H^+, M^+)|M$ is

$$E = \mathfrak{F}^{-1}(\Phi^{Pt} - \Phi^M) + \mathfrak{F}^{-1}(\tfrac{1}{2}D_{H_2} + I_H - \Phi^{Pt} + A^{H^+})$$

$$- (z_+\mathfrak{F})^{-1}\left(S^M + \sum I^M - z_+\Phi^M + A^{M^{z+}}\right)$$

$$= (z_+\mathfrak{F})^{-1}\left\{\left[z_+(\tfrac{1}{2}D_{H_2} + I_H) - \left(S^M + \sum I^M\right)\right] + zA^{H^+} - A^{M^{z+}}\right\}$$

The term in square brackets involves properties of the electrode material M (and H_2, but not Pt, in the case of $H_2|H^+$). The remaining (hydration) terms show the influence of the solvent. If the solvent only is changed, the square-bracketed expression remains the same. Hydration energies involve Born-type terms, $(z^2e^2N_{Av}/2R)(1 - \epsilon^{-1})$ (see Chapter 6, Section C) plus others for specific interactions. The value of the dielectric constant is thus very important, but it is not the whole story.

C. METAL–MOLTEN-ELECTROLYTE INTERFACE

Because of the absence of solvent, theoretical treatment of molten electrolytes may be simpler than for solutions. Electrocapillary curves have been determined for a number of metal–molten-electrolyte interfaces, although there are many experimental difficulties, and measurements of differential capacities have also been performed (Frumkin, Petrii, and Damaskin 1980, Section 5.6.3). Good agreement has been obtained in some cases between experimental electrocapillary curves and theoretical curves obtained by double integration of capacity as a function of cell potential. The capacity shows a minimum as a function of cell potential, usually quite close to the electrocapillary maximum (ecm). No theoretical justification for the coincidence of the ecm and the capacity minimum is yet available. The values of the minimum capacities are about the same size as for aqueous solutions, and pzc's also are often comparable to those for aqueous solutions. Of course, such features as the capacitance hump, which is related to solvent dipoles, are absent.

These similarities between the melts and the solutions are remarkable. Frumkin, Petrii, and Damaskin (1980, Chapter 5, pp. 279–282) compare the pzc of a melt and that of the corresponding aqueous solution. Noting that the difference in temperature is very large, the agreement between the two for a variety of electrode materials is striking. However, the capacitance of a metal-melt interface increases with temperature instead of decreasing (March and Tosi 1984, Section 8.6d).

In general, the capacitance varies approximately parabolically as a function of potential. The potential of minimum capacitance, which closely coincides with the pzc (ecm), depends on the cation, but is usually independent of the anion. The value of the minimum capacitance increases with temperature for alkali halides, but not for nitrates (Inman and Lovering 1983, Section 3.2).

Theories of the Gouy–Chapman type are certainly incapable of describing these systems, given the high densities of charged particles. It has been suggested that the structure of the double layer consists of alternating layers of charge (Ukshe et al. 1964). A theoretical description in terms of the distribution functions (Chapter 3, Section A) for bulk fluid (Dogonadze and Chizmadev 1964, Goodisman and Amokrane 1982) predicted a damped oscillating distribution of charge and a capacitance proportional to the inverse square of the Debye length. Other theoretical work produced a similar charge distribution (Inman and Lovering 1983, Section 3.2). None of the theories could account for the changes of capacitance with temperature and potential.

It seems that the electrodes used (molten Pb | molten alkali halide) are not at all ideally polarizable; the potential range for ideal polarizability may depend on temperature. The liquid lead electrode may even be exhibiting ideally reversible behavior for some of the potential range investigated (Inman and Lovering 1983, Section 3.3). The frequency dependence of the capacitance indicates that the capacitance is due to adsorption as well as to double-layer charging. If the discharge of adsorbed ions is very fast compared to the measuring frequency, one may still find that the Lippman equation is verified: The coincidence of the curve of surface tension vs. potential with the doubly integrated capacitance–potential curve is no proof that the capacitance is a true double-layer capacitance.

The difficulties in making experimental measurements on molten salts are balanced somewhat by the theoretical simplification, as compared to electrolyte solutions. Thus it has been possible to perform molecular dynamics simulations for a simple molten salt (KCl) in the presence of its vapor (Heyes and Clarke 1979) and in the presence of a rigid boundary and an electric field, which is a first approximation to the metal–molten salt interface (Heyes and Clarke 1981). In these simulations, the interionic interactions were taken as

$$u_{ij}(r) = \frac{q_i q_j}{r_{ij}} + A_{ij} e_{ij}^{-br} + C_{ij}(r_{ij})^{-6} + D_{ij}(r_{ij})^{-8}$$

where the terms express Coulombic interactions, short-range repulsions, and van der Waals interactions. The interaction of an ion with the wall was $u_w(z) = A(bz + 2) e^{-bz}$, where z is the distance from the wall. The simulation could follow the motions of 288 ions. It proved convenient to consider a film of molten KCl between two walls.

For uncharged walls, the one-particle distributions or density profiles for K^+ and Cl^- were essentially the same (the ions are about the same size). They showed strong oscillations in the vicinity of the walls, just as did the profiles for the free surface (in contact with vapor). These oscillations are associated with the hard-core repulsion; their wavelength is close to 3 Å, which is the nearest-neighbor separation in the crystal. The effect of an electric field is to build up K^+ concentration and decrease Cl^- concentration at the negatively charged wall and vice versa. The peaks in the density profile $\rho(z)$ now split into positive and negative components, and the spacing between like-ion peaks increases to 4 Å. In the presence of charged walls, the oscillations persist farther away from the wall into the melt than for the neutral system; this supports the idea that multilayers of charge exist in the electrified interface. Interestingly, the average of the positive- and negative-ion profiles was identical to $\rho(z)$ for the unperturbed (zero-field) case, so that the structure is always determined by packing considerations (hard-core interactions), with the applied field altering cation and anion distributions slightly in opposite directions (Heyes and Clarke 1981). The distribution functions in the direction tangential to the interface showed substantial clustering.

The tangential component of pressure (Chapter 3, Sections B and C) was calculated and resolved into contributions of the different parts of u_{ij}. The Coulombic and repulsive terms were each more than 10 times the total, but almost exactly canceled each other, so the other short-range terms played a major role. The surface tension at the rigid wall was similar to that of the free (liquid–vapor) surface, but the surface excess entropy was opposite in sign for the two cases, reflecting the quasi-crystalline ordering at the wall. Capacitances were calculated and were of the same order of magnitude as those measured for molten alkali halides.

D. WATER LAYER ON METAL

Central to any model of the electrode–solution interface is the molecular structure of the layer of water at the metal surface. The electrode–solution potential difference, and how it varies with the metal and the state of charge of the electrode, has been discussed largely in terms of the dielectric properties of water molecules in the compact layer.

The compact layer in the absence of specific adsorption consists of a layer of water molecules, which are physisorbed or chemisorbed on the electrode surface.

The properties of such a layer are expected to differ from the properties of bulk water. As already noted, the size of the inner-layer capacitance for water on mercury is about 30 $\mu F/cm^2$. An ideal capacitor with this capacitance would have $\epsilon/d = 3.4 \times 10^8$ cm^{-1}, corresponding to a dielectric constant of about 10 and a thickness (3 Å) of a monolayer of water. We have also noted that the capacitance of the diffuse layer becomes much larger than measured capacitances as soon as q exceeds a few microcoulombs per square centimeter, so that the potential drop across the diffuse layer is negligible compared to that across the compact layer. Thus the field through the compact layer is essentially the field at the electrode, which, for a surface charge of 10 $\mu C/cm^2$ and $\epsilon = 15$ is 6×10^5 V/cm. This is strong enough to have great effects on the properties of water in such a layer.

The dielectric constant of water in the compact layer differs appreciably from that of bulk water; this is expected because the large dielectric constant of water comes from the large dipole moments (connected with the association of water by hydrogen bonding) that are oriented by an electric field. The hydrogen bonding is lessened for the water near the electrode since (1) interactions with the electrode make the dipole less readily orientable and (2) the hydrogen-bond structure of water is broken at the electrode. According to the Onsager model (Chapter 3, Section D), the dielectric constant would only be about 30 if it were individual water molecules whose dipole moments were being oriented. Furthermore, it is possible that, under the high electric fields that can be produced in the compact layer, the polarization may cease to be proportional to electric field. The Langevin function $\mathcal{L}(x)$ may be approximated by $x/3$ only if x is small compared to 1, and (cf. Chapter 3, equation [33]) this requires $3\epsilon(2\epsilon + 1)^{-1} \mu E/kT$ to be small: For water at 300 K and a field of 10^4 esu/cm, which we have seen is quite possible, $\mu E/kT = 0.43$. Lastly, the thickness of the compact layer could vary with field, due to the electrostatic pressure term $\epsilon E^2/8\pi$.

As discussed in Chapter 4, Section A, the measurable surface excess entropy gives information on the solvent dipoles at the interface. After subtracting the contribution of the ions, one finds a solvent contribution that is roughly parabolic as a function of electrode charge. Its maximum occurs at a surface charge of 4–6 $\mu C/cm^2$, to the negative side of the pzc. This surface charge density thus produces the maximum disorder or minimum orientation of water molecules. The entropy of formation of the mercury–solution interface, $\Delta S^{M/S}$, also goes through a maximum when q^M is about 4–6 $\mu C/cm^2$, negative to the pzc.

The possibility that the difference between the pzc and the point of maximum entropy is related to the differing interactions of the anions and cations of the solution with the metal's ions and electron cloud is dismissed as follows: These interactions would change the surface excess of the metal (Hg) and would lead to an increase in $\Delta S^{M/S}$ for anodic potentials compared to cathodic potentials, which is the opposite to what is observed. The change in the solvent contribution to interfacial entropy could be due to a change in the amount of water in the surface layer, but the variation in the number of water molecules required to explain the changes is much larger (of the order of 50% of the total number) than the variations that are believed to occur. Diffuse-layer contributions can also be calculated and are found to be small. Therefore, the charge dependence of the surface entropy must be due to changes in the orientation of adsorbed water dipoles due to electric field. Further evidence that water orientation is at a minimum at the point of maximum entropy comes from the dependence of adsorption of organic molecules as a function of electrode charge.

The fact that the entropy maximum is at a potential negative to the pzc is consistent with the preferential adsorption of water molecules with their oxygen end toward the electrode at the pzc. A negative potential is necessary to overcome this preference and give equal populations of different orientations. Another explanation is provided by the electron density cloud of the metal. To the extent that the tail of this density extends past the water molecules in the first adsorbed layer, these molecules see a net field, corresponding to a positive charge on the metal, even at the pzc. If the electron tail resembles that of the metal in a vacuum, the electron density behaves something like $n_0 e^{-r/a}$, where n_0 is the bulk electron density, and a is about 1 Bohr radius. For a bulk electron density of $0.02 e/(a_0)^3$, the total electronic charge density beyond a plane at 1.5×10^{-8} cm from the electrode, where the centers of the water molecules are supposed to be located, is $n_0 a \exp(-1.5 \times 10^{-8}/a) = 6.7 \, \mu C/cm^2$. Thus, to have zero field at the plane of the water molecules, the charge on the metal should be about $-6.7 \, \mu C/cm^2$.

One way of estimating the surface potential of water molecules at a metal surface is from the shift of the pzc brought about by adsorption of organic molecules. For a given metal,

$$E_{pzc}^{(H_2O)} - E_{pzc}^{(org)} = -g^S(dip)^{(H_2O)} + g^S(dip)^{(org)}$$

Assuming $g^S(dip)^{(org)} = 0$, one obtains a value of 0.07 V for $g^S(dip)^{(H_2O)}$ on mercury; a lower value is obtained if $g^S(dip)^{(org)}$ cannot be neglected (and is in the same direction as for water). Trasatti (1979) discusses problems with this method. It is suggested that the organic molecules do in fact contribute to the surface potential, but that their moments produce extra orientation of the water molecules. It is shown that one can obtain, from shifts in pzc with adsorption of organics at the *free* surface of water, a value of -0.06 V for $g^{H_2O}(dip) - \chi^{H_2O}$; with the value of ~ 0.13 V for χ^{H_2O}, the surface potential at the free water surface, one again gets $g^{H_2O} = 0.07$ V.

The degree of orientation of water molecules at a metal surface is reflected in the contribution of the water to the metal–water dipolar potential difference, $\chi^S + \delta\chi_M^S$, where χ^S is the dipolar potential at a free water surface and $\delta\chi_M^S$ is the change due to the metal M. Thus $\delta\chi_M^S$ depends on the strength of the interaction of water with M, which should in turn depend on the electronegativity of the metal. Equation [13] may be written

$$E_{pzc} = \Phi^M/e + \delta\chi_S^M - \delta\chi_M^S + \text{constant}$$

If $\delta\chi_S^M$ is negligible, this says that, for metals of the same electronegativity, the pzc E_{pzc} equals Φ^M/e plus a constant. In plots of E_{pzc} vs. Φ^M/e, metals of the same electronegativity in fact fall on the same line of unit slope. From the lines corresponding to different electronegativities, one can deduce the degree of orientation of water on the metal (Reeves 1980, Section 3.3) and hence the relative strength of metal–solvent interaction. Then one deduces that for Au and Cu, with electronegativity ~2.0, there is no net orientation; for transition metals with electronegativity ~1.5, water is completely oriented; and in going from zero orientation to complete orientation there is a change in potential of 0.4 V.

Values of $g^{H_2O}(\text{dip})$ at the pzc for various metals thus vary between 0 and about 40 mV, always positive, corresponding to preferential adsorption with the oxygen end of the molecule toward the metal. It is generally thought that there is some extra attraction for the oxygen end of the molecule by the metal. Some authors interpret this as an enhanced dispersion interaction and others interpret it as a chemical or hydrophilic interaction. Surfaces that bind water by acting as proton donors are called *hydrophilic*, whereas surfaces with exposed oxygen groups are often *hydrophobic*. It has been proposed that metals for which $g^{H_2O}(\text{dip}) > \chi^{H_2O}$ should be classified as hydrophilic, and those for which $g^{H_2O}(\text{dip}) < \chi^{H_2O}$ should be classified as hydrophobic. Only Hg and Au are hydrophobic by this criterion. According to this, $g^{H_2O}(\text{dip})$ should correlate with the standard enthalpy of formation for the metal oxide, which also measures an affinity for oxygen, and such a correlation does exist (Trasatti 1979, pp. 110–112).

The linear correlation between $g^S(\text{dip})$ on M and the heat of formation of the metal oxide works well for many metals. Even if $\delta\chi_S^M$ is important, its variation from metal to metal probably parallels that of $g^S(\text{dip})_0$. The conclusion that the preferential orientation of water molecules at the interface increases as the affinity of the metal for oxygen increases is safe. But the correlation does not work well for d metals, only for s–p metals: $\delta\chi_S^M - g^S(\text{dip})_0$ is constant as the heat of oxygen adsorption changes among the d metals. Either $\delta\chi_S^M$ and $g^S(\text{dip})_0$ compensate each other, or $\delta\chi_S^M$ is small and $g^S(\text{dip})_0$ does not change, perhaps because the water molecules are always maximally oriented. Estimates of the strength of the interaction of water with s–p metals indicate it varies only by 4 kJ/mol from Hg to the most hydrophilic, Ga (Trasatti 1983). Although strong adsorption of H_2O on Ga has been much discussed, it seems that the bond between water and s–p metals should not be regarded as a chemisorptive bond, but something weaker, whose energy depends on surface electronegativity. On the contrary, a true chemisorptive bond is very likely to exist between water molecules and the surfaces of d metals. For such metals, bond strengths are of the order of tens of kilojoules.

A strongly hydrophilic surface should have a large effect on the structure of water, possibly orienting the water molecules beyond the first monolayer. It has been suggested that such an orienting effect is responsible for the inability of Gouy–Chapman theory to describe the diffuse layer on Ga electrodes with the usual value of the dielectric constant for bulk water. An appreciably higher value, about 120, is required. The dielectric constant of water molecules that are constrained from free orientation by the chemical forces exerted by the Ga electrode becomes anisotropic. The molecules are less able than bulk water to shield ion–ion interactions perpendicular to the metal surface, corresponding to a lower dielectric constant, but perhaps are more effective in shielding ion–ion interactions along the surface. Alternatively, the existence of an intermediate layer of water between the adsorbed layer and bulk water has been invoked. The molecules of this intermediate layer, which represents the transition from one kind of ordering to another, are completely disordered, and hence orient more readily in a field, corresponding to a higher dielectric constant (Trasatti 1979, pp. 112–113).

Given a value for $g^S(\text{dip})$, one can rewrite equation [13] in terms of the measurable work function $\Phi = -\mathcal{F}^{-1}\mu_e^M + \chi^M$,

$$E_{\text{pzc}} = -4.31 \text{ V} + \Phi + \delta\chi_S^M - g^S(\text{dip}) \qquad [17]$$

and derive values for $\delta\chi_S^M$. For Cd and Zn, values of -0.48 V and -0.40 V were found, but there were uncertainties of some tenths of volts in the values used for μ_e^M. On the other hand, if $\delta\chi_S^M$ is really much smaller than a volt, like $g^S(\text{dip})$, equation [17] shows that E_{pzc} should be about 4.3 V lower than Φ for all metals. In fact, relationships of the form $E_{\text{pzc}} = a\Phi - b$ have been proposed on the basis of experimental data. It appears that, with $a = 1$, such an equation works well if one uses different constants b for different groups of metals (Trasatti 1980) although experimental uncertainties in E_{pzc}, and its dependence on crystal face, cloud the picture somewhat. Trasatti has suggested that the necessity to use different values of b is related to changes in $g^S(\text{dip})$, that is, in water orientation, so that one could write

$$E_{\text{pzc}} = \Phi - 4.31 - 0.40\alpha \qquad [18]$$

where α is the degree of orientation of water molecules at the pzc. The value of α is correlated with the electronegativity of the metal. The metal dependence in this correlation could also arise from changed $\delta\chi_S^M$, whose value also is likely to correlate with the electronegativity of the metal.

The inner-layer capacitance at the pzc is about 30 $\mu F/cm^2$ for the mercury–aqueous-solution interface, and about the same for a number of other metals, but 53 $\mu F/cm^2$ for Cd, 80 for In, and 135 for Ga. The rate at which the capacitance increases with surface charge near the pzc changes from metal to metal like the value of the capacitance itself. The differences between metals are usually attributed to the behavior of the adsorbed water: The more hydrophilic the metal surface, the larger the electric fields that are required to reorient the water molecules. Then for a given change in the surface charge, there will be less change in the dipole potential of the adsorbed water, which means a larger contribution of water orientation to the dielectric constant and a larger capacity. It may be noted that the electron densities of the metals parallel the hydrophilicity scale so that the trends in capacitance could also be explained by invoking a contribution of the metal conduction electrons to capacitance.

In Figure 9 we showed the compact-layer capacitance as a function of surface charge, extracted by Grahame from capacitance measurements for mercury in NaF, neither ion of which is surface-active. The first point one should notice about the capacities is the fact that there is no symmetry about the ecm or point of zero charge ($q^M = 0$). Except at the lowest temperature, the capacitance increases with increasing q^M near the point of zero charge. Correspondingly, the minimum in the capacity occurs for q^M about $-10 \ \mu C/cm^2$. This is consistent with the fact that zero net orientation of water occurs when q^M is not zero, but slightly negative. Other evidence for this comes from measurements of the surface entropy (see above), from measurements of the adsorption of organic molecules, and from the temperature dependence of the point of zero charge. If one goes far from the point of zero charge in either direction, the capacity increases, but the increase is much more rapid on the positive-q side. At low temperatures, the capacity as a function of q actually involves two minima, separated by a maximum. Explaining this "capacitance hump" has been a preoccupation of theorists for a long time and has led to a lot of controversy. For negative electrode charges, the capacitance seems to become independent of temperature and does not vary much. Another feature to explain is the fact that the capacitance-charge curves all seem to intersect at one point.

At potentials far from the pzc, one may assume the orientation of water molecules is total, independent of the metal of the electrode. Then, comparing cell potentials for different metals M^1 and M^2 at the same surface charge q^M, we get

$$E^{M^1} - E^{M^2} = \mathcal{F}^{-1}(-\mu_e^{M^1} + \mu_e^{M^2}) + g^{M^1}(\text{dip}) - g^{M^2}(\text{dip})$$
$$-\delta \chi_{M^1}^S + \delta \chi_{M^2}^S = \Phi^{M^1} - \Phi^{M^2} + \delta \chi_S^{M^1} - \delta \chi_S^{M^2} \quad [19]$$

If the difference between $\delta\chi_S^{M^1}$ and $\delta\chi_S^{M^2}$ and the difference between $\delta\chi_{M^1}^S$ and $\delta\chi_{M^2}^S$ change little as q^M is changed far from the pzc, plots of E vs. q^M for different metals should be parallel. Their difference for a particular q^M value should be the difference in work functions. Using the variation of q^M with E obtained from capacitance data, this is found to be the case (Habib 1977, pp. 176-177).

The behavior of the potential drop across the electrode-solution double layer in the region of negative surface charge densities, and in the absence of specific adsorption, is easily explained by a model that neglects any contribution of the metal structure and considers only the effect of oriented solvent dipoles. The total potential difference for a surface charge density q^M, relative to its value when the charge density is zero (pzc), is $\Delta\Delta\phi = g(\text{ion}) - \Delta g^S(\text{dip})$. The former term is the "ideal-capacitor" term (cf. Chapter 2, equation [71]), due to the layers of free charge in the metal and at the outer Helmholtz plane:

$$g(\text{ion}) = \frac{q^M}{K_{\text{ion}}}$$

The value of the constant K_{ion}, a capacitance, is dependent on the dielectric constant between the charge layers and on the distance between the charge layers. The potential difference due to a dipole layer (Chapter 2, equation [78]) is $4\pi\epsilon^{-1}N\mu$, where μ is the component of the dipole moment in the field direction and there are

N dipoles per unit area. For $q^M < 0$, $g^S(\text{dip})$ will be less than 0, its value increasing in size with $|q^M|$, but, as discussed in relation to the orientational contribution to the dielectric constant (Chapter 3, equation [28]), it reaches a saturation value, corresponding to total orientation of the dipole moments in the electric field.

Thus a plot of $g^S(\text{dip})$ as a function of q^M must look like one of the curves of Figure 14a, the solid curve if $g^S(\text{dip}) = 0$ for $q^M = 0$ and the broken curve if there is a preferential orientation of the oxygen end of the orienting molecules at the pzc, so that $g^S(\text{dip}) > 0$ at the pzc, becoming 0 at a more negative potential. The free-charge contribution $g(\text{ion})$ as a function of q^M is simply a straight line passing through the origin (shown in Figure 14a). Subtracting $g^S(\text{dip})$ from $g(\text{ion})$, one sees that $\Delta\Delta\phi$ as a function of q^M must resemble one of the plots in Figure 14b. One commonly plots $-q^M$ vs. $-(E - E_{\text{pzc}})$ (which is the same as $\Delta\Delta\phi$), as shown in Figure 14c. The slope of this curve is the capacitance. Curves like this one are obtained experimentally for the mercury–alkali fluoride electrode. From the slope of the linear region of the plot (large negative q^M), one gets $1/K_{\text{ion}}$; subtracting q^M/K_{ion} from $E - E_{\text{pzc}}$, one gets $g^S(\text{dip})$ relative to its value at the pzc. For a series of metals, it is found that K_{ion} is the same, so that K_{ion} is a property of the solvent layer.

The value of K_{ion} is found to be 17 $\mu\text{F}/\text{cm}^2$ for NaF solution and is only slightly different for other cations. If this represents the capacitance of a water monolayer, ϵ/d, where d is the thickness of a water molecule (3.3 Å), then $\epsilon = 6.3$, which is a reasonable value for the high-frequency (electronic) dielectric constant. One could argue, however, that the thickness should be larger to take into account the size of the ions in the solution. If values much larger are used, though, ϵ becomes much larger than what is generally accepted. In fact (Trasatti 1979), recent measurements of ϵ^∞ yield 4.5; the corresponding value of d, 2.3 Å, seems unacceptably low.

Since the linear region of a q^M vs. E curve corresponds to saturation of the dipole-orientation contribution $g^S(\text{dip})$, comparison between different metals gives information on $g^S(\text{dip})$ at their points of zero charge: According to equation [13],

$$E_{\text{pzc}}^{(M_1)} - E_{\text{pzc}}^{(M_2)} = \mathfrak{F}^{-1}(-\mu_e^{M_1} + \mu_e^{M_2}) + g^{M_1,0}(\text{dip}) - g^{M_2,0}(\text{dip})$$
$$- g_{M_1}^{S,0}(\text{dip}) + g_{M_2}^{S,0}(\text{dip}) \qquad [20]$$

for two metals M_1 and M_2; $g_{M_1}^{S,0}(\text{dip})$ is the solvent dipole potential for solvent against metal 1 at the pzc for this metal. At the same very negative value of q^M for the two metals,

$$E_{\text{neg}}^{(M_1)} - E_{\text{neg}}^{(M_2)} = \mathfrak{F}^{-1}(-\mu_e^{M_1} + \mu_e^{M_2}) + g^{M_1,-}(\text{dip}) - g^{M_2,-}(\text{dip})$$
$$- g_{M_1}^{S,-}(\text{dip}) + g_{M_2}^{S,-}(\text{dip}) \qquad [21]$$

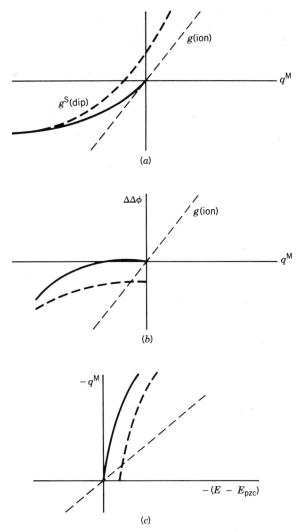

FIGURE 14. Dipolar, free-charge, and total potential drops across interface as a function of surface charge.

where the superscript minus sign indicates the same negative electrode charge; the free-charge contribution $g(\text{ion})$ is the same for both metals and the solvent dipole potentials have the same saturation values for the two metals. Now one can reasonably expect the change in the metal dipole potential on charging, $g^{M_1,0}(\text{dip}) - g^{M_1,-}(\text{dip})$, to be the same for all metals. In fact, $E_{\text{neg}}^{(M_1)} - E_{\text{neg}}^{(M_2)}$ is equal to the difference in work functions of the metals, plus the difference in the changes in the metal surface potentials due to charging, and, experimentally, $E_{\text{neg}}^{(M_1)} -$

$E_{neg}^{(M_2)}$ is found to be essentially equal to $\Phi_{M_1} - \Phi_{M_2}$ (Trasatti 1979, p. 97). Thus, subtracting equation [21] from equation [20] and taking $g^{M_1,0}(\text{dip}) - g^{M_1,-}(\text{dip})$ equal to $g^{M_2,0}(\text{dip}) - g^{M_2,-}(\text{dip})$, we have

$$[E_{pzc}^{(M_1)} - E_{pzc}^{(M_2)}] - [E_{neg}^{(M_1)} - E_{neg}^{(M_2)}]$$
$$= -[g_{M_1}^{S,0}(\text{dip}) - g_{M_2}^{S,0}(\text{dip})] \quad [22]$$

Using equation [22] one finds that $g_M^{S,0}$ is more negative by about -0.05 V for M = Au and Tl than for Hg. Other metals such as Pb, Bi, and In have $g_M^{S,0}(\text{dip})$ more positive than for Hg by similar amounts, and Ga makes the dipole-orientation potential at the pzc greater than that for Hg by 0.25 V or more. The actual value of $g_{Hg}^S(\text{dip})$ needed to convert these differences to actual dipole potentials seems to be (see above) about 0.07 V.

E. ADSORBED WATER STRUCTURE

Trasatti (1979) reviews much of what has been found experimentally about adsorbed water in the electrochemical interface. Successively more complicated theories have been adduced through the years to explain these results. These have been reviewed by a number of authors (Barlow 1970, Habib 1977, Reeves 1980, Trasatti 1979).

First let us consider what happens if there are adsorbed water molecules, with only two orientations possible. Let N_+ be the number of water dipoles per unit area oriented in the positive direction and let N_- be the number in the negative direction, so that the polarization (dipole moment per unit volume) is $(N_+ - N_-)m/x_2$, where m is the magnitude of the dipole and x_2 is the thickness of the compact layer. The potential difference across such a dipole layer is (Chapter 2, equation [78]) $4\pi(N_+ - N_-)m/\epsilon$, where the dielectric constant ϵ (due to electronic polarisation) is assumed uniform.

In the absence of the permanent dipoles, a surface charge density q^M produces an electric field $4\pi q^M/\epsilon$ and a potential difference across the interface of $\Delta\phi = 4\pi q^M x_2/\epsilon$, so that the capacitance is $C_{ion} = \epsilon/4\pi x_2$, assuming ϵ and x_2 are field-independent. Adding the potential difference due to permanent dipoles, we have

$$\Delta\phi = \frac{4\pi q^M x_2}{\epsilon} - \frac{4\pi(N_+ - N_-)m}{\epsilon} \quad [23]$$

and the capacitance $C_i = dq^M/d\Delta\phi$ obeys

$$(C_1)^{-1} = \frac{d[(4\pi q^M x_2/\epsilon) - 4\pi(N_+ - N_-)m/\epsilon]}{dq^M} = (C_{ion})^{-1} - (C_{dip})^{-1}$$

$$[24]$$

Here,

$$(C_{\text{dip}})^{-1} = \frac{4\pi m}{\epsilon} \frac{d(N_+ - N_-)}{dq^M} \quad [25]$$

and the problem becomes the determination of how $N_+ - N_-$ varies with q^M.

If there is only physisorption, one calculates $N_+ - N_-$ using the Boltzmann distribution for the two-state system:

$$\frac{N_\pm}{N} = \frac{e^{\pm mE/kT}}{e^{-mE/kT} + e^{mE/kT}} \quad [26]$$

where the electric field E is taken as $4\pi q^M/\epsilon$. Then

$$\frac{d(N_+ - N_-)}{dq^M} = \frac{Nd[\tanh(mE/kT)]}{dq^M} = \frac{4\pi Nm}{kT\epsilon} \text{sech}^2 \frac{mE}{kT}$$

and

$$(C_{\text{dip}})^{-1} = \frac{16\pi^2 m^2 N}{\epsilon^2 kT} \text{sech}^2 \frac{4\pi q^M m}{\epsilon} \quad [27]$$

The contribution of the permanent dipoles is largest for $q^M = 0$ (at the pzc) and decreases on either side of this value in a symmetrical fashion. The symmetry around $q^M = 0$ is incorrect.

The orientational entropy of adsorbed water molecules is easily calculated for this two-state model. Consider N surface sites, each occupied by a water molecule with up or down orientation. The number of arrangements, if N_+ molecules are up and N_- are down, is $N!/(N_+!N_-!)$ so that the entropy is

$$k \ln \frac{N!}{N_+!N_-!} = k[(N_+ + N_-) \ln N - N_+ \ln N_+ - N_- \ln N_-] \quad [28]$$

where Stirling's approximation and $N = N_+ + N_-$ have been used. For one mole of molecules,

$$S_{\text{or}} = -R(\theta_+ \ln \theta_+ + \theta_- \ln \theta_-)$$

where $\theta_+ = N_+/N$, $\theta_- = N_-/N$. Defining the degree of orientation Z by

$$Z = \frac{N_+ - N_-}{N} = \theta_+ - \theta_-$$

we have $\theta_+ = (1 + Z)/2$ and $\theta_- = (1 - Z)/2$ so

$$S_{or} = \tfrac{1}{2}R[(1 + Z)\ln(1 + Z) + (1 - Z)\ln(1 - Z) + 2\ln 2] \quad [29]$$

This quantity is a minimum for $Z = 0$, which occurs for zero field.

The incorrect symmetry around $q^M = 0$ can be removed by postulating an image-type or chemisorptive interaction between the dipoles and the electrode which differs for the two orientations. Then

$$\frac{N_+}{N} = e^{(mE + 2I)/kT} / \left(e^{(mE + 2I)/kT} + e^{-mE/kT} \right)$$

where $2I$ is the difference in the interaction between the positive and negative orientations. Then $N_+ - N_-$ becomes $\tanh[(mE + I/kT]$ and

$$(C_{\text{dip}})^{-1} = \frac{16\pi^2 m^2 N}{\epsilon^2 kT} \operatorname{sech}^2 \frac{mE + I}{kT}$$

The capacitance is now symmetric about $q^M = -\epsilon I/4\pi m$.

Another error in the formulas above is that the polarizing field E includes the effect of the charge q^M and the induced polarization of ϵ, but not the field of the permanent dipoles themselves. It has been suggested (Reeves 1980, p. 127) that this dipole–dipole interaction can be taken into account by using $\Delta\phi/x_2$ for E, where $\Delta\phi$ is given in equation [23]. Now

$$(C_i)^{-1} = \frac{d\Delta\phi}{dq^M} = (C_{\text{ion}})^{-1} - \frac{(4\pi Nm/\epsilon)\, d\{\tanh[(mE + I)/kT]\}}{dq^M}$$

$$= (C_{\text{ion}})^{-1} - \frac{4\pi Nm^2}{\epsilon kT x_2} \frac{d\Delta\phi}{dq^M} \operatorname{sech}^2 \frac{mE + I}{kT}$$

or, solving for $(d\Delta\phi/dq^M)$ and rearranging,

$$(C_i)^{-1} = (C_{\text{ion}})^{-1} \left(1 + \frac{4\pi Nm^2}{\epsilon kT x_2} \operatorname{sech}^2 \frac{mE + I}{kT} \right)^{-1} \quad [30]$$

If the second term is small compared to 1, we get

$$\frac{C_{\text{ion}}}{C_i} \sim 1 - \frac{4\pi Nm^2}{\epsilon kT x_2} \operatorname{sech}^2 \frac{mE + I}{kT}$$

If the molecules in the layer are completely oriented, the potential difference of the layer, in electrostatic units, is $4\pi Nm/\epsilon$. The area of a water molecule is about

12.5 Å² and the dielectric constant ϵ should be about 6.0 corresponding to electronic and vibrational polarizability only. Then the 0.4 V potential difference, which is supposed to correspond to complete orientation (see Section D), implies a dipole moment per molecule of 0.8 D. Since the permanent dipole moment of water is 1.85 D, it is implied that the molecules are oriented at an angle of 64° to the normal (cos 64° = 0.80/1.85). The existence of such an angle can be explained in terms of the structure of water (hydrogen bonds between water molecules): Individual molecules are not free to orient; on the other hand, tetramers or other units *are* free to orient. In a cluster of n molecules, the dipole moment is not n times 1.85 D, since the dipole moments of the molecules in the cluster are not parallel.

One knows, of course, that chemisorptive forces act to orient water on a metal surface. Chemisorption in the form of aggregates, with the energy of the chemisorption depending linearly on the electric field or electrode charge density, gives a monotonic variation of capacitance with q^M: The potential drop across a layer of chemisorbed molecules is proportional to N', their number per unit area, so that the contribution of these molecules to the inverse capacity is proportional to dN'/dq^M. If N' varies proportionally with $\exp(bq^M/kT)$, then dN'/dq^M is also proportional to $\exp(bq^M/kT)$, which is monotonic in q^M. Such a monotonic dependence is found for Cd in KF at 0°C. Only the number, and not the orientation, of the chemisorbed molecules is supposed to change with q^M; the energy of chemisorption is so large that only one orientation (with oxygen toward the metal) is permitted.

A macroscopic model that takes these effects into account was given by Macdonald and Barlow (Barlow 1970, pp. 222–226). Parameters in this macroscopic model were chosen to get the best fit of the capacitance for NaF solutions. It was possible to get a good fit to the capacitance as a function of potential and concentration for negative potentials, but there was a problem for surface charge densities greater than 2 μC/cm² on the positive side: The curves showed a maximum even at high temperatures, falling off with higher charge densities instead of rising sharply in value as the experimental capacitances do. A molecular model was also presented (Barlow 1970, pp. 226–231). The polarization near the electrode surface was separated into electronic (molecular) and orientational parts. The former is estimated from ϵ^∞ (Chapter 3, equation [63])

$$P_1 = n'\alpha E_{local} = n' \frac{3}{4\pi n} \frac{\epsilon^\infty - 1}{\epsilon^\infty + 2} E_{local}$$

where n' is the density of polarizable molecules in the surface region and E_{local} is the field that acts to polarize a water molecule (Chapter 3, equation [59]). The orientational part is calculated as

$$P_2 = n'\bar{\mu}$$

where $\bar{\mu}$ is the average component of the dipole moment along the direction of the field **E**. It is assumed that the water molecules in the compact layer are arranged in hexagonal close-packing with an interparticle spacing or lattice constant of r_1, so that each hexagon has an area of $3(r_1)^2 \, (\frac{3}{4})^{1/2}$ and contains three molecules (one at the center and six at the corners, each of which is shared between three hexagons). Therefore there are $(\frac{4}{3})^{1/2}/(r_1)^2$ molecules per unit area and $n' = (\frac{4}{3})^{1/2}(r_1^2 \, d)$, where d is the thickness of the layer. The hexagonal array of dipolar molecules is also used to calculate the local field at a water molecule in terms of $\bar{\mu}$ and P_1. The effective field is $4\pi q^M - (\frac{3}{4})^{3/4} \sigma n \, d(\bar{\mu} + P_1/n)$, where $\sigma = 11.034$ (from summation over nearest neighbors), P_1 is the polarization due to induced dipoles ($n\alpha E_{\text{local}}$), $\bar{\mu}$ is the average value of the component of a permanent molecular dipole in the field direction. If one neglects induced dipoles and keeps only permanent dipoles, and if the fractional surface coverage by ions is low,

$$E_{\text{local}} = 2\pi q^1 - (\tfrac{3}{4})^{3/4} \sigma (nd)^{3/2} \bar{\mu}$$

at the ecm ($q^M = 0$) and $(\frac{3}{4})^{3/4}\sigma$ is approximately 9 (Barlow 1970, pp. 234–235).

The preference of one orientation of the dipole or the other, due to interaction of water molecules with the electrode surface, is represented by a "natural field" E_n, so that

$$\bar{\mu} = \mu \mathcal{L}\left(\frac{\mu(E_{\text{local}} + E_n)}{kT}\right) \sim \frac{\mu^2(E_{\text{local}} + E_n)}{3kT}$$

Since $E_{\text{local}} = 2\pi q^1 - 9(nd)^{3/2}\bar{\mu}$, one can solve for $\bar{\mu}$:

$$\bar{\mu} = \frac{\mu^2}{3kT}\left(2\pi q^1 - 9(nd)^{3/2}\bar{\mu} + E_n\right)$$

$$\bar{\mu} = \frac{(\mu^2/3kT)(2\pi q^1 + E_n)}{1 + 9(nd)^{3/2}\mu^2/3kT}$$

The denominator is about 13 at room temperature. Then one can get P_1 and E_{local} as functions of the electrode charge. It appeared that, with proper choices of the parameters in this theory, one could reproduce experimental capacitances well.

A more careful consideration of the effect of dipole–dipole interaction was done by Bockris, Devanathan, and Muller (1963; also see Bockris and Reddy 1970, Section 7.4). They argued that because of the r^{-3} dependence of the dipole–dipole interaction, the interaction energy between one dipole and all the others concerned only the nearer neighbors, and that, if a fraction N_+/N of the dipoles were pointed in the positive direction, the interaction energy between a particular dipole and its neighbors would be $\pm U(N_+ - N_-)$. Here, U is a constant depending on the distance between neighboring dipoles and the sign depends on the orientation of the central dipole. (This is a mean-field theory, replacing instantaneous correlations between dipoles by the influence of an average field: See Chapter 3, Section D on theories for the dielectric constant.) Then

$$\frac{N_+}{N} = \frac{\exp\{[mE + 2I - U(N_+ - N_-)]/kT\}}{Q} \qquad [31]$$

where m is the magnitude of the dipole and $2I$ the difference in interaction energy with the electrode for the two dipole orientations.

$$Q = \exp\left\{\frac{[mE + 2I - U(N_+ - N_-)]}{kT}\right\} - \exp\left\{\frac{-[mE - U(N_+ - N_-)]}{kT}\right\}$$

so that

$$N_+ - N_- = N \tanh \frac{mE + I - U(N_+ - N_-)}{kT} \equiv N \tanh y \qquad [32]$$

Differentiating, we get

$$\frac{d(N_+ - N_-)}{dq^M} = \frac{N}{kT}(\operatorname{sech}^2 y)\left[m \frac{dE}{dq^M} - \frac{U d(N_+ - N_-)}{dq^M}\right]$$

or, since $E = \Delta\phi/x_2$,

$$\frac{d(N_+ - N_-)}{dq^M} = \frac{Nm}{x_2 kT}(\operatorname{sech}^2 y)\frac{d\Delta\phi}{dq^M}\left[1 + \frac{NU}{kT}\operatorname{sech}^2 y\right]^{-1}$$

This can then be used in differentiating equation [23] with respect to q^M to get the capacitance:

$$(C_i)^{-1} = \frac{d\Delta\phi}{dq^M} = (C_{\text{ion}})^{-1}$$

$$- \frac{4\pi Nm^2}{\epsilon x_2 kT}(\operatorname{sech}^2 y)\frac{d\Delta\phi}{dq^M}\left(1 + \frac{NU}{kT}\operatorname{sech}^2 y\right)^{-1}$$

Solving for $d\Delta\phi/dq^M$, we get

$$(C_i)^{-1} = \frac{(C_{\text{ion}})^{-1}}{1 + (4\pi Nm^2/\epsilon x_2 kT)\operatorname{sech}^2 y\left[1 + (NU/kT)\operatorname{sech}^2 y\right]^{-1}} \qquad [33]$$

Setting this equal to $(C_{\text{ion}})^{-1} - (C_{\text{dip}})^{-1}$ we obtain

$$(C_{\text{dip}})^{-1} = (C_{\text{ion}})^{-1} \frac{(4\pi Nm^2/\epsilon x_2 kT)(\text{sech}^2 y)\left[1 + (NU/kT)\text{sech}^2 y\right]^{-1}}{1 + (4\pi Nm^2/\epsilon x_2 kT)(\text{sech}^2 y)\left[1 + (NU/kT)\text{sech}^2 y\right]^{-1}}$$

$$= \frac{(16\pi^2 Nm^2/\epsilon kT)\,\text{sech}^2 y}{1 + (NU/kT)\text{sech}^2 y + (4\pi Nm^2/\epsilon x_2 kT)}$$

Bockris and Reddy (1970, Section 7.4.32) show that C_{dip} is typically considerably larger than experimental capacities, so its contribution is small, and it cannot explain the interesting experimental capacity curves.

Note also that the symmetry about $E = I$ is maintained. According to equation [33], changing the sign of y does not change the capacitance, and equation [29] shows that y, as well as $N_+ - N_-$, changes sign when $E - I/m$ changes sign. Thus the models presented so far are insufficient to give unsymmetric capacity–charge curves. Other models are reviewed by Reeves (1980, Section 3.6) and by Habib (1977).

Frumkin and co-workers (Damaskin and Frumkin 1974; Frumkin, Damaskin, and Petrii 1975) suggested that both free water molecules and aggregates of low dipole moment can be adsorbed. The existence of aggregates is implied by the small effective moment needed to explain the difference of only 0.4 V between the potential drops across layers of oriented and random water molecules and the small change in potential when adsorbed water is replaced by organic molecules.

The Bockris–Habib three-state model (Habib 1977, pp. 151–161) supposes that the N sites on the electrode surface are occupied by N_d nonpolar dimers, each occupying two sites, and N_m monomers, each occupying one site, which are adsorbed in one of two possible orientations. The dimers are not dissociated by the field, so their concentration does not change. The degree of orientation of the monomers, taking into account interactions between dipoles and preferential adsorption for the positive orientation (oxygen toward the metal), is given by

$$Z = \frac{\exp\left[-(\Delta G_+ - x)/kT\right] - \exp\left[-(\Delta G_- + x)/kT\right]}{\exp\left[-(\Delta G_+ - x)/kT\right] + \exp\left[-(\Delta G_- + x)/kT\right]} \qquad [34]$$

where ΔG_+ and ΔG_-, both negative, are free energies of adsorption for the two orientations. The difference $\Delta G_+ - \Delta G_-$ was made up of a difference in image and dispersion interaction energies with the metal, combining to give a value of -2.72 kJ/mol. The quantity x is defined as

$$x = \mu E - UcZ$$

where μ is the effective normal component of the dipole moment of a water molecule, c is the number of nearest neighbors of a water molecule, and U is a measure

E. ADSORBED WATER STRUCTURE

of the dipole–dipole interaction energy between adjacent molecules: cZ is the number of neighboring moments in the up direction, which determines the average field at a water molecule due to its neighbors. This field is subtracted from the double-layer field E, thus accounting for lateral interactions between dipolar molecules.

The capacitance expression is derived by differentiating the expression for the surface potential,

$$\frac{4\pi\mu}{\epsilon}(N_+ - N_-) = \frac{4\pi\mu N_m}{\epsilon} Z = \frac{4\pi\mu N_m}{\epsilon} \frac{\exp\left[(2x + \Delta G_- - \Delta G_+)/kT\right] + 1}{\exp\left[(2x + \Delta G_- - \Delta G_+)/kT\right] - 1}$$

with respect to q^M. Note that x depends on E and hence on q^M. The explanation of the capacity hump is in terms of the way in which the concentration of specifically adsorbed molecules increases with q^M. The way in which the hump depends on temperature is well described, but other features of the capacity curves are less well accounted for (Trasatti 1979, Section V.2).

According to equation [34], zero net orientation ($Z = 0$) occurs when $\Delta G_+ - x$ is equal to $\Delta G_- + x$ or

$$\Delta G_+ - \Delta G_- = 2x = 2\mu E \qquad [35]$$

The interaction energy parameters U and c do not enter here because $Z = 0$. Since $E = q^M/4\pi\epsilon_0$,

$$q^M = \frac{(\Delta G_+ - \Delta G_-)\epsilon_0}{8\pi\mu}$$

With $\Delta G_+ - \Delta G_- = -2.72$ kJ/mol, the maximum entropy ($Z = 0$) occurs at $q^M = -3.5$ μC/cm^2. In addition to the configurational entropy, one has to consider entropy associated with librations and internal vibrations of the adsorbed water molecules. These contributions, combined with the orientational contribution, give a value for the solvent excess entropy which agrees well with experiment.

A model proposed by Damaskin and Frumkin (1974) included water as chemisorbed individual molecules (with adsorption energy depending linearly on the electrode charge) and as physisorbed aggregates that are oriented by the electric field. The energies and entropies of adsorption of both species are parameters in the model. The two-dimensional aggregates have two possible orientations relative to the electrode, with relative probabilities depending on the electric field according to equation [26]. The potential drop across the water layer is a sum of two dipolar contributions, for the aggregates and for the chemisorbed molecules, and the ionic contribution $4\pi q^M x_2/\epsilon$. Thus C^{-1} is a sum of three terms, $(C_{ion})^{-1} - (C_1)^{-1} - (C_2)^{-1}$, where C_1 refers to the aggregates and C_2 refers to the single molecules. The contribution of the aggregates to the potential is $-4\pi(N_+ - N_-)m/\epsilon$, with N_+ and N_- given by equation [26], so $(C_1)^{-1}$ is of the form of equation [27], with appropriate values for m and N. The chemisorbed molecules contribute

$-4\pi N'm'/\epsilon$ to the potential drop, with N' governed by a Boltzmann factor, exp $[-(\Delta U - bq^M)/kT]$. Since these molecules are supposed to be chemisorbed with the oxygen and toward the metal, increasing q^M increases adsorption. The contribution to the inverse capacity, $(C_2)^{-1}$, is thus proportional to $\exp(bq^M/kT)$.

The total capacity then involves a symmetrical (sech^2) and an unsymmetrical (exp) function of q^M. Their combination leads to calculated capacity–charge curves that look like cubics, which resembles what one finds experimentally for Hg in NaF at 0°C (see Figure 9). The capacity rise for positive q^M comes from increased chemisorption. In passing from Hg to the more hydrophilic Cd, adsorption increases and the capacity rise is correctly predicted to be steeper.

Parsons (Parsons 1975) pointed out that Frumkin's model does not conserve the total number of water molecules in the compact layer. With the exponential dependence of N' on q^M, the number of chemisorbed molecules increases without limit as q^M becomes more positive. Parsons proposed a similar model free from this defect, in which there are four states of water on the surface, and there can be interconversion of aggregates and single molecules.

Parsons' model considers single water molecules of dipole moment μ and aggregates of dipole moment $\rho\mu$ per water molecule. The field at the electrode surface is $q^M/\epsilon = -q^S/\epsilon$, where ϵ is the dielectric constant in the compact layer. The single molecules are chemisorbed in either of two orientations: A molecule with dipole pointing toward the metal has an energy of $-\mu q^S/\epsilon + U_{b+}$; if the dipole points in the other direction, the energy is $\mu q^S/\epsilon + U_{b-}$. U_{b+} and U_{b-} represent the nonelectrostatic parts of the adsorption energies. The aggregates are physisorbed, so the energy per water molecule for an aggregate adsorbed with dipole pointing toward the metal is $-\rho\mu q^S/\epsilon$ and the energy per molecule for an aggregate with dipole pointing toward the solution is $\rho\mu q^S/\epsilon$. The value of μ is that of an individual water molecule, leaving ϵ, ρ, U_{b+}, U_{b-}, and ρ_T (the total number of water molecules per unit area of compact layer) as parameters of the model.

Letting the numbers of chemisorbed water molecules per unit area in the two orientations be n_+ and n_- and letting the numbers adsorbed in aggregates in the two orientations be n_{a+} and n_{a-}, one has $\rho_T = n_+ + n_- + n_{a+} + n_{a-}$. Then the number of water molecules in each state is given by a Boltzmann distribution:

$$\frac{n_+}{\rho_T} = Q^{-1} \exp\left(-\frac{-\mu q^S \epsilon^{-1} + U_{b+}}{kT}\right)$$

$$\frac{n_-}{\rho_T} = Q^{-1} \exp\left(-\frac{\mu q^S \epsilon^{-1} + U_{b-}}{kT}\right)$$

$$\frac{n_{a+}}{\rho_T} = Q^{-1} \exp\left(-\frac{-\rho\mu q^S}{\epsilon kT}\right)$$

$$\frac{n_{a-}}{\rho_T} = Q^{-1} \exp\left(-\frac{\rho\mu q^S}{\epsilon kT}\right) \qquad [36]$$

where

$$Q = \exp\left(\frac{\mu q^S \epsilon^{-1} - U_{b+}}{kT}\right) + \exp\left(\frac{-\mu q^S \epsilon^{-1} - U_{b-}}{kT}\right)$$
$$+ \exp\left(\frac{\rho \mu q^S}{\epsilon kT}\right) + \exp\left(-\frac{\rho \mu q^S}{\epsilon kT}\right)$$

The average dipole moment per unit area in the solution → metal direction is $n_+ \mu - n_- \mu + n_{a+} \rho \mu - n_{a-} \rho \mu$, so that the potential drop across the compact layer is

$$V_{s \to m} = 4\pi \epsilon^{-1}(n_+ - n_- + n_{a+}\rho - n_{a-}\rho)\mu$$

The expression for the differential capacity is given by Parsons (1975). The value of ρ, which is assumed independent of temperature, turns out to be 0.29; other parameters depend on temperature.

Capacity–potential curves for Hg in NaF at different temperatures (Figure 9) are well described by this model. Above room temperature one finds curves with a single minimum; below room temperature the curves show two minima with a maximum ("capacitance hump") between them, near the point of zero charge. In this model, the capacitance hump comes from the combination of the reorientation of clusters (which would make the effective dielectric constant a maximum at $q^M = 0$) and the increase in the effective dielectric constant associated with free molecules. The explanation of the capacity hump as a maximum in the effective dielectric constant is common to almost all recent models. The rise in capacity with increasingly negative q^M is due to orientation of single molecules and conversion of clusters to single molecules. All the capacity–charge curves for different temperatures intersect at two values of q^S, that is, for a certain potential more positive than the pzc the capacity is independent of temperature. This was observed in early measurements.

Fawcett (1978) improved the treatment of the clusters by introducing polymer statistics. The number of parameters is smaller by one than in Parsons' model (four), and the experimental data is fitted significantly better. A further advantage is that the values needed for the chemisorption energies U_{b+} and U_{b-} are only 6.9 and 0.9 kJ/mol, respectively, whereas Parsons' model required 20.9 and 10.5, which seem rather large. Other improvements, such as molecular rotation, multilayer adsorption, solvent compressibility, and dielectric saturation, have been suggested and used in models of other authors (Reeves 1980, pp. 130–131).

The various models have been compared (Trasatti 1979, Section IV.2). All the successful models have in common the necessity to invoke a structure of water at the interface which is more complicated than physisorbed dipole moments that can adopt positive or negative orientations. Some or all of the water molecules are supposed to be involved in clusters or other units. The free water molecules are analogous to the interstitial water molecules in some of the more successful theories of bulk water, whereas the clusters are analogous to the continuous network. The existence of clusters means that the effective dipole moment perpendicular to the surface is much less than the moment of a single water molecule. This corresponds to the fact that $g^{H_2O}(\text{dip})$ for a number of metals, derived by subtracting $g(\text{ion})$ from the measured electrode–solution potential difference, approaches about -400 mV for large negative q^M (the limiting value is approached for $q^M < -10\ \mu\text{C}/\text{cm}^2$). This corresponds to a dipole moment of about 0.8 D per molecule.

All the recent models predict a positive surface potential at the point of zero charge (preferential adsorption of water with oxygen toward the metal) of some hundredths of a volt or less, considerably smaller than that estimated from experiment (0.10 V for Hg, 0.25 V for Cd). None include the effect of the metal electrons, which itself can explain this in terms of the extension of the electronic tail into the solution: At the point of zero charge, the adsorbed water molecules are in a region where there is a positive electric field outward from the metal.

Corresponding to the positive field at the pzc, a negative surface charge is required to produce zero net orientation of water dipoles. Its value can be derived from a plot of $-q^M$ vs. $-E$ (E = electrode potential relative to a standard electrode). Given a value for the dipole potential of water on a metal electrode at the point of zero charge (derived from the value 0.07 V for mercury, which is based on a variety of experimental evidence), one can convert E values to values of the electrode–solution potential difference $\Delta_M^S\phi$. The free-charge potential drop is q^M/K_{ion}, where, from the slope of the charge–potential curves at large negative q^M or from the minimum in a curve of capacity vs. q^M, the value of K_{ion} is 17 $\mu\text{F}/\text{cm}^2$. Then q^M/K_{ion} may be subtracted from $\Delta_M^S\phi$ to yield $g^{H_2O}(\text{dip})$ as a function of q^M. The value of q^M which makes $g^{H_2O}(\text{dip})$ vanish is $-1.6\ \mu\text{C}/\text{cm}^2$ for Tl and -3.4 for Hg and increases with increasing hydrophilicity of the metal to about -7 for Cd and Ga. Except for the three-state Bockris–Habib model, the cluster models predict values of q^M significantly too low in size and perhaps of the wrong sign (Habib 1977).

The change in the shape of the capacity–charge curve with temperature, from one involving a maximum (near $q^M = 0$) between two minima to one with a single minimum at about $-10\ \mu\text{C}/\text{cm}^2$ for mercury, is well-reproduced by the models discussed. The measured temperature coefficient of the pzc is 0.57 mV/K. According to equation [13], this is $d\Phi/dT + \partial(\delta\chi_S^M)/\partial T - \partial g^S(\text{dip})/\partial T$. The derivative $\partial E/\partial T$ for other values of surface charge includes $\partial g(\text{ion})/\partial T$ as well. However, $\partial g(\text{ion})/\partial T$ and $\partial g^S(\text{dip})/\partial T$ probably vanish for large negative q^M (total orientation of dipoles). Correcting for $d\Phi/dT$, one can then (Trasatti 1979) superpose the linear portions of the q^M–E curves for different temperatures and interpret the differences in the nonlinear part of the curves as differences in $g^S(\text{dip})$. The temperature coefficient of $g^{H_2O}(\text{dip})$ is found to be negative at the point of zero surface charge density, that is, $g^{H_2O}(\text{dip})$ becomes less negative with increasing temperature. At more negative surface charge densities, the size of the temperature derivative decreases and it becomes zero below about $-9\ \mu\text{C}/\text{cm}^2$ (the value of $g^{H_2O}(\text{dip})$ is 0.07 V for 0–85°C). In the cluster models, the derivative of

g^{H_2O}(dip) with temperature is positive, in disagreement with what one infers from experiment.

In general, the inner-layer thickness d and/or the inner-layer dielectric constant ϵ are parameters in these models, corresponding to the value of the inner-layer capacitance $K_{ion} = \epsilon/4\pi d$. With $K_{ion} = 17~\mu F/cm^2$, there is a consensus that ϵ should be 5–6 and d should be 3–4 Å (size of a water molecule). The same value of ϵ is usually used in calculating the potential drop across the dipole layer, $4\pi N\mu/\epsilon$, although careful consideration suggests problems with this (Trasatti 1979, Section IV.2.iv). The value of μ, the effective component of the dipole moment perpendicular to the metal surface, is another important parameter. In the simplest two-state physisorption model, the value for individual water molecules, 1.85, is appropriate, but other values have been used for chemisorbed molecules. The value of N, the number of water molecules per unit area, is usually calculated by assuming a hexagonal close-packed layer, but this assumption, too, may be unjustified when chemisorption is important.

Surface potentials for various solvents have been measured relative to that of water. If χ^{H_2O} is taken as 0.13 V, the surface potentials for various alcohols, amides, and other compounds are all negative by several tenths of a volt, except possibly for formamide. Potentials of zero charge are also available for various metals in nonaqueous solvents. From these, plus a certain number of assumptions, one can derive g^S(dip) and other properties (Trasatti 1979, Section 2.VIII.2). Capacity–charge curves have also been measured for nonaqueous solvents. Sometimes, as for amides, these curves resemble those for water. In other cases, the capacity shows a single maximum or a single minimum in the range of potential investigated.

The four-state model of Parsons has been extended to other solvents (Parsons 1976). With proper choice of parameters, one can fit experimental results, but, in the case of curves unlike those found for water, the values of the parameters required are sometimes physically unreasonable. The results and the attempts to fit them to a model are reviewed by Trasatti (1979, Section 2.VIII.)

For unassociated solvents, Fawcett (1978) presented a three-state model in which the electrode is covered by a hexagonal close-packed monolayer of molecules, each of which can have its dipole moment oriented perpendicular or parallel to the surface. The perpendicularly oriented moments are either "up" (positive end toward the metal, designated "+" here) or "down" (negative end toward the metal, designated "−"), so that

$$N = N_+ + N_- + N_0$$

where N is fixed by the packing and the average permanent moment perpendicular to the surface is $(N_+ - N_-)\mu/N$. No aggregates on the surface are considered for solvents that do not associate in the bulk.

In addition to the permanent moment, a molecule has an induced moment, proportional to the electric field at the molecule (Chapter 3, Section D). This field is the external field E due to q^M, plus the mean field X due to surrounding dipoles, and is perpendicular to the surface. X is calculated in terms of the distance between molecules and the net polarization due to permanent and induced moments. Then the induced moment for a molecule whose dipole is up or down is $\alpha_1(E + X)$ and that for a molecule oriented parallel to the surface is $\alpha_2(E + X)$, where α_1 and α_2 are polarizabilities along and across the principal symmetry axis of a molecule. The free energy of the monolayer of dipoles is written as a sum of electrostatic terms (dipole–dipole interactions, induced dipole interactions proportional to $\alpha(E + X)^2$, dipole-field interactions proportional to q^M) and nonelectrostatic metal–solvent interactions, $N_+U_+ + N_-U_- + N_0U_0$.

The configurational entropy, assuming random mixing of the dipoles, is (see equation [28])

$$S = k \ln \frac{N!}{N_+!N_-!N_0!} = -k\left(N_+ \ln \frac{N_+}{N} + N_- \ln \frac{N_-}{N} + N_0 \ln \frac{N_0}{N}\right)$$

The free energy $U - TS$ is then minimized to obtain N_+, N_-, and N_0 given values for the parameters d (nearest-neighbor distance), μ, α_1, α_2, U_+, U_-, and U_0. Thus the potential drop across the monolayer is given by

$$\Delta\phi = 4\pi q^M d + 4\pi N(m_p + m_i) \quad [37]$$

where m_p and m_i represent the net electric polarization per site due to permanent and induced moments: $m_p = (N_+ - N_-)\mu/N$ and

$$m_i = \frac{[\alpha_1(N_+ + N_-) + \alpha_2 N_0](E + X)}{N} \quad [38]$$

Because the molecules are close-packed, the nearest-neighbor distance d is also the thickness of the layer. The moment $m_p + m_i$ is shown to be equal to $d^3 X/c_e$, where c_e is an effective coordination number that takes into account interactions of dipoles with their images in the metal. The capacitance of the layer, C_i, is $\partial q^M/\partial \Delta\phi$ so that

$$(C_i)^{-1} = \frac{\partial \Delta\phi}{\partial q^M} = 4\pi d + \frac{4\pi N d^3}{c_e}\frac{\partial X}{\partial q^M} \quad [39]$$

The derivative $(\partial X/\partial q^M)$ is calculated from the dependence of N_+, N_-, and N_0 on q^M or E.

The capacity curves for solvents with low dielectric constants generally have either a single maximum or a single minimum in the accessible range of potential (solvents such as water show a central maximum with minima on either side). These kinds of curves can be accounted for by Fawcett's model (Fawcett 1978). Curves with a capacity maximum correspond to predominance of molecules in the up and down orientations (see equation [23] and thereafter). A capacity minimum will be observed when molecules with dipoles oriented parallel to the surface predominate, since increased positive or negative fields will convert such molecules to down or up orientation, increasing solvent polarization and inner-layer capacity. In fitting this model to experimental data, the same value was used for α_1 and α_2, and the values of $U_+ - U_-$ and $U_0 - U_+$ were obtained from the inner-layer capacity and surface charge density at the extremum in capacity. Then d and α were chosen to give the best fit. Good results were obtained for the mercury–ethylene carbonate and the mercury–methanol interfaces, which correspond to the two kinds of capacitance curves mentioned above. This model, using fewer adjustable parameters, worked better than the previously proposed models for these systems, but could not reasonably be applied to the mercury–water interface and others for which association of molecules is important.

F. SPECIFIC ADSORPTION

The compact layer with specific adsorption of ions may be a more common situation than one without it. The precise definition of specific adsorption is not always easy (Habib and Bockris 1980, Section 4.2). Specific adsorption is supposed to refer to adsorption other than that required to balance the electrode charge. Thus, one says that there is specific adsorption at the point of zero charge if there is a positive Gibbs surface excess for any ionic species, and, at more negative (positive) potentials, if there is a positive Gibbs surface excess for any anionic (cationic) species.

Plots of surface tension vs. potential for the mercury–aqueous-solution interface tend to be independent of electrolyte for potentials more negative than the point of zero charge for NaF. For potentials more positive than this, the surface-tension–potential plots differ for different electrolytes, having their maxima at different potentials and with different heights. Since the derivative of surface tension with ion activity at constant electrode potential gives the surface excess, different surface tensions indicate different adsorptions for different electrolytes and hence specific adsorption. For negative potentials, one expects preferential adsorption of cations, so the surface-tension–potential curves suggest all cations are adsorbed in the same way, that is, nonspecifically.

Habib and Bockris (1980, Sections 6.3, 6.4, 6.5) review the history of the concept, the phenomenology of specific adsorption, and the methods for its measurement. As general rules, they list the following: (1) The amount of specific adsorption for any positive ion increases as the electrode charge becomes more negative; for negative ions, specific adsorption increases with increasingly positive electrode charge. (2) Anions generally are more likely to be adsorbed specifically than cations. (3) The greater the size of the ion, the greater its specific adsorption, so that certain large cations may have significant specific adsorption. (4) Specific adsorption is more favored for ions with weaker primary solvent sheaths: In aqueous solution, strongly hydrated anions adsorb less than weakly hydrated ones. (5) Specific adsorption increases with solution concentration at constant temperature. (6)

In most cases investigated, specific adsorption decreases with increasing temperature.

Much less theoretical work has been done to model specific adsorption than to model the layer of adsorbed water. Specifically adsorbed ions may be included in some of the models mentioned above for the compact layer. Of course, a general theory of specific adsorption may be a contradiction in terms.

G. SEMICONDUCTOR–ELECTROLYTE INTERFACE

As noted in Chapter 4, Section H, the diffuse layer at a semiconductor surface results from distributions of mobile charges (electrons and holes) and immobile charges (ionized impurity atoms). At the semiconductor–electrolyte interface, one may expect at least three regions of nonzero charge density: diffuse or Gouy layers in semiconductor and electrolyte, and a Helmholtz layer between them. Because of the low number density of charges, screening lengths in a semiconductor are typically much larger than in all but the most dilute electrolytic solutions.

For example, consider an n-type semiconductor in which the electrical potential is small. The relation between the electrical field E and the electrical potential ϕ ($\phi = 0$ in bulk) was given in equation [72] of Chapter 4. If the quantity ϕ/kT is small enough to write $\cosh[(p + e\phi)/kT]$ as a power series in ϕ/kT and truncate it after the quadratic terms, we obtain

$$E = -\left(\frac{16\pi e n_0}{\epsilon}\right)^{1/2} \left[\left(\phi \sinh \frac{p}{kT}\right) + \frac{e}{2kT}\phi^2 \cosh \frac{p}{kT} - \phi \sinh \frac{q}{kT}\right]^{1/2} \quad [40]$$

For an n-type semiconductor, p and q are equal and large compared to kT. Then equation [40] becomes

$$E = -\left(\frac{4\pi e^2}{\epsilon_{sc} kT}\right)^{1/2} \phi (n_0 e^{q/kT})^{1/2}$$

$$= -\left(\frac{4\pi e^2 n_e^0}{\epsilon_{sc} kT}\right)^{1/2} \phi$$

where n_e^0 is the concentration of electrons or ionized donors in bulk and the dielectric constant has been written ϵ_{sc}. We may write this as

$$\frac{d\phi}{dz} = \kappa_{sc} \phi \quad [41]$$

where $(\kappa_{sc})^{-1}$, a Debye screening length, is the length parameter characterizing the (exponential) decay of ϕ in space. In particular, the field and potential at the surface are related by $E_0 = \kappa_{sc}\phi_0$, corresponding to a plane capacitor of plate-to-plate thickness $(\kappa_{sc})^{-1}$.

In the interface, the potential difference between bulk electrolyte and bulk semiconductor is $\Delta_{el}^{sc}\phi = \Delta_{el}^{2}\phi + \Delta_{2}^{0}\phi + \Delta_{0}^{sc}\phi$, where the 2 refers to the outer Helmholtz plane and the 0 refers to the semiconductor surface. If there are no surface states or adsorbed ions, $\Delta_{2}^{0}\phi = E_H(z_2 - z_0)$. Since $\mathbf{D} = \epsilon\mathbf{E}$ is continuous (Chapter 2, equation [75] and thereafter), the field in the Helmholtz layer is related to E_0 by

$$\epsilon_{sc}E_0 = \epsilon_H E_H = \epsilon_{el}E_2 \qquad [42]$$

where ϵ_{el} is the dielectric constant in the diffuse layer and E_2 is the field at the outer Helmholtz plane. For small potentials, $E_2 = \kappa\Delta_{el}^{2}\phi$, analogously to equation [41], so

$$\Delta_{el}^{sc}\phi = \frac{E_2}{\kappa} + E_H(z_2 - z_0) + \frac{E_0}{\kappa_{sc}}$$

If ϵ_{sc} and ϵ_{el} are not of different orders of magnitude, equation [42] shows that E_0 and E_2 are about the same size. Then E_2/κ is much smaller than E_0/κ_{sc} because $\kappa \gg \kappa_{sc}$ and it is often possible to neglect E_2/κ and write

$$\Delta_{el}^{sc}\phi = (z_2 - z_0)\frac{\epsilon_{sc}}{\epsilon_H}E_0 + \frac{E_0}{\kappa_{sc}}$$

Since κ_{sc} is several orders of magnitude larger than $z_2 - z_0$ (which corresponds to one or two monolayers of solvent) the potential drop across the interface is essentially all in the semiconductor phase.

In general, the relation between E_0 and $\Delta_0^{sc}\phi$ is given by equation [72] of Chapter 4:

$$-E_0 = \left(\frac{16\pi n_0 kT}{\epsilon_{sc}}\right)^{1/2}\left[\cosh\left(Y + \frac{p}{kT}\right) - \cosh\frac{p}{kT} - Y\sinh\frac{q}{kT}\right]^{1/2}$$

where $Y = e\phi_0/kT = e\Delta_0^{sc}\phi/kT$, the dimensionless potential drop across the space-charge region. The quantity $e^{p/kT}$, the square root of which gives the ratio of electron-to-hole concentration in bulk (Chapter 4, equations [69] and [70]) is often denoted by λ^{-1} and is identical to $e^{q/kT}$ for a doped semiconductor, so that

$$E_0 = \mp\left(\frac{8\pi n_0 kT}{\epsilon_{sc}}\right)^{1/2}[\lambda(e^{-Y} - 1) + \lambda^{-1}(e^Y - 1) + (\lambda - \lambda^{-1})Y]^{1/2}$$

For $Y > 0$, the potential at the surface is higher than in bulk semiconductor and E_0 is negative, but for $Y < 0$ the surface potential is lower than in the bulk and one has to take the positive (lower) sign. With this more complicated expression, the potential difference across a semiconductor–electrolyte interface is still almost always dominated by the space-charge potential difference $\Delta_0^{sc}\phi$. The capacitance of the diffuse layer is $dq^S/d\phi_0$, where $q^S = -\epsilon_{sc}E_0/4\pi$, that is,

$$C_{sc} = \pm \left(\frac{\epsilon_{sc}n_0e^2}{2\pi kT}\right)^{1/2} \frac{d}{dY}\{\lambda(e^{-Y}-1) + \lambda^{-1}(e^Y-1) + (\lambda - \lambda^{-1})Y\}^{1/2}$$

$$= \epsilon_{sc}\pi^{-1}\kappa_{sc}(32)^{-1/2}\{[\lambda(e^{-Y}-1) + \lambda^{-1}(e^Y-1) + (\lambda-\lambda^{-1})Y]^{-1/2}$$

$$\cdot |-\lambda e^{-Y} + \lambda^{-1}e^Y + \lambda - \lambda^{-1}|\} \quad [43]$$

For an intrinsic semiconductor ($\lambda = 1$) the quantity in curly braces becomes $[e^{-Y} - 2 + e^Y]^{-1/2}|-e^{-Y} + e^Y| = e^{-Y/2} + e^{Y/2}$ so that the capacitance is $\epsilon_{sc}\pi^{-1}\kappa_{sc}8^{-1/2}\cosh(Y/2)$, a symmetric function of $Y = e\phi_0/kT$ with minimum at $Y = 0$ (like the diffuse-layer capacitance of an electrolyte). For example, for an intrinsic germanium semiconductor, with $\epsilon_{sc} = 5$ and $n_e^0 = 2 \times 10^{13}$ cm^{-3}, the capacitance at $Y = 0$ is 9.4×10^3 esu or 0.01 μF/cm^2. The Helmholtz-layer capacitance for a thickness of 2 Å and a dielectric constant of 5 is 10^7 esu or 11 μF/cm^2, so the capacitance of the interface is dominated by the capacitance of the semiconductor. It should be noted that for large $|Y|$ the capacity rises less sharply with $|Y|$ than indicated by the cosh function because the use of Boltzmann statistics for the concentrations of the carriers is not correct when these concentrations get large.

For an extrinsic semiconductor ($\lambda \neq 1$), C_{sc} is not symmetric about $Y = 0$ and the minimum in C_{sc} is not as sharp. This corresponds to a depletion layer in which the concentration of majority carriers is reduced by the potential and the concentration of minority carriers is still not appreciably large. Thus, if $\lambda^{-1} \gg \lambda$ (strong n-type doping) and $\lambda e^{-Y} \ll \lambda^{-1}$ (potential not negative enough for the hole concentration to approach the electron concentration at $z = 0$) we have, from equation [43],

$$C_{sc} = \epsilon_{sc}\pi^{-1}\kappa_{sc}(32\lambda)^{-1/2}[e^Y - 1 - Y]^{-1/2}[e^Y - 1]$$

or for small Y,

$$(C_{sc})^{-2} = \frac{8\pi\lambda kT}{\epsilon_{sc}n_e^0e^2}\left[Y + \frac{1}{2}Y^2\ldots\right]^{-2}\left[\frac{1}{2}Y^2 + \frac{1}{6}Y^3\ldots\right]$$

$$= \frac{4\pi\lambda kT}{\epsilon_{sc}n_e^0e^2}[1 - Y\ldots]\left[1 + \frac{1}{3}Y\ldots\right] \quad [44]$$

Thus $(C_{sc})^{-2}$ is linear in Y or $e\Delta_0^{sc}\phi/kT$.

G. SEMICONDUCTOR-ELECTROLYTE INTERFACE

If there are surface states (Chapter 4, Section I), of energy ϵ_s and with concentration N_{ss}, the number of electrons in these states is

$$n_s = \frac{N_{ss}}{1 + \exp\left[(\epsilon_s - \epsilon_F)/kT\right]}$$

where ϵ_s depends on the surface potential according to $\epsilon_s = \epsilon_s^0 - e\Delta_0^{sc}\phi$. These states may be electrically neutral when empty and make a negative contribution to the surface charge when occupied, or they may be donor levels, positively charged when empty and neutral when occupied.

In the latter case, the contribution of the surface states to the surface charge is

$$q^{ss} = eN_{ss} - \frac{eN_{ss}}{1 + \exp\left[(\epsilon_s - \epsilon_F)/kT\right]}$$

$$= N_{ss}\left[\exp\left(\frac{\epsilon_F - \epsilon_s^0 + e\Delta_0^{sc}\phi}{kT}\right) + 1\right]^{-1} \quad [45]$$

Instead of equation [42] one now has

$$\epsilon_H E_H = \epsilon_{sc} E_0 + 4\pi q^{ss} \quad [46]$$

For large enough q^{ss} we can neglect $\epsilon_{sc}E_0$ so that, combining equations [45] and [46], the potential drop in the Helmholtz layer is written as a function of the potential drop $\Delta_0^{sc}\phi$ across the diffuse space-charge layer:

$$E_H[z_2 - z_0] = 4\pi eN_{ss}(z_2 - z_0)(\epsilon_H)^{-1}\left[\exp\left(Y + \frac{\epsilon_F - \epsilon_s^0}{kT}\right) + 1\right]^{-1}$$

where the abbreviation $Y = e(\Delta_0^{sc}\phi)/kT$ has again been used. Differentiating, we find

$$\frac{d[E_H(z_2 - z_0)]}{dY} = \frac{-4\pi eN_{ss}(z_2 - z_0)(\epsilon_H)^{-1}\exp\left[Y + (\epsilon_F - \epsilon_s^0)/kT\right]}{\left\{\exp\left[Y + (\epsilon_F - \epsilon_s^0)/kT\right] + 1\right\}^2}$$

If the donor levels are weakly ionized, the exponential is large compared to unity so that the last expression becomes

$$-4\pi eN_{ss}(z_2 - z_0)(\epsilon_H)^{-1}\exp\left[-Y - \frac{\epsilon_F - \epsilon_s^0}{kT}\right] = -4\pi(z_2 - z_0)(\epsilon_H)^{-1}q^{ss}$$

The derivative of $E_H(z_2 - z_0)$, the Helmholtz-layer potential drop, with respect to space-charge-layer potential drop Y is of size unity when q^{ss} is about $(kT\epsilon_H/4\pi e)$

($z_2 - z_0$). Then the Helmholtz layer potential drop is of comparable size to that across the space-charge layer. This occurs when N_{ss} is about 10^{13} cm^{-2}.

The capacitance of the surface states is obtained from equation [45]:

$$C_{ss} = \frac{dq^{ss}}{d(\Delta_0^{sc}\phi)} = \frac{e^2 N_{ss}}{kT} \exp\left(Y + \frac{\epsilon_F - \epsilon_s^0}{kT}\right)\left[1 + \exp\left(Y + \frac{\epsilon_F - \epsilon_s^0}{kT}\right)\right]^{-2}$$

As a function of Y, this capacity has a single maximum, at $Y = (\epsilon_s^0 - \epsilon_F)/kT$. Its value is $e^2 N_{ss}/4kT$, which for $N_{ss} = 10^{14}$ cm^{-2} (a high value, representing about 10% of the surface atoms) is 1.3×10^8 cm^{-1} or 1.2×10^2 μF/cm^2. For large positive values of Y, C_{ss} becomes $(e^2 N_{ss}/kT)\,e^{-Y}$; for Y large and negative it approaches $(e^2 N_{ss}/kT)\,e^Y$.

The capacitance C_{ss} is in parallel with that of the space charge, C_{sc}, because the total surface charge density of the semiconductor is the sum of q^{ss} and q^s. The total capacity is then $C_{ss} + C_{sc}$. The total charge $q^{ss} + q^s$ is balanced, in the interface, by the charge in the electrolyte and the charge of specifically adsorbed ions. Dipoles in the electrolyte phase contribute to the potential drop across the interface, but not to the charge density.

Pleskov (1980, Section 6.4) discusses the experimental methods available for study of the structure of the double layer at a semiconductor electrode. In Section 6.3 he discusses the dynamic properties of the semiconductor–electrolyte interface (relaxation phenomena and the effect of irradiation).

6

FURTHER DEVELOPMENTS

A. IMPROVEMENTS ON GOUY–CHAPMAN THEORY

The Gouy–Chapman theory is almost universally used to describe the diffuse layer of an electrolyte at an electrode. Since contributions of the diffuse layer to surface charge, capacitance, and so on, calculated by the Gouy–Chapman theory, are subtracted from measured quantities to give compact-layer contributions, errors in the Gouy–Chapman theory could have serious consequences for our understanding of the electrochemical interface. Furthermore, it is obvious that the Gouy–Chapman theory becomes incorrect for large potentials and large ion densities.

According to equation [18] of Chapter 4, the surface excess of an ion i is proportional to $P_i/(1 + P_i)$, where $P_i = \tanh(q_i p/4kT)$ and p is the potential drop across the diffuse layer. Thus as p increases the surface excess increases without limit, which is impossible. According to the Boltzmann expression (equation [6] of Chapter 4) the concentration of an ion of charge q, at a point where the electrostatic potential is ϕ, becomes 100 times its concentration in bulk solution when $q\phi \sim 4.6kT$. The potential near the electrode is $4\pi q^M/\epsilon\kappa$, where q^M is the surface charge density on the metal, so, with $\kappa = 2.9 \times 10^7$ cm^{-1}, $q = e$, $T = 300$ K, and $\epsilon = 79$ this will occur when q^M is $(4.6kT/e)(\epsilon\kappa/4\pi) = 7.2 \times 10^4$ esu/cm^2 = 24 μC/cm^2, which is quite experimentally accessible. Of course, the presence of the compact layer makes the potential at the outer Helmholtz plane, where the diffuse layer begins, less than $4\pi q^M/\epsilon\kappa$.

What prevents ion concentrations from increasing without limit is of course the short-range repulsion between ions. Instead of writing $n_i/n_i(\infty) = \exp(-q_i\phi/kT)$, one could write $n_i/n_i(\infty) = \exp[(-q_i\phi + f)/kT]$, where f is a potential representing interionic repulsion. This would replace the interionic correlation by a mean repulsive field, which would itself have to depend on the local ion density. A better way of taking the repulsions into account is by giving each ion a hard-sphere core such that overlap of ionic volumes is forbidden (Reeves 1980, p. 118).

Since solvent molecules are no smaller than ions, one should give a size to molecules of solvent as well, rather than treating the solvent as a continuous medium characterized only by a dielectric constant. Treating the solvent as a continuous medium seems appropriate only when the solvent molecules are much smaller than the ions. Furthermore, even if a continuous medium is used, the value of the dielectric constant should differ from that appropriate to bulk solvent, since the field near the electrode may be several times 10^6 V/cm. For such fields, the dielectric constant is not even a constant, that is, it depends on field. Since the ions also exert very large fields, the dielectric "constant" should depend on ion concentration.

A number of theories attempting to take such effects as these into account have been proposed and are reviewed by Reeves (1980, p. 119). In many cases only small differences from the results of Gouy-Chapman theory were found. In fact, some of the corrections appear to cancel each other. The theories are quite involved and necessarily require approximations and assumptions to produce results to compare with experiment, so that the Gouy-Chapman theory is not considered to be invalidated. For example, for concentrations of 10^{-2} M and lower, and for a potential drop of 100 mV across the diffuse layer, the Gouy-Chapman results were changed only by a few percent (Reeves 1980, p. 120).

Clearly, there will also be situations where the Gouy-Chapman theory is inadequate. Because of its central role in theories of the electrochemical interface, it is important to understand why the theory works when it does and how it may be improved when it does not.

The Gouy-Chapman theory for ions near an electrode is closely analogous to the Debye-Hückel theory for ions in bulk solution. The interaction between an ion and the ions around it (the ion atmosphere) is responsible for the nonideal behavior of ions in solution. The solvent in Debye-Hückel theory is taken as a continuous dielectric medium, and the interaction between ions is reduced to the long-range Coulombic interaction, except for a hard-sphere repulsion between a central ion and the ions of its ion atmosphere (which means imposing a minimum distance of approach): There is no hard-sphere interaction between ions of the atmosphere. The total charge density at a distance r from an ion of type a is written $e \Sigma_b z_b \rho_b g_{ab}(r)$, where ρ_b is the bulk density of ions of type b, of charge $z_b e$, and g_{ab} is the two-particle distribution $n_{ab}(\mathbf{r}_1, \mathbf{r}_2)$ divided by the one-particle distribution $n_a(\mathbf{r}_1)$. The Debye-Hückel theory, like the Gouy-Chapman theory, uses the Poisson-Boltzmann equation, which comes about by replacement of the potential of mean force (Chapter 3, equation [11]) by the mean potential, that is, the Poisson equation is used to relate the charge density at a distance r from a central ion to the average electrostatic potential at r due to the central ion and the atmosphere, while each g_{ab} is assumed to obey a Boltzmann distribution in this average potential. Furthermore, as in some forms of the Gouy-Chapman theory, the Poisson-Boltzmann equation is linearized.

The linearization of the Poisson-Boltzmann equation requires that ion-ion interaction energies be small compared to kT, which is certainly not the case for small interionic distances. The Debye-Hückel theory succeeds at low ion concentrations because the average interionic distance is large, and the probability of ions approaching each other closely is small. For large interionic distance, solvent between the ions may more reasonably be approximated as a continuous medium, short-range forces become less important, and linearization becomes a better approximation. At small distance, all three aspects lead to problems, and the Debye–Hückel theory becomes quite incorrect for concentrations of 0.1 M or so.

Improvements in the Debye–Hückel theory are reviewed by Andersen (1975). It is pointed out that the McMillan–Mayer solution theory (McMillan and Mayer 1945) shows that replacement of the solvent by a continuum is not an approximation, provided that a correct choice is made for the short-range part of the interionic interaction. The correct effective ion–ion interaction at short distances incorporates the interactions with the molecules of the solvent as well as the direct short-range interionic repulsion.

This comes about as follows: In terms of the potential of mean force U,

$$g_{ab}(r) = \exp\left[\frac{U_{ab}(r)}{kT}\right]$$

(Chapter 3, equation [12]). For a low-density gas, for which only pair correlations are important, one can show (Chapter 3, equations [11] and [12]) that

$$g_{ab}(r) = \exp\left[\frac{-\phi_{ab}(r)}{kT}\right] \qquad [1]$$

where $\phi_{ab}(r)$ is the interaction potential between gas molecules. For an ionic solution, suppose that the pair correlation function when only two ions, one of type a and one of type b, are present (infinite dilution) is $g_{ab}^0(r)$. The potential of mean force at infinite dilution is

$$U_{ab}^0(r) = -kT \ln g_{ab}^0(r) \qquad [2]$$

and U_{ab}^0 includes the effect of the solvent on the ion–ion interaction. Now, if $U_{ab}^0(r)$ is taken as the effective potential, $g_{ab}(r)$ is given exactly by the dilute-gas formula (equation [1]), a not very useful tautology.

However, the McMillan–Mayer theory describes the conditions that would allow $U_{ab}^0(r)$ to be used for a system of finite (>0) concentration. Pair correlation functions $g_{ab}(r)$ for an ionic solution at finite concentration can be calculated using the dilute-gas formula and $U_{ab}^0(r)$ for an infinitely dilute solution at the same temperature, such that the chemical potential of the solvent is the same in the infinitely dilute solution and the actual solution. The equality of chemical potentials requires that the pressure of the infinitely dilute solution be lower than the pressure of the actual solution: the lowering is just the osmotic pressure. For each solute concentration, the required pressure for the reference (infinitely dilute) solution would be different, and the effective interionic potentials would be different. An alternative is to take the reference solution at 1 atm, obtain a single set of effective potentials for each choice of a and b, and use them to calculate properties of the actual solutions at appropriate pressures. Then corrections to 1 atm pressure are made on the calculated properties by thermodynamic methods.

This procedure justifies models for ionic solutions which ignore the solvent, except insofar as it modifies the interionic potentials. Now one can assume that the effective interionic potentials $w_{ab}^0(r)$ become $z_a z_b e^2/\epsilon r$ for large r. The validity of using a dielectric constant to represent the effect of the solvent in the limit of large r is shown by molecular theories of the dielectric constant and by the success of the Debye–Hückel theory when interionic distances are large. Of course, the behavior of the effective potentials $w_{ab}^0(r)$ for values of r decreasing to several molecular diameters is not known, although one can assume $w_{ab}^0(r)$ is a continuous function and approaches $z_a z_b e^2/\epsilon r$ smoothly as r becomes large. Most important, because of the McMillan–Mayer theory, one can be assured that the effective potential, involving the ions only, exists.

Given the effective potential, one needs to calculate properties of an assembly of ions interacting with this potential (and no solvent) that is, a plasma. Some statistical mechanical methods for calculating the thermodynamic properties of a plasma are reviewed by Andersen (1975, pp. 23–27): They include (Chapter 3, Section E) cluster theories and integral-equation methods. According to statistical mechanics (Chapter 3, equations [5]–[7]), the pair correlation function for an isotropic homogeneous system is

$$g_{ab}(r_{12}) = N_a N_b \frac{\int d\mathbf{r}_3, \ldots, d\mathbf{r}_N \exp\left[-\Phi(\mathbf{r}_1, \ldots, \mathbf{r}_N)/kT\right]}{\int d\mathbf{r}_1, \ldots, d\mathbf{r}_N \exp\left[-\Phi(\mathbf{r}_1, \ldots, \mathbf{r}_N)/kT\right]}$$

where Φ is the total interaction energy of all particles. The Monte Carlo method is one way to evaluate the integrals. By applying the Monte Carlo method to a simple model (e.g., the primitive model) one has results that can be used to check other statistical mechanical approaches. Such theories may also show clearly and with more understanding than numerical methods how the interaction potentials determine the properties of the system.

Another approach is to infer features of the effective potential from experimental data on solutions. This is complicated by the fact that, in even the simplest solution, there are at least two kinds of ions and hence at least three effective potentials. One can assume functional forms for the potentials, with more or less theoretical justification, and choose parameters in them to fit calculated properties to experimentally measured properties. Andersen (1975, pp. 16–19) reviews such choices.

Perhaps the simplest choice corresponds to the "primitive model" of an electrolyte, that is,

A. IMPROVEMENTS ON GOUY-CHAPMAN THEORY

$$w_{ab}^0(r) = \frac{z_a z_b e^2}{\epsilon r}, \qquad r > r_a + r_b$$
$$w_{ab}^0(r) = \infty, \qquad r < r_a + r_b \qquad [3]$$

The core parameters r_a and r_b are chosen to fit experiment. This choice for w_{ab}^0 is not adequate to describe solutions of simple salts for concentrations up to 1 M. Improvements on this model involve more parameters, with the number of parameters increasing rapidly with the complexity of the model. Calculation of the properties of a system given the pair potential becomes difficult for a complicated model, and there is another problem. With many adjustable parameters, one of course can fit all the experimental data, but several quite different potentials work equally well, so not much is learned by the fitting procedure. Part of the problem is due to the fact that the experimental data are thermodynamic properties, which are not very sensitive to the details of the interionic potentials (as seen in Chapter 3, they are averages over ionic distribution functions). One would like measurable properties that depend more directly on the pair correlation function.

The problems with the Gouy-Chapman theory have been traced to the neglect of short-range repulsions between ions and the replacement of a potential of mean force by an average potential (neglect of instantaneous correlations or fluctuations). Since the primitive model includes short-range repulsions, it has often served as the testing ground for improvements on the Poisson-Boltzmann theory or Gouy-Chapman theory. However, the primitive model necessarily leaves out effects of solvent structure and is not appropriate for describing the compact layer. When one models the electrochemical interface by primitive-model ions at a charged wall, the charged wall thus represents the compact layer as well as the metal of the interface, and the distance of closest approach is the outer Helmholtz plane.

Let the concentration of ions of species i at a distance x from the wall be

$$n_i(x) = n_i^0 g_{0i}(x)$$

where n_i^0 is the bulk concentration and g_{0i} is written to make the wall-ion distribution function analogous to an ion-ion distribution function. It can be shown (Outhwaite, Bhuiyan, and Levine 1980) that

$$g_{0i}(x) = v_i \exp\left(\frac{-e_i \phi}{kT}\right) \exp\left(\frac{w_i}{kT}\right) \qquad [4]$$

where $v_i(x)$ takes into account the exclusion volume and w_i is the effect of instantaneous correlations. The Poisson equation is

$$\frac{d^2\phi}{dx^2} = -\frac{4\pi}{\epsilon} \sum e_i n_i^0 g_{0i}(x) \qquad [5]$$

In the Poisson–Boltzmann theory, $v_i(x) = 1$ and $w_i = 0$. Given $v_i(x)$ and w_i, the combination of equation [4] and [5] produces a modified Poisson–Boltzmann (MPB) equation to be solved numerically for ϕ. One then obtains the $g_{0i}(x)$, and properties of the interface can be calculated.

In general, the results of different MPB equations (corresponding to different v_i and w_i) resemble each other; they differ importantly from the results of Gouy–Chapman theory by predicting that $\phi(x)$ is nonmonotonic. The Gouy–Chapman theory predicts that ϕ and the density of each ion vary monotonically with distance from the electrode: According to equation [6] and [17] of Chapter 4,

$$n_i = n_i(\infty) \left(\frac{1 + Pe^{-\kappa x}}{1 - Pe^{-\kappa x}} \right)^{-1/2}$$

for a symmetric electrolyte, with $P = \tanh(qp/4kT)$; $(1 + Pe^{-\kappa x})/(1 - Pe^{-\kappa x})$ is a monotonically increasing function of x since $-1 \leq P \leq 1$. The density profile of an ion in the linearized Gouy–Chapman theory for the primitive model (charged hard spheres in a uniform dielectric medium) is

$$g_i(x) = g_0(x) + \frac{z_i eE}{\epsilon \kappa kT} e^{-\kappa(x - \sigma/2)}, \quad x > \frac{\sigma}{2}$$

where g_i is the ion density divided by its bulk value and g_0 is the density profile for hard spheres near a hard wall (Henderson, Blum, and Lozada-Cassou 1983). For large ion concentrations (~ 2 M), the modified Poisson–Boltzmann theories give a potential ϕ that decreases rapidly from its (positive) value at $x = 0$, passes through a minimum for x between one and two ionic diameters, and rises to its asymptotic value by three ionic diameters. The g_{0i} more closely resemble those for the Gouy–Chapman theory than does the potential, but also show nonmonotonic behavior in some cases. The oscillations in density profiles lead to oscillations in the charge density, that is, charge layering near the electrode surface.

The possibility of charge layering at a charged wall was predicted by early theories that improved on Gouy–Chapman theory. Stillinger and Kirkwood (1960; Martynov, 1963, 1973), considering a hard-sphere ionic system, showed that the charge density profile becomes oscillatory when the Debye length is comparable to ionic diameters. Since $\kappa^{-1} = 3.4 \times 10^{-7}$ cm for 0.01 M solution in water at 300 K, the Debye length will be 5 Å at a concentration of $(34/5)^2(0.01) = 0.5$ M, or at half this concentration for doubly charged ions. The stratification of charge near a charged wall comes about because, around each ion, there is an excess of ions of opposite charge, so that an ion plus its atmosphere is neutral. At high enough concentrations (small average interionic distances) the ion atmospheres overlap, instantaneous correlations between ions become important, and g_{ab} becomes oscillatory. Because of the charge on the wall, this leads to ordering perpendicular to the wall. Thus the ordering results from an interplay between short-range and long-range forces. For molten-salt systems, one can expect the effect of layering to be very important.

The primitive model can also be treated by the hypernetted chain (HNC) and other approximations. These theories improve on the Gouy-Chapman theory by taking into account hard-core interactions between ions. Torrie and Valleau (1980, 1982) have performed Monte Carlo computer simulations using the primitive model for the double layer, to check the validity of the Gouy-Chapman theory and other theories. It appears that the HNC represents the actual (simulation) results well for 1-1 and 2-2 electrolytes with surface charges up to 0.3 C/m^2. The Gouy-Chapman theory markedly overestimates the double-layer thickness, as well as the potential difference between electrode and bulk. For 2-2 electrolytes, this potential difference actually passes through a maximum at about 0.15 C/m^2, which the Gouy-Chapman theory of course cannot reproduce. The ionic density profiles $g_i(x)$ seem to become oscillatory, and even nonmonotonic, at these charge densities, in the simulations and in the HNC. Although the Gouy-Chapman theory cannot reproduce the oscillations, it does give reasonable values of total ionic adsorptions (which are integrals of g_i) because they obey the electroneutrality condition,

$$\sum z_i e \rho_i^b \int_0^\infty g_i(x)\, dx = \frac{E}{4\pi}$$

and the sum is dominated by the counterion.

A striking result of the Monte Carlo simulations is that there may be an extremum in a plot of diffuse-layer potential drop vs. surface charge. It occurs at negative electrode charges q^M when the cations are divalent. This is an electrostatic effect rather than a result of the excluded volumes of ions. The diffuse-layer potential becomes nearly independent of surface charge over a substantial region, although Gouy-Chapman theory predicts the usual exponential dependence. The modified Poisson-Boltzmann theories agree qualitatively, but not always quantitatively, with the results of the Monte Carlo simulations.

Within Gouy-Chapman theory itself, a more realistic treatment of ion sizes also produces interesting effects. The ion sizes enter only in determining the distances of closest approach of ions to the electrode ("outer Helmholtz plane"), and these should be different for different ions. The effect of this should be greatest at small surface charges, for which concentrations of cations and anions near the electrode are of comparable size. In particular, at the point of zero surface charge density the different distributions of cations and anions will produce a potential difference.

The Gouy-Chapman theory for different ion sizes has recently been elucidated (Valleau and Torrie 1982). Consider ions of charges $\pm q$ and sizes r_1 and r_2. For $z > r_1$ the Poisson-Boltzmann equation (Chapter 4, equation [12]) is

$$\frac{d^2\phi}{dz^2} = \frac{8\pi \rho_b q}{\epsilon} \sinh \frac{q\phi}{kT} \qquad [6]$$

but for $r_1 > z > r_2$ there is only one kind of ion present and the Poisson–Boltzmann equation is

$$\frac{d^2\phi}{dz^2} = \frac{4\pi\rho_b q}{\epsilon} \exp\left(\frac{q\phi}{kT}\right) \quad [7]$$

The solution to equation [6] (Chapter 4, Section A) is to be matched to a solution to equation [7] at $z = r_1$, with ϕ and $d\phi/dz$ continuous. Results were obtained (Valleau and Torrie 1982) assuming anions could get closer to the metal than cations by about one water-molecule diameter—this actually simulates specific adsorption. At large surface charges, where one ion or the other is present in excess, they resemble the results of the conventional Gouy–Chapman theory. At small surface charges, there are differences because of the different distances of closest approach. The capacitance is not a symmetric function of the surface charge. Similarly, a plot of surface tension vs. potential, obtained by integrating the Lippman equation with calculated surface charge as a function of potential, is not symmetric about the maximum. Because there is a potential drop across the diffuse layer even at zero surface charge density, due to the different locations of cations and anions, the pzc now depends on the ion concentration c. Unlike the actual Esin–Markov coefficient, the point of zero charge is not linear with $\ln c$, but approximately linear with $c^{1/2}$.

Monte Carlo calculations have been performed for a system of ions of different sizes at a charged wall (primitive model), including polarization effects (image forces—see Section B) as well. The existence of a diffuse-layer potential at zero surface charge was confirmed, as well as the dependence of the point of zero charge on ion concentration. The image forces enhance this dependence substantially, since they have a larger effect on the ion that can come closer to the surface. Nonmonotonic behavior of the electrostatic potential and ion densities, as well as charge inversion with a reduction of diffuse-layer potential, are also enhanced by the image forces.

B. IMPROVED MODELS FOR THE INTERFACE

Since we are now adding to our knowledge of the diffuse layer, the theory for which was worked out to almost general satisfaction many years ago, it is not surprising that there is much theoretical work being done on the rest of the electrochemical interface. Most past theories neglect or oversimplify one part of the double layer in order to give a realistic description of another. The metal has been particularly slighted in this regard, often being reduced to a source of electrical potential and an impenetrable boundary for the electrolyte. Recently, theories for the metal have begun to appear, as well as models that attempt to model all parts of the interface realistically.

Badiali, Rosinberg, and Goodisman (1981) used a jellium model and a variational calculation of the conduction electron density profile to obtain work functions and surface potentials for liquid metals. The effect of the solvent was represented by a repulsive potential, either a hard-wall barrier or a δ-function. The change in the surface potential of the metal caused by the solvent, $\delta\chi_S^M$, was calculated for a number of simple metals. It was always negative, and more negative for metals of higher electron density, that is, -0.24 V for Hg and about -1 V for Al. The metals with higher electron density also have smaller ion size so that, in this model, the molecules of the solvent approach the metal more closely, which also makes the magnitude of $\delta\chi_S^M$ larger. Subsequently, the variational calculation was replaced by a self-consistent calculation of the electron density profile, and the step-function ion profile of the jellium model was replaced by an oscillatory ion profile, which was shown to be more realistic for a liquid metal (Goodisman 1985).

The jellium model was extended to calculation of the capacitance of the metal in the cases of Hg and Ga (Badiali, Rosinberg, and Goodisman 1983a). This capacitance is due to the change in the electron density profile of the metal as it is charged. Although the value of the capacitance depends on the parameters used to characterize the water-conduction electron interaction, $C^M(\text{dip})$ was always significantly larger for Ga, with higher electron density, than for Hg. The actual differential capacity of the interface is given by

$$C^{-1} = [C(\text{ion})]^{-1} + [C^M(\text{dip})]^{-1} + [C^S(\text{dip})]^{-1} \qquad [8]$$

where the three terms refer to the contributions of free charges, of the metal surface, and of the water surface. Using $C(\text{ion}) = 17$ $\mu\text{F}/\text{cm}^2$ $(4\pi(d_2 - d_1)/\epsilon_1, \epsilon_1 = 6, d_2 - d_1 = 0.3$ nm), and $C^S(\text{dip}) = 25$ $\mu\text{F}/\text{cm}^2$, one can extract the contribution of the metal. It appears that the jellium model explains the values qualitatively, as well as the differences in capacitance from one metal to another, although uncertainties in the parameters of the model make it impossible to give quantitative predictions. The difference between the Hg/H$_2$O and Ga/H$_2$O interfaces can even be qualitatively predicted by this model without invoking adsorption of the solvent on the metal; probably, both adsorption and modification of the metal electron density profile should be considered.

The assumption that the interface behaves as several capacitances in series, used in these calculations, implies that only the field of the free charges was used in calculating the electronic distribution, just as only the field extending from the metal surface is normally used in calculating the behavior of solvent dipoles. The fact that the electronic tail penetrates the region of the solvent dipoles means two things: (1) The electrostatic potential felt by the electrons is not just that corresponding to a uniform field, but has a contribution from the dipoles. (2) The electric field due to free charges on the metal and at the outer Helmholtz plane is not constant in the region of the dipoles. To represent the effect of the dipoles on the metal electrons, the electrostatic potential of a dipole layer was added to the electronic energy functional. It was found that the effect on the electronic distribution was small, so that it is legitimate to calculate the electronic profiles without considering the solvent polarization.

On the other hand, the effect of the extension of the electronic tail into the region of the solvent dipoles is that the field to be used in model calculations of dipole behavior should not be $(4\pi/\epsilon_1)\, q^M$, but $(4\pi/\epsilon_1)(q^M + \delta q)$, where δq is the electronic charge that is found on the solution side of the dipoles. In particular, there is an electric field at the position of the solvent dipoles even at the potential of zero charge (pzc). Correspondingly, there should be some orientation of dipoles at the point of zero charge. For mercury, zero field at the dipoles occurs for $q^M = -0.4\ \mu C/cm^2$, which agrees with some estimates made from experiment. A problem that has been raised (Badiali, Rosinberg and Goodisman 1983a, b) is that the position of the solvent dipoles relative to the jellium (in general, the ionic profile of the metal) should be determined by a calculation including both metal and solution species; in particular, its value could depend on the charge q^M and such dependence could have a major effect on the capacity of the metal. Models treating both metal and solution phase in a realistic way are required.

The jellium model for a metal has been used to calculate the rate of electron transfer from metal to a proton in solution. The proton was represented (Halley, Pratt, and Johnson 1983) as a point charge in a dielectric medium of fixed dielectric constant ϵ, taken to be the high-frequency dielectric constant. The structure of the electrolyte phase is thus ignored, reversing the usual procedure of including the structure of the electrolyte phase but ignoring or idealizing the electronic structure of the metal phase. The energies of the system, metal plus proton, and the system, metal plus neutral atom, are calculated using a local density functional. The one-electron states obtained by solution of the self-consistent problem should include a bound state localized on the proton as well as a continuum. This bound state is taken as filled for H and empty for H^+. Then, energies for the two systems as a function of the separation of the proton from the jellium edge, $E_H(z)$ and $E_{H^+}(z)$, can be computed (in fact a number of approximations were made so that the actual calculation of the electron density in the field of semi-infinite jellium plus a proton was not carried out). The effect of a potential drop E across the interface was taken to be a displacement of E_H, relative to E_{H^+}, by a constant energy eE (Chapter 1, Section L). The barrier to electron transfer is the difference in energy between the minimum in $E_{H^+}(z)$ and the intersection of $E_H(z) + eE$ and $E_{H^+}(z)$. Assuming that the free energy of activation can be equated to the energy of activation, that is, the barrier, one then gets the transfer coefficient α from the slope of a plot of barrier height vs. E.

If the dielectric constant of the metal differs from that of the region adjoining it, charges in the latter will polarize the charge density of the metal, inducing charges on the metal surface. Considering a charge q at some $z > 0$, one has to find the electrical potential $\phi(\mathbf{r})$ such that the Poisson equation is satisfied, with the normal component of the electrical displacement $D_z = -\epsilon(z)(d\phi/dz)$ and the parallel component of electric field $E_z = -(d\phi/dz)$ continuous across the surface $z = 0$; the solution corresponds to a fictitious charge for $z \le 0$. The interaction of the induced charge with the true charge is an *image force*. Let the dielectric constant for $z > 0$ be ϵ, and let that for $z < 0$ be ϵ'. Then the interaction between two real charges i and j is $q_i q_j/\epsilon r_{ij}$ plus the image interaction

B. IMPROVED MODELS FOR THE INTERFACE

$$\frac{\epsilon - \epsilon'}{\epsilon + \epsilon'}(q_i q_j \epsilon^{-1})[(x_{ij})^2 + (y_{ij})^2 + (z_i + z_j)^2]^{-1/2}$$

and the interaction of a charge with the surface includes a term

$$\frac{[(\epsilon - \epsilon')/(\epsilon + \epsilon')](q_i)^2}{4\epsilon z_i}$$

and a modification of the interaction with the true surface charge.

Torrie, Valleau, and Patey (1982) performed Monte Carlo calculations on the primitive model to investigate the effect of image forces on the diffuse layer. The ions are of diameter d, so their centers are limited to the region $z > d/2$, with the region $z < 0$ being occupied by the metal and compact layer. They considered the cases $\epsilon' = \infty$, a conducting medium for which each ion generates an equal and opposite image charge, and $\epsilon' = 1$, like vacuum, to represent electrodes with low dielectric constant, since $\epsilon = 78.5$. There are great effects of the image forces on the distribution of counterions (ions with charge opposite to the electrode surface charge). With $\epsilon' = 1$ and low surface charges, there is a maximum in their density profile near $z = d$ because of the repulsive image interactions, and co-ion (same charge as electrode) concentrations seem to increase for distances less than d. The electrostatic potential is much less affected. The modified Poisson–Boltzmann theory reproduces these effects well.

The HNC integral equation theory could not be applied to the system with the correct form of the image forces given above, but could be applied to a "screened self-image" model that replaces all these terms by the single-particle interaction

$$u_i(z) = \frac{\epsilon - \epsilon'}{\epsilon + \epsilon'} \frac{(q_i)^2}{4\epsilon z} e^{-2\kappa z}$$

Monte Carlo calculations were performed for this model as well and also showed a pronounced maximum in the counterion density between $z = d$ and $z = 2d$; the HNC theory reproduced this behavior.

Within the electrolyte part of the interface, the distinction between compact (Stern) and diffuse layers seems very artificial when one looks at the solvent from a molecular point of view. The water molecules are treated as a uniform dielectric medium when discussing the outer (diffuse) layer and are treated as a collection of orientable dipoles, or as a medium of different dielectric constant from the outer layer, when discussing the inner (compact) layer close to the electrode. In fact, all water molecules are the same and are subject to the same forces. Thus, it is of interest to ascertain whether the distinction between inner and outer layers, useful as it is in representing many of the properties of the double layer, occurs naturally (Carnie and Chan 1980). The water molecules should be treated as orientable dipoles having a finite size, just like the ions; one knows that the interaction of charges in a medium containing orientable dipole moments will resemble the Coulombic interaction divided by the dielectric constant when the charges are far enough apart.

The simplest model of an electrolyte which treats both the ions and solvent molecules as discrete is a mixture of charged and dipolar hard spheres, all of the same diameter σ. Some of the spheres have charges $z_i e$ at their centers, others have dipole moments μ. Higher-order moments of the water molecules are neglected, as is their nonspherical shape and short-range interactions other than the hard-sphere interactions. These simplifications, necessary to allow analytical solutions for the distribution functions of the ions and molecules, clearly leave out important features of the real system. On the other hand, no dielectric constant is put into the model: The interaction between ions and dipoles is unscreened. If the effect of the solvent dipoles is to screen Coulombic interactions with a dielectric constant ϵ, this will have to come out of the calculations.

Further simplifications to make the model tractable relate to the description of the electrode: It is represented as a hard flat wall bearing a surface charge, that is, its only interactions with the electrolyte phase are an infinite repulsion on contact and the electrostatic interaction due to the field of an ideal charged plane. There are no image forces or non-Coulombic short-range interactions. With the above assumptions, the model can be treated by the mean spherical approximation (Carnie and Chan 1980, Blum and Henderson 1980), which, since it is a linearized theory, is expected to be valid only near the point of zero charge.

The analytic solution is complicated, but can be transformed to give explicit results for low concentrations ($<0.01\ M$). The theory then gives expressions for the density profiles of the ions, $g_i(x)$, and for the density profiles of the dipoles, $g_s(x, \theta)$, where x is the distance from the electrode and θ is the angle between the direction of a dipole and the normal to the surface. The actual density of ions of species i at x is $\rho_i^b g_i(x)$, where ρ_i^b is the bulk density ($x \to \infty$) of this species, and the density of dipolar species at x is $\rho_s^b \int d\theta \sin \theta\ g_s(x, \theta)$, where ρ_s^b is the bulk density of solvent molecules.

Analogously to the Debye length, the quantity κ is defined by $\kappa^2 = (4\pi e^2/\epsilon kT) \sum (z_i)^2 \rho_i^b$, where ϵ is now a complicated function of the dipole moment μ and the density ρ_s^b. For low values of κ, the potential drop across the double layer becomes

$$\Delta V = \frac{E}{\epsilon\kappa} + \frac{E\sigma}{2\epsilon} + (1 - \epsilon^{-1})\frac{E\sigma}{2\lambda} \qquad [9]$$

where the surface charge on the electrode is $E/4\pi$. Like ϵ, λ is a function of μ and ρ_s^b; it is related to ϵ by $16\epsilon = \lambda^2(1 + \lambda)^4$. Gratifyingly, ΔV looks like the sum of a diffuse-layer term $E/\epsilon\kappa$ (which becomes small for high ion concentration) and a compact-layer term proportional to σ. Using the correct value for ϵ, 78.4, one finds for the compact-layer term $E\sigma/2\epsilon'$ with $\epsilon' = 4.8$. Then, with $\sigma = 2.76$ A, one finds capacitances at the point of zero charge which, as a function of concentration, behave qualitatively like those found from experiment. With $\epsilon' = 6.7$ the capacitances agree with experiment within experimental error.

However, in spite of the resemblance of the two parts of ΔV, $E/\epsilon\kappa$ and $E\sigma/2\epsilon'$, to diffuse- and compact-layer potential drops, these terms do not represent potential drops across separate regions of x which would correspond to compact and diffuse layers. It is found that the contribution of oriented dipoles to the potential, though largest at $x = \frac{1}{2}\sigma$ (distance of closest approach), does not go to zero at the first layer. The conventional separation of the solvent layer into inner (compact) and outer (diffuse) parts implies two distinct regions, in one of which $g_s(x, \theta)$ differs importantly from 1, and in the other of which $g_i(x)$ differs importantly from 1, with the former region lying closer to the electrode than the latter. This is decidedly not the case in the mean-spherical model results, but it must be noted that the separation between layers is supposed to result from the fact that ions, being hydrated, cannot approach as closely to the electrode as water molecules, and the mean spherical model gives ions and dipoles hard spheres of the same size.

Vericat, Blum, and Henderson (1983) have applied the Generalized Mean Spherical Approximation (GMSA) to the nonprimitive electrolyte (charged and dipolar hard spheres) at a charged wall. This improvement to the Mean Spherical Approximation (MSA) does not give results that are linear in the charge density of the electrode, so the GMSA may be useful further from the point of zero charge than the MSA. The ionic density profiles calculated from the GMSA, like those from other improvements on the Gouy–Chapman theory, also show layering of sheets of alternating charge density near the electrode. In fact, oscillations of this type are more pronounced in the GMSA than in the MSA.

The theoretical models all take electrical potential and ion densities as functions of a single variable, the distance from the electrode, thus ignoring inhomogeneities in directions parallel to the electrode surface. Although such inhomogeneities may not affect properties such as the capacitance, they can be probed by electroreflectance and photoemission studies (Liu 1983), and theoretical models that take inhomogeneities into account are required. The values of the Esin–Markov coefficient also rule out theories that consider only variation perpendicular to the interface, and require replacement of the average electrostatic potential in the compact layer by a micropotential that depends on the positions of the ions and molecules. For the diffuse layer, it has been suggested that the necessity to use a very high value of the dielectric constant to describe the diffuse layer on Ga electrodes is due to the anisotropic shielding and anisotropic dielectric constant for water molecules that are strongly oriented by the chemical forces from the electrode. The oriented water molecules are less able than bulk water to shield ion–ion interactions perpendicular to the metal surface (lower effective dielectric constant) but shield these interactions more effectively than bulk water in directions parallel to the metal surface (higher effective dielectric constant) (Trasatti 1979, pp. 112–113). Finally, the value of the inner-layer capacity K_{ion}, whose value is 16.8 $\mu F/cm^2$ for water on liquid mercury for surface charges less than -10 $\mu F/cm^2$, apparently varies from one crystal face of a solid metal to another. This is explained as a change in inner-layer thickness due to penetration of water molecules into interatomic depressions on the surface. Calculations, again involving theoretical models that take into account inhomogeneities parallel to the electrode surface, show that a change in d

of several hundredths of an angstrom can result from such packing. The resulting variation in K_{ion} from one metal to another is about 5%. An increase of about the same size is predicted in going from the liquid to the solid phase of the same metal, assuming that the liquid surface is flat and does not allow penetration of the solvent molecules (Trasatti 1983).

In the simplest model for the dipoles in the compact layer, the effective field acting on a dipole is $E_{eff} = E - \lambda P$, where P is the polarization, depending on E_{eff} through the Langevin function, and λ is a parameter whose value expresses the influence of surrounding dipoles on a central dipole. Since dipole–dipole interactions imply a tendency of nearest-neighbor dipoles to be antiparallel, a minus sign has been used in E_{eff}, so that λ is positive. For low fields, the Langevin function can be linearized to yield

$$P = \frac{N\mu^2 E_{eff}}{3kT}$$

where N is the number density of dipoles. Then one can solve for P and for the dielectric susceptibility $\chi = P/E$:

$$\chi = \frac{N\mu^2}{3kT}\left(1 + \lambda \frac{N\mu^2}{3kT}\right)$$

The dielectric constant $\epsilon = 1 + 4\pi\chi$ and the dipole-layer capacitance $C_d = \epsilon/4\pi d$. The value of λ can then be derived. If $d = 0.3$ nm and $C_d = 26$ μF/cm^2 (as measured for Hg in saturated NaF), one finds $\epsilon = 8.8$. Then, with $N = 3.35 \times 10^{22}$ molecules/cm^3 and $\mu = 1.87 \times 10^{-18}$ esu, one finds $\lambda = 1.69$; its value is practically independent of the electrolyte (Liu 1983).

The average dipole moment in the direction perpendicular to the electrode surface is $\langle \mu_n \rangle = \mu \mathcal{L}(x)$, where $x = \mu E_{eff}/kT$. The mean-square dipole moment in this direction is $\Delta\mu_n = (\langle \mu_n^2 \rangle - \langle \mu_n \rangle^2)^{1/2} = \mu(1 + x^{-1} - \coth^2 x)$, which, for a wide range of potentials and values of λ, is about $\mu/3^{1/2}$. The average value of the dipole moment in a direction parallel to the electrode, $\langle \mu_P \rangle$, is of course zero, but the mean-square value is

$$(\Delta\mu_P)^2 = \langle (\mu_P)^2 \rangle - 0 = \tfrac{1}{2}\left(\mu^2 - \langle (\mu_n)^2 \rangle\right) \neq 0$$

The factor of $\tfrac{1}{2}$ is here because there are two directions parallel to the electrode surface so that $2\langle (\mu_P)^2 \rangle + \langle (\mu_n)^2 \rangle = \mu^2$. In photoemission into an electrolyte, the parallel component of electronic momentum is not always conserved, and the random potential associated with $\Delta\mu_P$ may help to explain this (Liu 1983).

Schmickler (1983) presented a model that simultaneously deals with metal, modeled as jellium, and the compact layer of solvent, modeled as an ensemble of point dipoles. The dipoles, which are assumed to adopt one of three orientations, were arranged in a hexagonal lattice. Nearest-neighbor interactions between dipoles were treated exactly, non-nearest-neighbor interactions were treated by a mean-field approximation. The electronic profile for the metal was taken as a one-parameter exponential, the parameter chosen to minimize the density-functional expression for the energy (Chapter 4, Section F) for a given surface charge density. The field at the position of the dipoles, due to the jellium background plus the conduction electrons, was used, together with the dipole–dipole interactions, to compute the average orientation of the dipoles, $\langle S \rangle$, and the mean field at a dipole. This involves a self-consistency condition, since the mean field is proportional to $\langle S \rangle$, and $\langle S \rangle$ is determined by considering a dipole in the field of the nearest neighbors plus the mean field. The difference of potential across the dipole layer was divided by the thickness of the compact layer to get the average field that acts on the metal electrons. The parameter of the electronic distribution was then recalculated and the process was repeated until self-consistency was attained (a second self-consistency condition).

The capacity is peaked near the point of zero charge, but the peak value increases considerably with the bulk electron density of the metal. Higher electron density means more charge that can be polarized by the applied field, and therefore means an increase in capacity. This was found in all the jellium calculations for the capacity of the liquid metal surface (Badiali, Rosinberg, and Goodisman 1983a).

C. HYDRATED ELECTRON

We have had occasion to refer to the electron in aqueous solution in connection with the definition of standard electrode potentials (Chapter 3, Section H). It was noted that electrons are not normally components of aqueous solutions. They can, however, be produced by ejection of electrons from a metal in contact with the solution, under the influence of light. The understanding of the emission process is relevant to the structure of the metal–solution interface, while theories for the solvated electron are relevant to theoretical study of the interaction of electrons of a metal with an aqueous solution. The emission of an electron from a metal to a solution is in a sense the simplest electrochemical reaction, although whether a hydrated electron can be produced electrochemically (without absorption of radiation) is a subject of current investigation.

Another kind of experiment may give more information on the molecular nature of the double layer and help theoretical and experimental progress toward understanding, namely, the study of the "emersed electrode." The double layer at a metal electrode can be studied in isolation from the solvent (Hansen 1983). It seems that, when the electrode is removed from its liquid phase, the double layer out to the shear plane, which includes virtually all the charge for concentrated electrolytes, remains with the electrode. Thus it can be studied by various spectroscopies and surface analysis techniques, which have produced so much information about clean surfaces. The charge that is lost during emersion can be monitored, and work functions can be measured.

In photoemission from a metal to an electrolyte solution, electrons are ejected from the metal by absorption of light quanta, pass through the metal–electrolyte interface, and are thermalized and solvated. The time required to form solvated electrons is of the order of 10^{-11} s. Subsequently, the solvated electrons react chemically with electron scavengers such as H_3O^+, NO_3^-, O_2, or many organic molecules; if none are present, the electrons rapidly return to the electrode. The photocurrent depends strongly on the state of the metal–electrolyte interface. In particular, the work function, or minimum energy required to eject electrons, depends linearly on $\Delta\phi$, the potential difference between metal and solution: $\Phi^{M/S}(\Delta\phi) = \Phi^{M/S}(\Delta\phi = 0) + e\Delta\phi$. If the electrode is part of an electrochemical cell, the electrode–solution potential difference differs from the cell potential by a constant, so

$$\Phi^{M/S}(\mathcal{E}) = \Phi^{M/S}(0) + e\mathcal{E} \qquad [10]$$

where $\Phi^{M/S}(0)$ is the work function when the cell potential is zero. Of course $\Phi^{M/S}$ differs from Φ (work function for emission to vacuum) because the final state of the electron after emission is very different in the two situations (Pleskov and Rotenberg 1978).

The photocurrent is zero if the frequency of the light is below a threshold frequency ν_0, where $h\nu_0$ is equal to the energy of the solvated electron minus the energy of the Fermi level in the interior of the metal. For frequencies above the threshold, the photocurrent varies as

$$I = A(h\nu - h\nu_0)^{5/2} \qquad [11]$$

where A is a constant. (This differs from the law for photoemission to vacuum, $I = A'(\nu - \nu_0)^2$, because the long-range image forces between emitted electron and electrode are screened off by the ions of the electrolyte solution.) Since the work function $\Phi^{M/S}$ is identical to $h\nu_0$, equation [11] may be combined with equation [10] to give

$$I = A(h\nu - h\nu_0^0 - e\mathcal{E})^{5/2} \qquad [12]$$

where ν_0^0 is the threshold frequency at $\mathcal{E} = 0$. If j is the photocurrent density, equation [12] shows that a plot of $j^{2/5}$ vs. $\mathcal{E}$ should be linear. This is observed to hold except when the thickness of the diffuse layer exceeds the De Broglie wavelength of the emitted electron or when the electrode reaction of the reduced scavenger is slow.

The threshold frequency for photoelectron emission from a metal to a vacuum depends on the metal, since different metals have different Fermi energies. For a metal in contact with a solution, the Fermi energy may be changed by changing the metal's potential, which is done by changing $\mathcal{E}$. In fact, two chemically different metals at the same potential relative to the same standard electrode have the same

Fermi energy relative to solution. In this sense, the photoemission threshold depends on the electrode potential, not on the nature of the metal. The difference between metals is reflected in the structure of the interface, but the interface affects the rate of photoemission (since it constitutes a barrier through which electrons must tunnel–see Chapter 9, Section C) and not the energetics. Thus plots of $j^{2/5}$ vs. $\mathcal{E}$ for different metals are straight lines that intersect at $\mathcal{E} = h(\nu - \nu_0^0)/e$. Even adsorption on the metal surface does not affect this result, since it affects only the structure of the interface.

For quanta of light with energy $h\nu = 3.4$ eV, the potential of the metal corresponding to threshold is -0.05 ± 0.05 V against a saturated calomel electrode (SCE). Thus $h\nu_0^0 = 3.4 \pm 0.1$ eV and the work function depends on the potential according to

$$\Phi^{M/S} = h\nu_0 = 3.4 \text{ eV} + e\mathcal{E}_{cal} \qquad [13]$$

where $\mathcal{E}_{cal}$ is the electrode potential against an SCE. Equation [13] holds for all metals in contact with aqueous solution. It allows one to obtain the work function for any potential. Then, for mercury at its pzc ($\mathcal{E}_{cal} = -0.43$ V), the work function for emission to the solution is 3.0 ± 0.1 eV. This is to be compared to the work function for emission of electrons from the (neutral) metal to vacuum, 4.5 eV. The emitted electron is thus stabilized by 1.5 eV in the solution relative to vacuum by the electron's interaction with the solvent, and by the lower potential of the solvent.

The potential of solvent relative to vacuum is about 0.15 V (surface potential of the free water surface) so the solvent–electron interaction is -1.35 eV, which is less than the electron hydration energy. The implication is that the electron is emitted into a state ("dry electron") that differs from the state of a fully solvated electron. Emitted electrons are supposed to be moving so fast that the electronic polarization of the medium, but not the orientational polarization, can react to them. As they become thermalized or slowed down, hydration becomes possible (Pleskov and Rotenberg 1978, Section 4.4).

For dilute solutions, the Debye length of the diffuse layer may become larger than the De Broglie wavelength of the emitted electron. If ψ', the potential across the diffuse layer, is negative, and if the emitted electron has energy less than $-e\psi'$, the electron cannot travel through the diffuse layer. Then the potential inside the diffuse layer (at the outer Helmholtz plane), rather than the potential in bulk solution outside the diffuse layer, determines the final state of the electron. Instead of equation [12], we have

$$I = A(h\nu - h\nu_0^0 - e\mathcal{E} + e\psi')^{5/2}$$

A comparison of photocurrent curves for concentrated and dilute solutions allows the direct determination of ψ'. Results of such measurements (actually, for the difference of ψ' values at solution concentrations of 0.01 and 0.001 N and for a range of potentials) were in good agreement with the Gouy–Chapman theory.

Another effect of negative ψ' is to retard the return of hydrated electrons to the electrode. As noted above, scavengers normally trap some or all of the electrons before their return to the electrode; how well they do this determines how large a photocurrent can be attained at steady state. Negative ψ' allows one to attain the limiting photocurrent with lower concentrations of scavengers. This effect also may be used to determine the absolute value of ψ'; results for acid solutions of 0.1–0.001 N agreed with Gouy–Chapman theory.

At the pzc, if there is no specific adsorption, $\psi' = 0$, and all effects of electrolyte concentration on photocurrent which work through ψ' should vanish. Thus the photocurrent curves ($j^{2/5}$ vs. $\mathcal{E}$) for different electrolyte concentrations should cross at $\mathcal{E}$ equal to the pzc, as is approximately the case. Values of the point of zero charge determined in this way agree with values determined from capacitance measurements; the precision of the two methods is the same. Specific adsorption changes ψ', but also has other effects: By covering the electrode surface, specifically adsorbed ions may decrease the photocurrent or perhaps may change the quantum yield. This makes it impossible to use the photoemission to measure the changes in ψ' with specific adsorption (Pleskov and Rotenberg 1978, Section 5.5).

In certain solvents, electrons do not react rapidly, so they can be stabilized, and exist at equilibrium. Then an electrochemical potential can be defined for them in the usual way. For an electrically neutral solvent phase, one has

$$\tilde{\mu}_{e^-}^S = \mu_{e^-}^S - e\chi^S \qquad [14]$$

since there is likely to be a surface potential χ^S. The chemical potential $\mu_{e^-}^S$ of a solvated electron may be written as a sum of several contributions (Trasatti 1980, Section 3.1):

$$\mu_{e^-}^S = \mu_{e(m)}^S + \Delta G_{in} - \Delta G_{or} \qquad [15]$$

to emphasize the fact that the electron orients the solvent molecules around it. Here, $\mu_{e(m)}$ includes the interaction of the electron with surrounding solvent only by electronic polarization, corresponding to mobile electrons moving too fast for nearby solvent molecules to reorient. The reorganization of the solvent molecules to form a potential well around the electron decreases the free energy by ΔG_{or}, and the interaction of the electron with the polarized solvent molecules increases the free energy by ΔG_{in}.

The chemical potential $\mu_{e^-}^S$ depends on the electronic activity like that of other solvated species:

$$\mu_{e^-}^S = \mu_{e^-}^{0S} + kT \ln a_e$$

The value of μ_e^{0S} for water is derived from a thermodynamic cycle involving the reaction

$$H(H_2O) \rightarrow H^+(H_2O) + e^-(H_2O)$$

taking into consideration the surface potential χ^S of 0.13 V. The result is that $\mu_e^{0S} = -1.44$ eV (Trasatti 1980; Pikaev 1971, Section 1.10). Since μ_e^{0S} refers to a standard state of 1 mole of electrons per liter, one may calculate μ_e^{*S} for transfer of a single electron into a large volume of solvent by adding the chemical potential of an ideal gas of electrons, $kT \ln[(2\pi mkT/h^2)^{3/2} V/N]$. The result is -1.34 eV, which, corrected for the surface potential, gives the equilibrium electron work function of the solution: $\Phi^S = -\mu_e^{*S} + e\chi^S = 1.47$ eV. For a delocalized (mobile) electron in solution, not interacting with a potential well of oriented solvent molecules, the work function would differ from Φ^S by $\Delta G_{or} - \Delta G_{in}$ and perhaps would have to be corrected for the kinetic energy of the mobile electrons. A third work function is measured by the photoelectric method, since an electron is ejected from an oriented-solvent potential well but the process is too fast for the solvent to relax. This nonequilibrium work function is

$$_{ne}\Phi^S = -\mu_e^{*S} + \Delta G_{or}^* + e\chi^S = -\mu_{e(m)}^{*S} - \Delta G_{in}^* + e\chi^S \quad [16]$$

where $\mu_{e(m)}$ is for mobile electrons. Here ΔG_{in}^* and ΔG_{or}^* refer to infinite dilution,

$$_{ne}\mu_{e(m)}^{*S} = \mu_e^S - \Delta G_{or}^*$$

where the ne stands for nonequilibrium and the m stands for mobile.

D. THEORIES OF THE HYDRATED ELECTRON

Electrons produced by irradiation of aqueous solutions may become hydrated, that is to say, stabilized or trapped in a cavity in the solution by their interaction with the polarized solvent outside the cavity. An early theoretical description of such an electron was given by Jortner (1959, 1964). Inside the cavity, the dielectric constant is unity; outside it has the value ϵ_s, the static dielectric constant, since the solvated electron is supposed to move slowly enough so that the solvent can polarize and reorient around it. The electron is supposed to be described by a one-electron wave function, which is the solution to a Schrödinger equation that includes the polarization interaction with the solvent in the potential energy function. The interaction potential between electron and solvent is then derived self-consistently.

Suppose one has a spherically symmetrical charge distribution $c(r)$ (which will eventually be $-e$ times $|\psi|^2$, where ψ is the electronic wave function) in a cavity of radius R. If the dielectric constant ϵ depends on r alone, Poisson's equation gives the electrostatic potential in electrostatic units as

$$\phi(r) = \phi(0) - 4\pi \int_0^r dr'(r')^{-2} \int_0^{r'} \frac{dr''(r'')^2 c(r'')}{\epsilon(r'')}$$

6. FURTHER DEVELOPMENTS

Inside the cavity ($r < R$), where $\epsilon = 1$,

$$\phi(r) = \phi(0) - 4\pi \int_0^r dr'(r')^{-2} \int_0^{r'} dr''(r'')^2 c(r'') \qquad [17]$$

and outside ($r \geq R$), where the dielectric constant has the constant value ϵ_s,

$$\phi(r) = \phi(0) - 4\pi \int_0^R dr'(r')^{-2} \int_0^{r'} dr''(r'')^2 c(r'')$$

$$- 4\pi \int_R^r dr'(r')^{-2} \left[\int_0^R dr''(r'')^2 c(r'') + \int_R^{r'} \frac{dr'' r''^2 c(r'')}{\epsilon_s} \right] \qquad [18]$$

where $\phi(r)$ includes the potential of $c(r)$ as well as the electrical potential of the polarized medium.

To find the latter, we may use the formula (end of Chapter 2, Section H) for the potential of a polarized body:

$$\phi^P = \int [\mathbf{P}(\mathbf{r}') \cdot (\mathbf{r} - \mathbf{r}')] |\mathbf{r}' - \mathbf{r}|^{-3} d^3r'$$

The polarization depends only on the distance from the origin and is zero for $r' < R$. Then, if θ is the angle between $\mathbf{r}$ and $\mathbf{r}'$,

$$\phi^P = 2\pi \int_R^\infty dr'(r')^2 P(r') \int \sin\theta \, d\theta (-r' + r\cos\theta) [(r')^2 + r^2 - 2rr'\cos\theta]^{-3/2}$$

$$= 2\pi \int_R^\infty dr' \, P(r') \left(-1 + \frac{r - r'}{|r - r'|} \right)$$

Here, $\mathbf{P} = \mathbf{D}(1 - (\epsilon_s)^{-1})/4\pi$ (Chapter 2, Section H, equation [71]), where the displacement $\mathbf{D}$ is just the field of the charge distribution, that is,

$$D(r') = 4\pi(r')^{-2} \int_0^{r'} dr''(r'')^2 c(r'')$$

Thus we have for the potential due to the medium

$$\phi^P = [1 - (\epsilon_s)^{-1}] 2\pi \int_R^\infty dr' \left(-1 + \frac{r - r'}{|r' - r|} \right) (r')^{-2} \int_0^{r'} dr''(r'')^2 c(r'')$$

Inside the cavity ($r < R$), we get

$$\phi^P = 4\pi[1 - (\epsilon_s)^{-1}] \int_R^\infty dr'[(-r')^2]^{-1} \int_0^{r'} dr''(r'')^2 c(r'')$$

$$= -4\pi[1 - (\epsilon_s)^{-1}] \left[R^{-1} \int_0^R dr'(r')^2 c(r') + \int_R^\infty dr' r' c(r') \right] \quad [19]$$

on integration by parts. Outside the cavity the potential due to the medium is

$$\phi^P = 2\pi[1 - (\epsilon_s)^{-1}] \int_r^\infty dr'(-2)(r')^{-2} \int_0^{r'} dr''(r'')^2 c(r'')$$

$$= -4\pi[1 - (\epsilon_s)^{-1}] \left[r^{-1} \int_0^r dr'(r')^2 c(r') + \int_r^\infty dr'' r'' c(r'') \right] \quad [20]$$

Because the potential ϕ^P is induced by the charge density $c(r)$, the interaction energy of $c(r)$ with the medium is $\frac{1}{2} \int \phi^P c d\tau$, as can be seen by imagining the charge of c to increase gradually from 0 to its full value. Thus, the effective potential $V(r)$ for the electronic Schrödinger equation is $\frac{1}{2}(-e) \phi^P$. Replacing $c(r)$ by $-e|\psi|^2$, we have from equations [19] and [20]

$$V(r) = -\frac{e^2}{2} 4\pi[1 - (\epsilon_s)^{-1}] \left[r^{-1} \int_0^r dr'(r')^2 |\psi(r')|^2 \right.$$

$$\left. + \int_r^\infty dr' r' |\psi(r')|^2 \right] \quad [21]$$

and $V(r) = V(R)$ for $r < R$. This potential enters the Schrödinger eigenvalue equation,

$$-\frac{\hbar^2}{2m} \nabla^2 \psi + V\psi = E\psi \quad [22]$$

which determines ψ. Since V depends on ψ, the wave function is to be consistent with the potential: One requires that the solution to equation [22] be identical to the wave function that is used to construct V (self-consistency). We note that $V(r)$ is always negative [since $(\epsilon_s)^{-1} < 1$] for all r, while

$$\frac{dV}{dr} = 2\pi e^2 [1 - (\epsilon_s)^{-1}] r^{-2} \int_0^r dr'(r')^2 |\psi(r')|^2 > 0 \quad [23]$$

for $r > R$ (and 0 for $r \leq R$). Thus $V(r)$ is an attractive potential. From equations

[21] and [23] it also appears that V becomes less attractive as ψ becomes more diffuse. The potential V tends to confine the electron to the cavity, while the kinetic energy, which is represented in the Schrödinger equation by the operator $T = -(\hbar^2/2m)\nabla^2$, tends to favor a more spread-out wave function.

One may find approximate solutions to equation [22] by variation, that is, one varies a trial function ψ_t to minimize $\int \psi_t^*(T + V)\psi_t d\tau$, with $\int |\psi_t|^2 d\tau = 1$ (normalization) and V given by equation [21]. The value of the integral of $(T + V)$ is an upper bound to the exact energy, E of equation [22], and may even be a reasonable approximation to E when ψ_t is not a good approximation to ψ.

For example, consider the trial function $\psi_a = N(r + a)^{-2}$, where a is a variational parameter and N is a normalization constant. This function leads to simple integrals, which are easily evaluated by integration by parts. To have $4\pi \int_0^\infty r^2(\psi_a)^2 dr = 1$ (normalization), N^2 must equal $3a/4\pi$. The kinetic energy is

$$-\frac{\hbar^2}{2m} 4\pi N^2 \int_0^\infty dr\, r^2(r + a)^{-2} r^{-2} \frac{d}{dr} r^2 \frac{d}{dr}(r + a)^{-2}$$

$$= -\frac{e^2 a_0}{2} 3a \int_0^\infty dr\,(r + a)^{-2}[2r(r + a)^{-3} - 3r^2(r + a)^{-4}] = \frac{e^2 a_0}{5a^2} \quad [24]$$

where we have introduced $a_0 = \hbar^2/me^2 = 0.529 \times 10^{-8}$ cm, the Bohr radius. The polarization potential corresponding to this wave function is

$$V(r) = -\tfrac{1}{2}e^2[1 - (\epsilon_s)^{-1}]4\pi\left[\frac{1}{r}\int_0^r dr'\frac{N^2(r')^2}{(r' + a)^4} + \int_r^\infty dr'\frac{N^2 r'}{(r' + a)^4}\right]$$

$$= -\tfrac{1}{4}e^2[1 - (\epsilon_s)^{-1}](2r + a)(r + a)^{-2} \quad \text{for } r \geq R$$

$$V(r) = -\tfrac{1}{4}e^2[1 - (\epsilon_s)^{-1}](2R + a)(R + a)^{-2} \quad \text{for } r < R$$

The expectation value of $V(r)$ over the wave function is

$$4\pi \int (\psi_a)^2 V r^2\, dr$$

$$= 4\pi N^2(-\tfrac{1}{4}e^2)[1 - (\epsilon_s)^{-1}][(2R + a)(R + a)^{-2}$$

$$\times \int_0^R r^2\,dr\,(r + a)^{-4} + \int_R^\infty r^2\,dr\,(2r + a)(r + a)^{-2}(r + a)^{-4}]$$

$$= -e^2[1 - (\epsilon_s)^{-1}](\tfrac{1}{2}R^4 + aR^3 + a^2R^2 + \tfrac{1}{2}a^3R + \tfrac{1}{10}a^4)(R + a)^{-5} \quad [25]$$

Consider first $R = 0$ (no cavity): The energy from equations [24] and [25] is $e^2 a_0/5a^2 - (e^2/10)[1 - (\epsilon_s)^{-1}]/a$. Minimizing this with respect to a we find

D. THEORIES OF THE HYDRATED ELECTRON

the optimum value of a is $4a_0[1 - (\epsilon_s)^{-1}]^{-1}$, which makes the minimum energy $(-1/80) [1 - (\epsilon_s)^{-1}]^2 (e^2/a_0)$. Since $e^2/a_0 = 27.21$ eV (1 atomic unit), this means an energy of -0.34 eV if $1 - (\epsilon_s)^{-1}$ is approximated by unity.

This is a first approximation to the hydration energy of the electron. From a thermodynamic cycle for the reaction $e^-(aq) + H^+(aq) \to \frac{1}{2}H_2(aq)$, a value of -1.7 eV has been derived for this quantity. Changing the value of ϵ_s would make our variational energy less negative and increase the discrepancy with the experimental hydration energy. Increasing the cavity radius R would do the same, as can be seen as follows: Rewrite the potential energy as $-\{(e^2/10a) [1 - (\epsilon_s)^{-1}]\} \{1 - [R/(R+a)]^5\}$. For a given value of a, the derivative with respect to R is then $(e^2/2a) [1 - (\epsilon_s)^{-1}] aR^4/(R+a)^6$, which is positive. Thus increasing the cavity radius can only raise the energy.

To get closer to the experimental hydration energy, one can use a better variational function. The form of ψ_a is not particularly suited to a problem involving an electrostatic potential; an exponential function of r, such as in atomic orbitals, is more appropriate. In fact, one may use the hydrogenic wave function $\psi_t = (a^3/\pi)^{1/2} e^{-ar}$, which obeys the normalization condition, and vary a. The integrals in equation [21] are straightforward, giving

$$V(r) = -\tfrac{1}{2} e^2 [1 - (\epsilon_s)^{-1}] [r^{-1} - e^{-2ar}(a + r^{-1})], \qquad r \geq R \quad [26]$$

$$V(r) = -\tfrac{1}{2} e^2 [1 - (\epsilon_s)^{-1}] \left[R^{-1} - e^{-2ar}(a + R^{-1}) \right], \qquad r < R$$

The expectation value of this potential is then

$$\int \psi_t V \psi_t \, d\tau = -\tfrac{1}{2} e^2 [1 - (\epsilon_s)^{-1}] \Bigg\{ 4\pi \int_0^R dr\, r^2 \frac{a^3}{\pi}$$

$$\cdot e^{-2ar} [R^{-1} - e^{-2ar}(a + R^{-1})] + 4\pi \int_R^\infty dr\, r^2 \frac{a^3}{\pi}$$

$$\cdot e^{-2ar} [r^{-1} - e^{-2ar}(a + r^{-1})] \Bigg\}$$

$$= -2e^2 [1 - (\epsilon_s)^{-1}] \left[\frac{1}{4R} + e^{-2aR} \left(\frac{-a}{2} + \frac{-1}{2R} \right) \right.$$

$$\left. + e^{-4aR} \left(\frac{R^2 a^3}{4} + \frac{5Ra^2}{8} + \frac{21a}{32} + \frac{1}{4R} \right) \right] \quad [27]$$

which is to be added to the expectation value of the kinetic energy, $\tfrac{1}{2}\hbar^2 a^2/m$, to get the total energy, which is then to be minimized with respect to a.

For $R = 0$ (no cavity) the calculation is simple. The energy is

$$-2e^2 [1 - (\epsilon_s)^{-1}] \frac{5a}{32} + \frac{\hbar^2 a^2}{2m} = \frac{-5e^2}{16} [1 - (\epsilon_s)^{-1}] a + \frac{e^2 a^2 a_0}{2}$$

Differentiating with respect to a and setting the expression equal to zero gives $a = 5[1 - (\epsilon_s)^{-1}] (16a_0)^{-1}$ and an energy equal to $-(25/512) [1 - (\epsilon_s)^{-1}]^2 (e^2/a_0)$. With $e^2/a_0 = 27.2$ eV this gives the interaction energy of the electron with the solvent as -1.33 eV if $(\epsilon_s)^{-1}$ is negligible compared to unity. Again, a finite value of ϵ_s would make the predicted interaction energy less negative.

If one now considers a cavity of radius R, the energy to be minimized with respect to a is

$$-e^2[1 - (\epsilon_s)^{-1}] \left[(2R)^{-1} - e^{-2aR}(a + R^{-1}) \right.$$
$$\left. + e^{-4aR} \left(\frac{R^2 a^3}{2} + \frac{5Ra^2}{4} + \frac{21a}{16} + \frac{1}{2R} \right) \right] + \frac{e^2 a^2 a_0}{2}$$

Because $dV/dr > 0$ for $r > R$ and $dV/dr = 0$ for $r < R$, one can anticipate that an increase in the cavity radius leads to a less negative energy. In fact, the occurrence of R in exponentials means that the energy is quite insensitive to R: For $R = 1a_0$, the best values of a and energy, are 0.315 and $-0.04875 e^2/a_0$, respectively (compare 0.3125 and -0.04883 for $R = 0$). One could look for better trial functions, with more than one variational parameter, but it turns out that the energy cannot be lowered much. The problem is in the model.

The same conclusion follows from consideration of the electronic excitation energy, calculated by supposing that the excited state has a wave function resembling a $2p$ atomic orbital, with the ground state resembling a $1s$ atomic orbital. The normalized $2pz$ wave function is $\psi_{2p} = (b^5/\pi)^{1/2} r e^{-br} \cos\theta$, where b is a variable parameter. Its energy is calculated as the kinetic energy $b^2 e^2/8a_0$ plus the potential energy $\int (\psi_{2p})^2 V\, d\tau$, where V is given by equation [26], that is, the potential is due to the solvent polarized by an electron in the *ground* state. This is in accordance with the Franck–Condon principle, which states that the heavy particles (solvent dipoles in the present case) do not move during an electronic transition. The potential energy of the excited state is then

$$\int (\psi_{2p})^2 V\, d\tau = -\tfrac{1}{2} e^2 [1 - (\epsilon_s)^{-1}] \frac{4\pi}{3} \frac{b^5}{\pi}$$
$$\cdot \left\{ \int_0^R dr\, r^2 [R^{-1} - e^{-2aR}(a + R^{-1})] r^2 e^{-2br} \right.$$
$$\left. + \int_R^\infty dr\, r^2 [r^{-1} - e^{-2ar}(a + r^{-1})] r^2 e^{-2br} \right\}$$

Minimizing the total energy with respect to b one obtains an excited state energy. The electronic transition frequency is the energy difference between ground and excited states, divided by h, and is too small.

D. THEORIES OF THE HYDRATED ELECTRON   249

One may argue that the *electrons* in the solvent are able to rearrange during the electronic transition, so that an additional term,

$$-\frac{1}{2} e^2 \frac{1 - (\epsilon_{op})^{-1}}{1 - (\epsilon_s)^{-1}} \left[\int (\psi_{2p})^2 V_{2p} \, d\tau - \int (\psi_{2p})^2 V_{2s} \, d\tau \right]$$

where ϵ_{op} is the optical dielectric constant (due to electronic polarization only), should be added to the energy of the $2p$ state. This term represents the energy change due to repolarization of the solvent electrons from a distribution appropriate for a $1s$ electron in the cavity to a distribution appropriate for a $2p$ electron. With this correction, a transition energy of 1.35 eV is found, which is about 20% lower than the experimental value. Increasing the cavity radius lowers the predicted transition energy.

To go beyond the continuum model for the solvent, one can treat the solvent molecules of the first solvation shell explicitly and use the continuum model for the molecules beyond this shell, somewhat as is done in the Onsager theory of the dielectric constant (Chapter 3, Section D). Such a theory was given by Fueki, Feng, and Kevan (1973). Each solvent molecule is supposed to have a dipole moment μ_d and an electronic polarizability α.

If R_d represents the distance from the first solvent shell to the center of the electron density, the total electronic charge within this shell is

$$C_d = -4\pi e \int_0^{R_d} |\psi_i|^2 r^2 \, dr \qquad [28]$$

where ψ_i is the electron's wave function. The electric field due to this charge is $E_d = C_d/(R_d)^2$, in the radial direction. A dipole moment αE_d is induced in the electron cloud of each solvent molecule in the shell, so that there is an interaction energy with the central electron cloud of $\frac{1}{2}\alpha(C_d)^2/(R_d)^4$. Similarly, the permanent dipole of each solvent molecule is oriented by the field of the central electron so that its magnitude in the radial direction is $\mu_d \langle \cos \theta \rangle$, where $\langle \cos \theta \rangle$, according to the Langevin equation, is $\coth \chi - \chi^{-1}$, where $\chi = \mu_d E_d/kT$. There is then an interaction energy between the central electron and the permanent dipole of $-\mu_d \langle \cos \theta \rangle C_d/(R_d)^2$. The two field–dipole interactions are to be added together and multiplied by N, the number of molecules in the solvation shell at R_d.

To these attractive energies, that stabilize the hydrated electron, one must add several short-range repulsive interactions. These interactions arise from exchange, or overlap between the central electron density and the electrons of the solvent molecules, from interaction between the dipoles on different molecules in the solvent shell, and from the rearrangement of solvent required to form the solvent-free cavity of size R_d. The last term is written as the surface tension multiplied by the difference in surface area between the cavity of radius R_d and a cavity the size of a solvent molecule: $\gamma[4\pi(R_d)^2 - 4\pi(R_s)^2]$. The exchange repulsion is represented very crudely, and involves an adjustable parameter, chosen so that the 1s–2p transition energy has its correct value. The solvent outside a radius $R_d + R_s$, that is, beyond the first solvation shell, is supposed to be represented by a continuous dielectric, characterized by the dielectric constants ϵ_d and ϵ_{op}. Its interaction energy with the central electron is then given by expressions such as those given above (equations [25]–[28]).

The total energy of the hydrated electron, calculated with a hydrogenlike 1s wave function, is then minimized with respect to the parameter in the wave function for a given value of R_d. Repeating the calculation for various values of R_d, one obtains the energy of an electron in the ground state as a function of R_d, so that the value of R_d is chosen by the model, as the value giving the lowest energy. The calculation is performed for the 2p excited state as well as for the 1s ground state and for N, the number of dipoles in the first solvation shell, equal to either 4 or 6. The cavity radius R_d comes out to be ~ 1.9 Å for $N = 4$ and ~ 2.5 Å for $N = 6$. The resulting electron hydration energies for the ground state are higher than the experimental value of 1.7 eV by several tenths of an electron volt. The discrepancy is explained in terms of the energy required to break hydrogen bonds in the first hydration shell, which the model does not include. The 1s–2p excitation energy is also a bit higher than the experimental value.

Several workers have given structural models for the hydrated electron, which avoid the use of the continuum entirely. In such a model, one considers explicitly the arrangement of the individual water molecules that coordinate the electron. The orientation and arrangement of these molecules result from hydrogen bonding and the presence of the electron that is solvated. Evidence from electron spin resonance is that the electron is within a polyhedron of oriented water molecules. Thus a possible model (Natori and Watanabe 1966) considers a tetrahedron of water molecules, with their oxygen atoms at the vertices of the tetrahedron, and one hydrogen atom on each water oriented toward the central electron. The potential in which the electron moves is taken as a sum of potentials due to the 12 atoms, each a function of the distance between the electron center and the atom and hence dependent on the orientation and position of the water molecules, plus a potential representing the electron's interaction with water molecules outside the solvation shell constituted by the first four molecules. Each of the atomic potentials is a sum of the electrostatic potentials due to the nucleus and the 1s, 2s, and 2p electrons, although this neglects (1) the change of the electron density due to bonding within water molecules and (2) the short-range exchange repulsion between the central electron and the electrons of water molecules.

The Schrödinger equation for an electron moving in the field described above was treated variationally by Natori and Watanabe (1966). The variational wave function was a linear combination of four 1s atomic orbitals, that is, $(Z^3/\pi) \exp(-z|r - r_i|/a_0)$, on the hydrogen atoms pointing toward the central electron, with Z, as well as the linear combination coefficients, being treated as variational parameters. This produces four solutions, each a combination of the four atomic orbitals. The solution of lowest energy represents the ground electronic state of the hydrated electron. The other three solutions (which are degenerate in energy be-

D. THEORIES OF THE HYDRATED ELECTRON 251

cause of the tetrahedral symmetry) represent the excited state, corresponding to the $2p$ orbital of previously mentioned treatments. The transition energy from this calculation is less than half of the experimental value, however. To estimate the hydration energy, the calculated electronic energy is corrected for electronic polarization of the water molecules and for hydrogen-bond breakage. The value of the hydration energy, too, is not in good agreement with experiment. However, with a better wave function for the hydrated electron, in particular, one orthogonalized to the wave functions of the water molecules, excellent agreement is obtained for the hydration energy (Natori 1964).

A calculation by Ray (1971) considered four water molecules at the corners of a tetrahedron and a fifth, with or without the extra electron (H_2O^-) at the center. Note that this calculation is based on a model in which the hydrated electron is no longer within a cavity, but instead associated with a water molecule. Only the electronic structure of the central molecule was considered explicitly, the others being represented by dipoles. The use of a dipole moment of 3.8 D for each of these (instead of 1.8 D for a free water molecule) was intended to simulate the effect of correlations between water molecules: We have noted (Chapter 2, Section F) that the value of the dielectric constant of water is consistent with orienting dipoles of much larger size than those of single water molecules, that is, to clusters moving together. The electronic Hamiltonian for the central H_2O or H_2O^- is that for an isolated H_2O or H_2O^- plus the electrostatic potential of four dipole moments μ_n. Thus, for the ith electron (of 10 electrons for H_2O, 11 for H_2O^-), the additional term is $-e \sum_j \mu_n \cos \theta_j |\mathbf{r}_j - \mathbf{r}_i|^2$, where j runs from 1 to 4, and θ_j is the angle between the direction of the jth dipole, located at $\mathbf{r}_j$, and the vector $\mathbf{r}_j - \mathbf{r}_i$. Self-consistent field molecular orbital calculations were carried out to determine five molecular orbitals for H_2O (each doubly occupied in the ground state) or six molecular orbitals for H_2O^- (one singly occupied). The largest difference between the energies of H_2O and H_2O^- was found with the surrounding dipoles oriented toward the center. The value of this energy difference, 1.79 eV, is quite close to the experimental binding energy of the hydrated electron, 1.7 eV.

One would like to have quantum mechanical calculations of electronic energy for a system of a large number of water molecules, with and without an extra electron, and for a series of arrangements of the nuclei. From the electronic energy as a function of nuclear configuration, one would obtain the configuration corresponding to equilibrium, as well as the force constants for distortions of this structure. From the energies with and without the extra electron, one would obtain the hydration energy. The full program is, however, too difficult to carry out at present, and compromise has been necessary. In some cases, semiempirical calculations, in which only valence electrons are considered and certain energy quantities are assigned values instead of being calculated, have been performed instead of ab initio calculations, in which all integrals are evaluated exactly. The number of water molecules has often been limited to four or six, and only a small number of molecular configurations are considered. Such calculations are reviewed by Bockris and Khan (1979, Section 11.4).

7

DIFFUSION

A. LAWS OF DIFFUSION

At equilibrium, the electrochemical potential of a substance must be the same throughout the system. Otherwise, there will be a tendency for the substance to move from regions of higher electrochemical potential to regions of lower electrochemical potential, although the rate at which this occurs may be very slow. One can expect the substance's flow velocity to be proportional to the gradient of the electrochemical potential. If the medium in which the substance is found is itself moving, there will also be a flow of the substance due to the motion of the medium. We write for $\mathbf{j}_i$, the material flux of substance i at a point in space,

$$\mathbf{j}_i = c_i \mathbf{v} - \frac{c_i D_i (\nabla \tilde{\mu}_i)}{RT} \qquad [1]$$

where $\mathbf{v}$ (a vector) is the velocity of motion of the fluid at the point, c_i is the concentration of i, D_i is the diffusion coefficient, and $\nabla \tilde{\mu}_i$ is the gradient of the electrochemical potential. Fick's first law of diffusion, $\mathbf{j}_i = -D_i \nabla c_i$, follows from equation [1] if (a) $\mathbf{v} = 0$, (b) the species i is uncharged, and (c) the solution is ideal so $\mu_i = \mu_i^0 + RT \ln c_i$.

Expressing the electrical part of $\tilde{\mu}_i$ in terms of the inner potential of the phase as

$$\tilde{\mu}_i = \mu_i + z_i \mathcal{F} \phi$$

A. LAWS OF DIFFUSION

and replacing $\nabla \mu_i$ by $(\partial \mu_i / \partial c_i) \nabla c_i$, we obtain

$$\mathbf{j}_i = c_i \mathbf{v} + \frac{D_i z_i \mathfrak{F} c_i}{RT}(-\nabla \phi) - \frac{c_i D_i (\partial \mu_i / \partial c_i) \nabla c_i}{RT} \qquad [2]$$

The first term represents the contribution of convection, the second term represents the contribution of migration ($-\nabla \phi$ is the electric field so

$$\frac{D_i |z_i| \mathfrak{F}}{RT} = u_i \qquad [3]$$

is the mobility), and the third term represents the contribution of diffusion. If $\mu_i = \mu_i^0 + RT \ln c_i$ as for an ideal solute, the third term is $-D_i \nabla c_i$. Of course diffusion is a result of random motions, and the net flux of particles is a consequence of the flux of particles from regions of high concentration exceeding the flux from regions of low concentration: There is no actual force on the particles. However, the effective diffusive force naturally appears, added to the electric field force.

The change per unit time of N_i^V, the number of particles in a volume V of electrolyte, is given by the negative of the surface integral $\int \mathbf{j}_i \cdot d\mathbf{S}$, where $d\mathbf{S}$ is a normal in the outward direction on the surface bounding V. Since $\int \mathbf{j}_i \cdot d\mathbf{S} = \int (\nabla \cdot \mathbf{j}_i) \, dV$ by Gauss' theorem,

$$-\int (\nabla \cdot \mathbf{j}_i) \, dV = \frac{dN_i^V}{dt} = \int \frac{\partial c_i}{\partial t} dV$$

or $\partial c_i / \partial t = -\nabla \cdot \mathbf{j}_i$. Using equation [2] for $\mathbf{j}_i$ we have

$$\frac{\partial c_i}{\partial t} = -\mathbf{v} \cdot \nabla c_i - c_i \nabla \cdot \mathbf{v} + \frac{z_i (c_i D_i \nabla^2 \phi - \mathbf{E} \cdot \nabla D_i c_i) \mathfrak{F}}{RT} + \frac{\nabla \cdot (c_i D_i \nabla \mu_i)}{RT}$$

For an incompressible fluid, $\nabla \cdot \mathbf{v}$ vanishes. If we further assume ideal solute behavior so that $\nabla D_i = 0$ and $\nabla \mu_i / RT = (c_i)^{-1} \nabla c_i$, the equation simplifies to

$$\frac{\partial c_i}{\partial t} = -\mathbf{v} \cdot \nabla c_i + \frac{z_i D_i (c_i \nabla^2 \phi - \mathbf{E} \cdot \nabla c_i) \mathfrak{F}}{RT} + D_i \nabla^2 c_i \qquad [4]$$

According to the Poisson equation, $\nabla^2 \phi = -\rho / \epsilon$ so that, if there is local electroneutrality, this term also vanishes.

The effective force for diffusion of an individual charged particle should be written as $-\nabla \tilde{\mu}_i / N_A$, dividing by Avogadro's number to get a force per particle.

7. DIFFUSION

In the presence of a field, the particle accelerates until the force of viscous drag, which is proportional to particle velocity, is equal and opposite to the force $-\nabla\tilde{\mu}_i/N_A$. Writing the viscous force as $-F_i\mathbf{v}_i$ to show its dependence on the particle velocity, we have

$$-N_A F_i \mathbf{v}_i = \nabla\tilde{\mu}_i$$

so that the flux due to diffusion and migration (equation [1] with zero velocity $\mathbf{v}$ of the fluid) is

$$\mathbf{j}_i = \frac{-c_i D_i(-N_A F_i \mathbf{v}_i)}{RT} = \frac{c_i D_i F_i \mathbf{v}_i}{kT}$$

Since $\mathbf{j}_i = c_i \mathbf{v}_i$, the diffusion coefficient D_i is kT/F_i. For example, if $F_i = 6\pi\eta r_i$, as for a spherical particle of radius r_i in a continuous medium of viscosity η (Stokes' law: Reif 1965, Section 15-6), we have $D_i = kT/6\pi\eta r_i$. This means that $D_i\eta$ for a particular ion in different solvents is a constant, or, since $u_i = D_i|z_i|\mathcal{F}/RT$, that $u_i\eta$ is the same for a particular ion in different solvents (Walden's rule: Koryta, Dvořák, and Boháčková 1978, p. 86). This is only approximately true because the effective radius of an ion differs in different solvents.

The electrical current density in an electrolyte is obtained as the sum of contributions of different ions:

$$\mathbf{i} = \sum z_i \mathcal{F} \mathbf{j}_i = \sum z_i \mathcal{F} c_i \mathbf{v} + \sum \frac{(z_i)^2 \mathcal{F}^2 D_i c_i \mathbf{E}}{RT}$$
$$- \sum \frac{z_i \mathcal{F} c_i D_i (\partial \mu_i / \partial c_i) \nabla c_i}{RT}$$

where $\mathbf{E}$ is the electrical field and equation [2] has been used. The second term on the right-hand side is the electric field multiplied by the electrolytic conductivity (see Chapter 1, Section J)

$$\kappa = \sum \frac{(z_i)^2 \mathcal{F}^2 D_i c_i}{RT} = \sum |z_i| \mathcal{F} c_i u_i \qquad [5]$$

If $\mathbf{v} = 0$ or if $\sum c_i z_i$ is approximately zero everywhere (local electroneutrality) we may write

$$\mathbf{i} + \sum \frac{z_i \mathcal{F} c_i D_i (\partial \mu_i / \partial c_i) \nabla c_i}{RT} = \kappa \mathbf{E} \qquad [6]$$

In an electrolyte solution, the electric field $-\nabla\phi$ includes the imposed electric field as well as fields due to the ions. Their different diffusion coefficients mean that different ions move at different velocities under a concentration gradient, which sets up regions of nonzero charge density and local electric fields. We see that $\mathbf{E}$ is the sum of $\mathbf{i}/\kappa$, the ohmic electrical field strength that is related to the external electrical field, and a term due to diffusion, which vanishes if there is local electroneutrality and all particles diffuse equally.

B. LIQUID JUNCTION POTENTIAL

At the boundary between two electrolyte solutions 1 and 2, concentration gradients will cause diffusion of different charged particles at unequal rates. This will lead to a transition layer or diffusion layer in which the separation of differently charged ions will produce an electric field. The second term in equation [6] represents the electric field that results:

$$\mathbf{E}_{\text{diff}} = -\nabla \phi_{\text{diff}} = \kappa^{-1} \sum \frac{z_i \mathcal{F} D_i c_i (\partial \mu_i / \partial c_i) \nabla c_i}{RT} \qquad [7]$$

Assuming planar symmetry so that ϕ_{diff} depends on z alone, we may integrate over the diffusion layer to get the liquid junction potential:

$$-(\phi^{(2)} - \phi^{(1)}) = \frac{\mathcal{F}}{RT} \sum z_i \int_1^2 D_i \kappa^{-1} c_i \frac{\partial \mu_i}{\partial c_i} \frac{dc_i}{dz} dz \qquad [8]$$

If the structure of the diffusion layer does not change much during the time that measurements are made on a cell it is reasonable to calculate the liquid junction potential in this way, even though an irreversible process is involved. Treatments based on the thermodynamics of irreversible processes also exist.

It is possible to obtain an explicit result for the potential difference across the junction between two solutions of the same electrolyte at different concentrations if certain assumptions hold. Expressing diffusion coefficients and conductivities in terms of ion mobilities, we have

$$\phi^{(2)} - \phi^{(1)} = -\sum_i \frac{z_i}{|z_i|} \int_1^2 u_i c_i \left(\mathcal{F} \sum c_j |z_j| u_j \right)^{-1} \frac{\partial \mu_i}{\partial z} dz$$

Since the transport number of ion i is $t_i = c_i |z_i| u_i / \sum c_j |z_j| u_j$

$$\phi^{(2)} - \phi^{(1)} = -\int_1^2 \sum_i \frac{t_i}{z_i \mathcal{F}} \frac{\partial \mu_i}{\partial z} dz \qquad [9]$$

The mobilities u_i are functions of concentration as are the transport numbers t_i, but the t_i may vary relatively little, especially if the electrolyte concentrations in the phases 1 and 2 are not too different. If the t_i are taken as constants, the liquid junction potential is simply

$$\phi^{(2)} - \phi^{(1)} = -\sum_i \frac{t_i}{z_i \mathcal{F}} (\mu_i^{(2)} - \mu_i^{(1)}) \qquad [10]$$

For a binary electrolyte, this becomes

$$\frac{RT}{\mathcal{F}} \left(\frac{t_+}{z_+} \ln \frac{a_+^{(2)}}{a_+^{(1)}} - \frac{t_-}{z_-} \ln \frac{a_-^{(2)}}{a_-^{(1)}} \right)$$

where differences of ionic chemical potentials have been expressed in terms of ionic activities. With $z_+ = z_- = 1$, replacing $a_+^{(2)}/a_+^{(1)}$ and $a_-^{(2)}/a_-^{(1)}$ by ratios of mean ionic activity coefficients yields the expression derived thermodynamically in Chapter 1, Section G. If $t_+ > t_-$, the solution of higher concentration has the lower electrical potential and vice versa; there is no liquid junction potential if $t_+ = t_- = \frac{1}{2}$.

In a more general expression, such as equation [9], one sees that, when solution 2 is more concentrated than solution 1, cations make a negative contribution to the potential difference whereas anions ($z_i < 0$) make a positive contribution. Diffusion of cations out of the more concentrated solution will tend to give it a lower electrical potential and vice versa for anions.

To integrate equation [9] in general, one must have information about how transport numbers and other properties vary through the interface (note that this is a quasithermodynamic approach, which assumes that one can define, at each value of z within the interface, values for properties like chemical potential and transport number, which really refer to bulk phases). For example, Henderson made the following assumptions (Koryta, Dvořák, and Boháčková 1970, pp. 161–162): (a) All the concentrations vary together, as if at each value of z one had a certain mixture of the two homogeneous solutions, (b) the chemical potentials obey the ideal solute law, and (c) ionic mobilities are the same throughout, so transport numbers vary only due to the changing concentrations. According to (a), all concentrations vary according to

$$c_i = (1 - x) c_i^{(1)} + x c_i^{(2)} \qquad [11]$$

so everything depends on the single variable x, whose value goes from 0 to 1 through the interface. The chemical potentials for a dilute or ideal solution are

$$\mu_i = \mu_i^0 + RT \ln c_i$$

and one has for the potential difference

$$\phi^{(2)} - \phi^{(1)} = -\frac{RT}{\mathcal{F}} \sum_i \frac{|z_i| u_i}{z_i} \int \frac{(dc_i/dz)\, dz}{\sum_j |z_j| u_j c_j}$$

$$= -\frac{RT}{\mathcal{F}} \sum_i \frac{|z_i| u_i}{z_i} \int_0^1 \frac{(c_i^{(2)} - c_i^{(1)})\, dx}{a + bx}$$

with $a = \sum |z_j| u_j c_j^{(1)}$ and $b = \sum |z_j| u_j (c_j^{(2)} - c_j^{(1)})$. The integration is easy and one finds

$$\phi^{(2)} - \phi^{(1)} = -\frac{RT}{\mathcal{F}} \frac{\sum |z_i| u_i (c_i^{(2)} - c_i^{(1)})/z_i}{\sum |z_j| u_j (c_j^{(2)} - c_j^{(1)})} \ln \frac{\sum |z_j| u_j c_j^{(2)}}{\sum |z_j| u_j c_j^{(1)}}$$

For a uni-univalent electrolyte on both sides of the junction, this potential difference is

$$\phi^{(2)} - \phi^{(1)} = -\frac{RT}{\mathcal{F}} \frac{u_+ - u_-}{u_+ + u_-} \ln \frac{c^{(2)}}{c^{(1)}} \qquad [12]$$

which is identical to equation [10] if the activity ratio is replaced by the concentration ratio; in this case assumption [11] is identical to the assumption of constant transport numbers through the interface.

C. DIFFUSION OF ELECTROLYTES

In the absence of convection ($v = 0$), equation [2] gives the flux of the ith component of an electrolyte. Writing $\mu_i^0 + RT \ln a_i$ for μ_i, we have

$$\mathbf{j}_i = -c_i D_i \left[z_i \mathcal{F}(RT)^{-1} \nabla \phi + \frac{\partial \ln a_i}{\partial c_i} \nabla c_i \right]$$

If there is no external electrical field, ϕ is the diffusion potential, introduced in equations [6] and [7]. Introducing the activity coefficient γ_i by $a_i = c_i \gamma_i$, we now have

$$\mathbf{j}_i = -c_i D_i z_i \mathcal{F}(RT)^{-1} \nabla \phi_{\text{diff}} - D_i \left(1 + \frac{\partial \ln \gamma_i}{\partial \ln c_i}\right) \nabla c_i \qquad [13]$$

Absence of external electrical field means absence of current; for a single electrolyte $C_{\nu_+}A_{\nu_-}$ this means

$$z_+ \mathbf{j}_+ + z_- \mathbf{j}_- = 0$$

In this case the electric field $\mathbf{E}_{\text{diff}}$ of equation [7] becomes

$$-\nabla\phi_{\text{diff}} = c\mathfrak{F}\kappa^{-1}\left[z_+ D_+(\nu_+)^2 \frac{\partial \mu_+}{\partial c_+} + z_- D_-(\nu_-)^2 \frac{\partial \mu_-}{\partial c_-}\right]\nabla c(RT)^{-1}$$

where c is the concentration of the electrolyte, $\nu_+ c$ is the concentration of the cation, and $\nu_- c$ is the concentration of the anion. Again introducing activities and using the expression (equation [5]) for the conductivity κ we have

$$-\nabla\phi_{\text{diff}} = \frac{RT[z_+ D_+ \nu_+ (\partial \ln a_+ / \partial \ln c) + z_- D_- \nu_- (\partial \ln a_- / \partial \ln c)]}{c\mathfrak{F}[(z_+)^2 D_+ \nu_+ + (z_-)^2 D_- \nu_-]} \nabla c$$

This is substituted into equation [13] to obtain the fluxes of cations and anions. The total flux of electrolyte, $\mathbf{j}$, is equal to $\mathbf{j}_+/\nu_+$ (or to $\mathbf{j}_-/\nu_-$), so

$$\mathbf{j} = -D_+\left[z_+ \mathfrak{F}c(RT)^{-1}\nabla\phi_{\text{diff}} + \nu_+\left(1 + \frac{\partial \ln \gamma_+}{\partial \ln c_+}\right)\nabla c\right]$$

$$= D_+ D_- z_+ z_-[(z_+)^2 D_+ \nu_+ + (z_-)^2 D_- \nu_-]^{-1}\left[\nu_-\left(1 + \frac{\partial \ln \gamma_-}{\partial \ln c}\right)\right.$$

$$\left. + \nu_+\left(1 + \frac{\partial \ln \gamma_+}{\partial \ln c}\right)\right]\nabla c \qquad [14]$$

after some algebra. We have used the condition

$$z_-\nu_- = -z_+\nu_+ \qquad [15]$$

to get the last form, symmetric in anion and cation quantities. According to equation [15], $(z_+)^2 D_+ \nu_+ + (z_-)^2 D_- \nu_-$ is $-z_+ D_+ z_- \nu_- - z_- D_- z_+ \nu_+$, so we have finally

$$\mathbf{j} = -D_+ D_-(\nu_- D_+ + \nu_+ D_-)^{-1}\left(1 + \frac{\partial \ln \gamma_\pm}{\partial \ln c}\right)\nabla c \qquad [16]$$

where

$$(\nu_+ + \nu_-)\ln \gamma_\pm = \nu_+ \ln \gamma_+ + \nu_- \ln \gamma_-$$

defines the mean ionic activity coefficient. The quantity multiplying ∇c in equation [16] is the negative of the effective diffusion coefficient D_{eff} for a salt.

D_{eff} is a mean of the diffusion coefficients for cation and anion and includes the effect of $\mathbf{E}_{\text{diff}}$. Because of the proportionality between mobilities and diffusion coefficients (equation [3]) we may write the effective diffusion coefficient as

$$D_{\text{eff}} = \frac{RT}{\mathcal{F}} \frac{u_+ u_- (\nu_+ + \nu_-)[1 + (\partial \ln \gamma_\pm / \partial \ln c)]}{\nu_+ z_+} \qquad [17]$$

If the solution is dilute enough for ions to migrate independently (see Chapter 1, Section J), we may replace the mobilities by their limiting values and write D_{eff} in terms of transport numbers and molar conductivities at infinite dilution:

$$D_{\text{eff}} = \frac{RT}{\mathcal{F}\nu_+ z_+} t_+^\infty t_-^\infty \Lambda^\infty (\nu_+ + \nu_-) \left(1 + \frac{\partial \ln \gamma_\pm}{\partial \ln c}\right)$$

If the ions of the electrolyte $C_{\nu_+} A_{\nu_-}$ are at low concentration, with another electrolyte present in excess, the diffusion potential drop is suppressed. This is seen in equation [7], since the conductivity now becomes large compared to the terms in the numerator referring to $C_{\nu_+} A_{\nu_-}$. Then the diffusion flux of each ion, from equation [13], is

$$\mathbf{j}_i = -D_i \left(1 + \frac{\partial \ln \gamma_i}{\partial \ln c_i}\right) \nabla c_i$$

Since the activity coefficients in this case depend essentially on the electrolyte present in high concentration, the partial derivative also vanishes, and one has just

$$\mathbf{j}_i = -D_i \nabla c_i \qquad [18]$$

which is Fick's first law.

D. INTEGRATION OF DIFFUSION EQUATION

In the absence of convection and migration, equation [4] becomes

$$\frac{\partial c}{\partial t} = D \nabla^2 c \qquad [19]$$

which is Fick's second law of diffusion. It is important to have solutions to this equation for various geometries and initial conditions. The simplest geometry is linear, with the concentration varying in one spatial direction (x) only and $\nabla^2 c$ becoming $\partial^2 c / \partial x^2$.

We consider first the case of a semi-infinite tube, in which at time $t = 0$ the concentration is uniform, $c = c_0$, for all $x > 0$. The concentration at $x = 0$ is maintained at zero for all $t > 0$. Diffusion will occur to the left for all $x > 0$ and will be greater where the gradient of c is greater, for smaller x. The solution to the equation [19] is then

$$c(x, t) = c_0 \, \text{erf}\left(\frac{x}{2D^{1/2}t^{1/2}}\right) \qquad [20]$$

Here, the error function is defined by

$$\text{erf}(z) = 2\pi^{-1/2} \int_0^z \exp(-u^2) \, du$$

so $d[\text{erf}(z)]/dz = 2\pi^{-1/2} \exp(-z^2)$ and $\text{erf}(0) = 0$, $\text{erf}(\infty) = 1$. In Figure 15, $c(x)$ is shown for various times t.

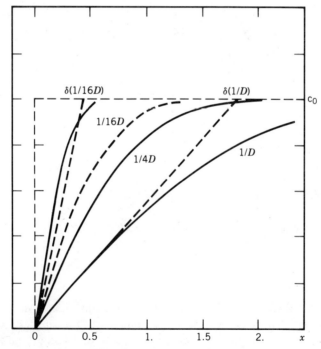

FIGURE 15. Concentration profiles $c(x)$ for various times ($1/16D$, $1/4D$, $1/D$) for an initial step-function concentration profile (broken line). Initial slopes and intercepts (thicknesses of diffusion layer) are shown for $1/16D$ and $1/D$.

According to equation [20], the concentration for all $x > 0$ decreases with t, as does the concentration gradient:

$$\frac{\partial c}{\partial x} = c_0 (\pi D t)^{-1/2} \exp\left(\frac{-x^2}{4Dt}\right) \qquad [21]$$

Thus the mass flux at $x = 0$ decreases inversely with the square root of the time:

$$-D\left(\frac{\partial c}{\partial x}\right)_{x=0} = -c_0 \left(\frac{D}{\pi t}\right)^{1/2} \qquad [22]$$

Since the slope at $x = 0$ is $c_0(\pi Dt)^{-1/2}$, a tangent to $c(x)$ at $x = 0$ intersects the horizontal $c = c_0$ at $x = \delta$, where

$$\delta = (\pi Dt)^{1/2}$$

This is shown in Figure 15 for $t = 1/D$ and for $t = 1/16D$. If the diffusion process is considered (Chapter 1, Section N) to form a diffusion layer in which the concentration gradient is constant, bounded by bulk solution with undisturbed concentration c_0, δ represents the thickness of the diffusion layer; it increases with the square root of time.

If one prescribes the value of the concentration gradient instead of the concentration at $x = 0$, the solution to equation [19] is

$$c = c_0 - 2G\left(\frac{Dt}{\pi}\right)^{1/2} \exp\left(\frac{-x^2}{4Dt}\right) + Gx \operatorname{erfc}\left[\frac{1}{2} x(Dt)^{-1/2}\right] \qquad [23]$$

where the complementary error function is

$$\operatorname{erfc}(z) = 1 - \operatorname{erf}(z)$$

and G is the value of the gradient $\partial c/\partial x$ at $x = 0$. Prescribing the value of G is appropriate for electrode processes in which the rate of reaction is fixed and is equal to the rate at which reactant is transported to the electrode surface. According to equation [23], the concentration at $x = 0$ is $c_0 - 2G(Dt/\pi)^{1/2}$, a monotonically decreasing function of t which becomes 0 at $t = \pi(c_0/2G)^2 D$. For t greater than this value, the diffusion cannot proceed under the conditions of this model. If the reaction rate at $x = 0$ (electrode surface) is proportional to the concentration of dissolved substance at this point, the boundary condition on the diffusion equation is

$$D\left(\frac{\partial c}{\partial x}\right)_{x=0} = kc_{x=0} \qquad [24]$$

where k is a rate constant.

Laplace transformation in time is a powerful technique for solving diffusion equations. Multiplying equation [19] by e^{-st} and integrating over t from 0 to ∞, we have

$$\int_0^\infty dt\, \frac{\partial c(x,t)}{\partial t} e^{-st} = D \int_0^\infty dt\, \frac{\partial^2 c(x,t)}{\partial x^2} e^{-st}$$

Let $\int_0^\infty dt\, c(x,t)\, e^{-st}$, the Laplace transform of c, be designated $\tilde{c}(x,s)$. Integration by parts in the first integral then produces

$$-c(x,0) + s\tilde{c}(x,s) = D \frac{\partial^2 \tilde{c}}{\partial x^2} \qquad [25]$$

If $c(x, 0) = c_0$, equation [25] is

$$D \frac{\partial^2 y}{\partial x^2} = sy$$

where $y = \tilde{c} - c_0/s$; since y should be finite for all x, it must be equal to a constant A multiplied by $\exp[-x(s/D)^{1/2}]$.

For the first problem of this section, $c(0, t) = 0$ for all t so $\tilde{c}(0, s) = 0$ and $y(0, s) = -c_0/s$. Thus A must be chosen as $-c_0/s$ and

$$\tilde{c} = \frac{c_0}{s} - \frac{c_0}{s} \exp\left[-x\left(\frac{s}{D}\right)^{1/2}\right]$$

which, according to a table of Laplace transforms, is the Laplace transform of

$$c_0 - c_0\, \mathrm{erfc}\left(\frac{xD^{-1/2}}{2t^{1/2}}\right) = c_0\, \mathrm{erf}\left[\frac{x}{2(Dt)^{1/2}}\right]$$

as given in equation [20]. The boundary condition of equation [24], on Laplace transformation, becomes $D(\partial \tilde{c}/\partial x)_{x=0} = k\tilde{c}_{x=0}$, which requires $-DA(s/D)^{1/2} = k[(c_0/s) + A]$. With

$$A = \frac{kc_0/s}{k + (sD)^{1/2}}$$

one can see that the solution is the inverse transform of

$$\frac{c_0}{s} - \frac{kc_0 \exp(-xs^{1/2}D^{-1/2})}{ks + D^{1/2}s^{3/2}}$$

so that $(\partial c/\partial x)_{x=0}$ is the inverse transform of $c_0 D^{-1/2}/(s^{1/2} + k^{-1}D^{1/2}s)$, that is

$$\left(\frac{\partial c}{\partial x}\right)_{z=0} = \frac{kc_0}{D} \exp\left(\frac{k^2 t}{D}\right) \text{erfc}\left(\frac{kt}{D^{1/2}}\right)$$

In this case, the gradient of the concentration at $x = 0$ is $kc_0 D^{-1}$ at $t = 0$ and decreases monotonically with time, approaching 0 as $t \to \infty$. Other solutions to the diffusion equation are derived in Koryta, Dvořák, and Boháčková (1970, Appendix A).

For an infinite tube, containing initially a solution of concentration c_1 for $x < 0$ and a solution of concentration c_2 for $x > 0$, one can solve the linear diffusion equation to obtain the relaxation of the concentration profile from its initial step. The concentrations at $x \to \pm\infty$ remain at their initial values and the concentration at $x = 0$ is always the average, $(c_1 + c_2)/2$. The solution to equation [19] [with $\nabla^2 \to (\partial^2/\partial x^2)$] is

$$c = \tfrac{1}{2}(c_1 + c_2) + \tfrac{1}{2}(c_2 - c_1) \, \text{erf}\left[\tfrac{1}{2} x(Dt)^{-1/2}\right] \qquad [26]$$

The concentration gradient is then

$$\frac{\partial c}{\partial x} = \frac{1}{2}(c_2 - c_1)(\pi Dt)^{-1/2} \exp\left(\frac{-x^2}{4Dt}\right)$$

which decreases as $t^{-1/2}$. One can introduce the notion of a transition layer at $x = 0$, within which the concentration gradient is constant at $\tfrac{1}{2}(c_2 - c_1)(\pi Dt)^{-1/2}$ and outside of which the concentrations have their original values. Then the width of the layer on the positive side of the origin is

$$\delta = \frac{c_2 - \tfrac{1}{2}(c_1 + c_2)}{\tfrac{1}{2}(c_2 - c_1)(\pi Dt)^{-1/2}} = (\pi Dt)^{1/2}$$

The width on the negative side of the origin is also $(\pi Dt)^{1/2}$.

Diffusion under conditions of spherical symmetry is often appropriate, as for a spherical mercury drop electrode. The diffusion equation (equation [19]) is then

$$\frac{\partial c}{\partial t} = D\left[\frac{1}{r^2}\frac{\partial}{\partial r}\left(r^2 \frac{\partial c}{\partial r}\right)\right] \qquad [27]$$

In the simplest case, one has initially $c = c_0$ for $r > r_0$ (r_0 = radius of the sphere) and $c = 0$ for $r = r_0$, $t > 0$. Of course c remains at c_0 at $r \to \infty$ for all t. This corresponds to a spherical electrode at which the diffusing substance is consumed as fast as it arrives. The solution to equation [27] in this case is

$$c = c_0 \left\{ 1 - \frac{r_0}{r} \operatorname{erfc} \left[\frac{r - r_0}{2(Dt)^{1/2}} \right] \right\}$$

The flux into the sphere at $r = r_0$ is

$$j_0 = D \left(\frac{\partial c}{\partial r} \right)_{r_0} = Dc_0 [(\pi Dt)^{-1/2} + (r_0)^{-1}] \qquad [28]$$

In the steady state, $(\partial c/\partial t) = 0$, one has from equation [27] that $r^2(\partial c/\partial r)$ is independent of r. Given the boundary condiitions, this means

$$c_{st} = c_0(1 - r_0 r^{-1})$$

and implies a flux at $r = r_0$ equal to $Dc_0 r_0^{-1}$. Thus equation [28] may be written

$$j_0 = j_{st} + j_{td}$$

where the time-dependent flux j_{td} is $c_0(D/\pi t)^{1/2}$, identical to the linear diffusion result (equation [22]). The value of j_{st} becomes smaller as r_0 becomes larger; $r_0 \to \infty$ corresponds to a planar electrode.

In polarography (Chapter 1, Section Q) it is usual to use a dropping mercury electrode. The growth of the mercury drop is equivalent to a radial motion of the solution toward the electrode surface. Consider a perfect sphere of radius r_0, growing in volume at a constant rate s with $r_0 = 0$ at $t = 0$, so that

$$\frac{4\pi (r_0)^3}{3} = st$$

Letting a be the radius of the sphere at $t = 1$ we have

$$(r_0)^3 = a^3 t$$

or $r_0 = at^{1/3}$. Thus the radial velocity at $r = r_0$ is

$$\frac{dr_0}{dt} = \frac{at^{-2/3}}{3} = \frac{a^3}{3(r_0)^2}$$

and the distance from a point at r to the surface of the sphere is

$$x = r - at^{1/3}$$

At a distance $r > r_0$, the outward movement of the electrode corresponds to a convective radial velocity of $a^3/3r^2$ ($\mathbf{v}$ in equation [1]).

Introducing the additional convective term into the spherical diffusion equation for a single species (equations [4] and [27]), we have

$$\frac{\partial c}{\partial t} = D\left[\frac{1}{r^2}\frac{\partial}{\partial r}\left(r^2\frac{\partial c}{\partial r}\right)\right] - \frac{a^3}{3r^2}\frac{\partial c}{\partial r}$$

It is assumed that the electrical terms are negligible or that there is local electroneutrality. The boundary conditions we consider for polarography are: $c = c_0$ for all r at $t = 0$, $c = c_0$ for $r \to \infty$ at all t, and $c(r_0) = c^*$ for $t > 0$ (previously we took $c^* = 0$). The equation must now be rewritten in terms of x. If the thickness of the diffusion layer is small compared to the size of the sphere, as is usually the case, there is some simplification; the resulting equation is discussed in Koryta, Dvořák, and Boháčková (1970, p. 137).

8

ELECTRODE KINETICS

A. INTRODUCTION

The overpotential or overvoltage at an electrode is the difference between the electrical potential of an electrode (relative to some reference electrode) when an electron transfer reaction is taking place and current is flowing, and its value when no current is flowing and there is equilibrium with respect to this electron transfer reaction. At equilibrium the net reaction rate is zero and there is no net current. The overpotential is obviously related to the rate at which electrochemical reaction occurs at the electrode but, since the reaction involves a series of processes, the rate of each of which can depend on the electrode potential, the overvoltage is considered to consist of components corresponding to each such process: charge transfer between the electrode and solution species, diffusive mass transport from bulk solution to the electrode or vice versa, chemical reaction either at the electrode or in bulk electrolyte, incorporation of a species into the electrode structure, and so on. Resistance of the circuit also appears as overvoltage, but normally is not considered along with the others, since it is unrelated to the electrode reaction.

As discussed in Chapter 1, Section L, the net current density can be written as the difference between cathodic and anodic current densities. The cathodic current density is proportional to the concentrations of oxidized species in bulk solution and to a factor $\exp(-\Delta G_r^{\ddagger}/RT)$, where $\Delta G_r^{\ddagger}$ is the activation free energy for the forward reaction (reduction), while the anodic current density is proportional to the concentrations of reduced species in bulk solution and to a factor $\exp(-\Delta G_o^{\ddagger}/RT)$, where $\Delta G_o^{\ddagger}$ is the activation free energy for the reverse reaction (oxidation). The free energies of activation, being differences of electrochemical potentials, can be separated into chemical and electrical parts, as can the free-energy change for any of the steps involved in the overall reaction. If the electrical part of the free-energy change for the charge-transfer step, written as a reduction, is $n\mathfrak{F}(\phi^M - \phi^P)$, it is assumed that the free energy of activation for the reduction reaction has as its electrical part $\alpha n \mathcal{F}(\phi^M - \phi^P)$, where α, the charge-transfer coefficient, is between 0 and 1. The electrical part of the free energy of activation for the oxidation reaction is thus $(\alpha - 1) n\mathcal{F}(\phi^M - \phi^P)$. Here n is the number of electrons transferred between solution and electrode, ϕ^M is the electrical potential in the electrode, and ϕ^P is the electrical potential at the position occupied by a charged species when electron transfer to or from the electrode occurs. In Chapter 1, Section L we took ϕ^P as the inner potential of the bulk solution phase, but some authors prefer to take ϕ^P as the potential at the outer Helmholtz plane or other position in the double layer, separating out the effect of the diffuse layer explicitly.

The current density now can be written in terms of the transfer coefficient α, the chemical part of the activation energies for the forward and reverse processes, and the bulk concentrations of oxidized and reduced species. It is also convenient to introduce the exchange current density j^0, which is the density of anodic or cathodic currents when the electrode potential is at its equilibrium value so that the two are equal. The value of j^0 depends on the concentrations of oxidized and reduced species. Methods of determining the exchange current density and the transfer coefficient α are reviewed in Bockris and Khan (1979), Chapter 2. Thus if the cathodic density is $Ke^{-\alpha n \mathcal{F} \Delta \phi / RT}$ and the anodic density is $Me^{(1-\alpha) n \mathcal{F} \Delta \phi / RT}$, where K and M are constants and $\Delta \phi = \phi^M - \phi^P$,

$$j^0 = Ke^{-\alpha n \mathcal{F} \Delta \phi' / RT} = Me^{(1-\alpha) n \mathcal{F} \Delta \phi' / RT}$$

where $\Delta \phi'$ is the equilibrium value of $\Delta \phi$, so that

$$j = j^0 e^{-\alpha n \mathcal{F} (\Delta \phi - \Delta \phi') / RT} - j^0 e^{(1-\alpha) n \mathcal{F} (\Delta \phi - \Delta \phi') / RT}$$

and $\Delta \phi - \Delta \phi'$ is the overpotential η if ϕ^P is in bulk solution.

If the reaction involves movement of an ion of charge $+ze$ from bulk solution to position P and movement of a reduced ion of charge $(z - n) e$ from P to bulk solution after reaction, the cathodic current should include a factor $\exp(-\alpha z \mathcal{F} \phi^P / RT)$ and the anodic current should include a factor $\exp[(1 - \alpha)(z - n) \mathcal{F} \phi^P / RT]$. The exchange current density includes corresponding factors, and (Andersen and Eyring 1970, pp. 250–256) one gets for the current density

$$j = j^0 \exp\left[\frac{(\alpha n - z)(\phi^P - \phi_e^P)}{RT}\right] \left\{ \exp\left(\frac{-\alpha n \mathcal{F} \eta}{RT}\right) - \exp\left[\frac{(1 - \alpha) n \mathcal{F} \eta}{RT}\right] \right\} \quad [1]$$

where ϕ_e^P is the value of ϕ^P at electrode equilibrium. It may be argued that the difference between ϕ^P and ϕ_e^P is usually small, so that

$$j = j^0 [e^{-\alpha n \mathcal{F} \eta / RT} - e^{(1 - \alpha) n \mathcal{F} \eta / RT}] \quad [2]$$

as in Chapter 1, equation [54]. The parameters j^0 and α are supposed to be independent of j and η.

As shown in Chapter 1, Section M, η is linear in $\ln |j|$ when $|\eta|$ is large enough to neglect one term in equation [2] compared to the other. Since the ratio of the reducing and oxidizing currents is $e^{n\mathcal{F}\eta}/RT$, and $RT/\mathcal{F} = 0.0258$ V at $T = 300$ K, the ratio of currents increases by a factor of 10 for an increase in $|\eta|$ of 59 mV, and the linearity of η in $\ln |j|$ is commonly observed. For low overpotentials, j is proportional to η; this is often expressed by defining the charge-transfer resistance as

$$R_c = -\left(\frac{d\eta}{dj}\right)_{j \to 0} = \frac{RT}{n\mathcal{F}j^0} \quad [3]$$

The expressions for current as a function of overpotential all assume a steady state: Concentrations of reactants and products, and double-layer structure, are supposed not to change with time. This means that all the current is involved with the electrochemical reaction, none going, for example, to charge the double layer (Andersen and Eyring, 1982, Section 3,II.D).

For the simple reaction $O + ne^- \to R$ with O and R in the solution and e^- in the electrode, a schematic free-energy diagram can be drawn to show why the electrical part of the activation free energy should be a fraction α of the overall free-energy change and what the value of α is likely to be. If the free energies of $O + ne^-$ and of R are drawn as functions of the reaction coordinate (Figure 16) their intersection represents the activated complex. The activation free energy

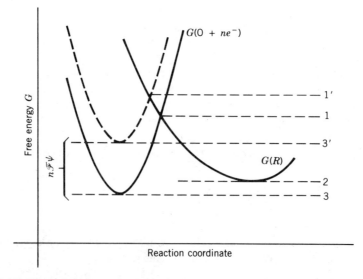

FIGURE 16. Effect of displacement on the intersection of free-energy curve of reduced species and free-energy curve of oxidized species.

$\Delta G_r^{\ddagger}$ is equal to $G_1 - G_3$, while the overall free-energy change for the reaction is $G_2 - G_3$. If the potential of the electrode relative to the solution is changed by $-\psi$, the curve for $O + ne^-$ is displaced upward by $n\mathfrak{F}\psi$ (broken curve in Figure 16). The overall free-energy change is now $G_2 - G_3'$, which is lower by $n\mathfrak{F}\psi$ than its value for $\psi = 0$. The activation free energy is now $G_1' - G_3'$, which differs from its value for $\psi = 0$ by $(G_1' - G_1) - (G_3' - G_3)$. Clearly, $G_3' - G_3 = n\mathfrak{F}\psi$ and $G_1' - G_1$ is a fraction of $n\mathfrak{F}\psi$.

Suppose the left branch of $G(R)$ in the region of the intersection is given approximately by $G_a - ax$ (x = reaction coordinate) and the right branch of $G(O + ne^-)$ is given by $G_b + bx$. The intersection point of the curves is at $x = (G_a - G_b)/(a + b)$ and the free energy at this point is $G_1 = G_b + b(G_a - G_b)/(a + b)$. If $G(O + ne^-)$ is increased by $n\mathfrak{F}\psi$, so that its right branch is now given by $G_b + n\mathfrak{F}\psi + bx$, the new intersection point will be at $x = (G_a - G_b - n\mathfrak{F}\psi)/(a + b)$ and the free energy at this point will be

$$G_1' = G_b + n\mathfrak{F}\psi + \frac{b(G_a - G_b - n\mathfrak{F}\psi)}{a + b}$$

Therefore

$$G_1' - G_1 = n\mathfrak{F}\psi\left(1 - \frac{b}{a + b}\right) = \frac{an\mathfrak{F}\psi}{a + b} \qquad [4]$$

and the transfer coefficient α is $a/(a + b)$. Its value depends on the relative sizes of the slopes of the curves $G(R)$ and $G(O + ne^-)$, being equal to $\frac{1}{2}$ if the magnitudes of a and b are equal. The term *symmetry factor* is often used for α because it depends on the relative slopes and hence the symmetry of the activation barrier.

For an electrode that is an electronic conductor in contact with an ionic conductor, one may note that a change in the electrical potential corresponds to a change in the Fermi level of electrons in the electronic conductor. At equilibrium, ϵ_F has some value, which differs from its value for the same electronic conductor in vacuum. An increased negative potential of the electrode (meaning increased negative potential of the electronic conductor relative to the other phase) corresponds to raising the Fermi level above its equilibrium value, which increases the rate of electron transfer out of the electrode and decreases the rate of electron transfer into the electrode, thus producing a net current flow into the electrode. The incorporation of these concepts into a theory to explain equation [1] and the values for j^0 and α will emerge in later sections.

B. REACTION AND MASS TRANSPORT

If the electrode reaction is rapid enough, the activity of a reactant in the region near the electrode from which it is supposed to react (P in Section A) will drop below its activity in bulk solution, because the rate of mass transport is not large enough to supply reactant as fast as it is consumed. Similarly, there may be a buildup of product at P because product is not transported away as rapidly as it is formed. Then, as discussed in chapter 1, Section O, the overpotential will include a contribution due to diffusion and migration. The importance of migration (motion under the influence of an electric field) will be low if a nonelectroactive supporting electrolyte is present in high concentration to carry most of the current. We assume this is the case for now.

According to equation [55] of Chapter 1,

$$j^0 = n\mathfrak{F}k_r^0 a_O^b e^{-\alpha z \mathfrak{F} E'/RT} = n\mathfrak{F}k_o^0 a_R^b e^{(1-\alpha) z \mathfrak{F} E'/RT} \quad [5]$$

where b refers to bulk solution and E' refers to the electrode–solution potential difference at open circuit. Then the current is

$$j = j^0 \left[\frac{a_O^P}{a_O^b} e^{-\alpha z \mathfrak{F} \eta/RT} - \frac{a_R^P}{a_R^b} e^{(1-\alpha) z \mathfrak{F} \eta/RT} \right] \quad [6]$$

which replaces equation [2]. Here, P refers to the region of solution from which reaction takes place. For large enough currents, a_O^P and a_R^P will start to deviate from a_O^b and a_R^b; the ratios a_O/a_O^b and a_R/a_R^b will be functions of current and hence of potential.

The diffusion-layer picture of Chapter 1, Section M is very commonly used to discuss mass transfer. By supposing that the activities of all species reach their bulk values at a distance δ from the electrode and that activities vary linearly between distances 0 and δ, it avoids the need for detailed description of concentration profiles. Values of the diffusion-layer thickness depend on the way in which the solution is stirred, as well as on the properties of the electrolyte phase (Andersen and Eyring 1970, Section III.B.2). The thickness is usually about 0.03 cm for unstirred solutions and one or two orders of magnitude smaller for stirred solutions, but always much larger than characteristic lengths for the double layer.

Assuming a diffusion layer, the concentration gradient of any species k is just the difference of its concentrations in bulk and at the electrode, $a_k^P - a_k^b$, divided by δ. The diffusive flow of a species is proportional to its concentration gradient, and the current density or rate of electrode reaction is proportional to the flow of a reactant to the electrode or of a product away from the electrode. Then, if δ may be taken as a constant, $a_k^P - a_k^b$ is proportional to the current density. The limiting cathodic current density (see Chapter 1, Section N) j_c^1 occurs when $a_k^P = 0$ for a reactant species. Therefore

$$\frac{j}{j_c^1} = \frac{a_k^b - a_k^P}{a_k^b}$$

if k is an oxidized (reactant) species. Similarly, the limiting anodic current density j_a^1 occurs when $a_j^P = 0$ for a reduced (product) species, so

$$\frac{-j}{j_a^1} = \frac{a_k^b - a_k^P}{a_k^b}$$

where k is a product (reduced) species (note that anodic currents are negative by convention). Then equation [6], which refers to the reaction $O + ne^- \rightarrow R$, becomes

$$j = j^0\left[\left(1 - \frac{j}{j_c^1}\right)e^{-\alpha z\mathcal{F}\eta/RT} - \left(1 + \frac{j}{j_a^1}\right)e^{(1-\alpha)z\mathcal{F}\eta/RT}\right] \quad [7]$$

which may be solved for j:

$$\frac{j}{j^0} = \frac{\exp[-\alpha z\mathcal{F}\eta/RT] - \exp[(1-\alpha)z\mathcal{F}\eta/RT]}{1 + (j^0/j_c^1)\exp[-\alpha z\mathcal{F}\eta/RT] + (j^0/j_a^1)\exp[(1-\alpha)z\mathcal{F}\eta/RT]} \quad [8]$$

If the exchange current j^0 is small compared to the limiting diffusion currents j_c^1 and j_a^1, only the electrode reaction rate matters, and equation [8] reduces to equation [2]. If j^0/j_c^1 and j^0/j_a^1 are large, the current density approaches

$$j = \frac{\exp[-\alpha z\mathcal{F}\eta/RT] - \exp[(1-\alpha)z\mathcal{F}\eta/RT]}{(1/j_c^1)\exp[-\alpha z\mathcal{F}\eta/RT] + (1/j_a^1)\exp[(1-\alpha)z\mathcal{F}\eta/RT]}$$

$$= [1 - e^{z\mathcal{F}\eta/RT}]/[(1/j_c^1) + (1/j_a^1)e^{z\mathcal{F}\eta/RT}] \quad [9]$$

so that, for a given j, η depends on the limiting diffusion currents, and not on j^0 or α, which characterize the electrode reaction. Equation [9] can be rearranged to

$$\frac{z\mathcal{F}\eta}{RT} = \ln\frac{1 - j/j_c^1}{1 + j/j_a^1} \quad [10]$$

showing the dependence of overpotential η on current density for a reaction whose rate is controlled by diffusion.

When both diffusion and electrode reaction are important, but $\eta > 50$ mV so that the second exponential is large compared to the first, equation [8] becomes

$$\exp\left[\frac{(1-\alpha)z\mathcal{F}\eta}{RT}\right] = \frac{-j/j^0}{1 + j/j_a^1} \quad [11]$$

(j is negative for an anodic current). Similarly, if $\eta < -50$ mV so that the first exponential in equation [8] dominates the second, we get

$$\exp\left[\frac{-\alpha z\mathcal{F}\eta}{RT}\right] = \frac{j/j^0}{1 - j/j_c^1} \quad [12]$$

In either case, $\ln[|j|^{-1} - (j^1)^{-1}]$ will be linear in the overpotential. From the slope of a plot of $\ln[|j|^{-1} - (j^1)^{-1}]$ vs. η, one can obtain α and, from the intercept, j^0.

Finally, consider low overpotentials so that the exponentials can be linearized ($e^x \sim 1 + x$). Equation [8] becomes

$$j = \frac{-z\mathcal{F}\eta}{RT}\left(\frac{1}{j^0} + \frac{1}{j_c^1} + \frac{1}{j_a^1}\right)^{-1} \qquad [13]$$

If j^0 is much smaller than j_c^1 and j_a^1, the slope $(d\eta/dj)_{j \to 0}$ is

$$-\frac{RT}{z\mathcal{F}j^0} = R_c$$

the charge-transfer resistance. If the limiting diffusion currents are much smaller than the exchange current,

$$\left(\frac{d\eta}{dj}\right)_{j \to 0} = -\frac{RT}{z\mathcal{F}}\left(\frac{1}{j_c^1} + \frac{1}{j_a^1}\right) \equiv R_d \qquad [14]$$

and R_d may be referred to as the diffusion resistance.

The equations so far have not considered the contribution of migration under the influence of the electric field in the diffusion layer. In general, consider the electrode reaction

$$u_1 O_1 + u_2 O_2 + \cdots + ne^- \to v_1 R_1 + v_2 R_2 + \cdots \qquad [15]$$

Overall electroneutrality requires

$$\sum_k z_k v_k = \sum_k z_k u_k - n$$

where z_k is the charge number of species k. The rate of production of the species R_k per unit area of electrode is $jv_k/n\mathcal{F}$; the rate of consumption of O_k is $ju_k/n\mathcal{F}$. At steady state, the rate of production or consumption of a species must equal the rate at which it is transported through the diffusion layer. Then, if D_k is the diffusion coefficient of species k,

$$\frac{jv_k}{n\mathcal{F}} = -D_k \frac{\partial a_k}{\partial x} - \frac{z_k a_k D_k \mathcal{F}}{RT}\frac{\partial \phi}{\partial x} \qquad [16]$$

$$\frac{ju_k}{n\mathcal{F}} = D_k \frac{\partial a_k}{\partial x} + \frac{z_k a_k D_k \mathcal{F}}{RT}\frac{\partial \phi}{\partial x} \qquad [17]$$

(Chapter 7, equation [2]), where x is the coordinate perpendicular to the electrode surface and $a_k(\sim c_k)$ is the activity of substance k. For a substance not involved in reaction [15],

$$D_k\left(\frac{\partial a_k}{\partial x} + \frac{z_k a_k \mathcal{F}}{RT}\frac{\partial \phi}{\partial x}\right) = 0 \qquad [18]$$

The current density is given by an equation like equation [6], with $a_O(a_R)$ replaced by a suitable combination of the activities of the $O_k(R_k)$.

Equation [6] as written refers to a reaction like reaction [15] in which there are only two electroactive species, an oxidized species and a reduced species, both in solution. If we go to an even simpler case, in which the reduced species is neutral and is either incorporated into the electrode or leaves as a gas (H_2), it is possible to solve the problem, taking into account migration, diffusion, and electrode reaction. The current density is given by equation [6] with the factor a_R^P/a_R^b replaced by unity. Equation [17] for species O with stoichiometric coefficient 1 is

$$\frac{j}{n\mathcal{F}} = \frac{D_O(a_O^b - a_O^P)}{\delta} + \frac{jz_O a_O D_O \mathcal{F}}{RT\kappa} \qquad [19]$$

In equation [19] the diffusion-layer thickness δ has been introduced and the field $E = -(\partial \phi/\partial x)$ has been written as $-j/\kappa$. The conductivity κ is given by equation [5] of Chapter 7, and the mobility of species i is written as $D_i|z_i|\mathcal{F}/RT$. Then equation [19] becomes

$$\frac{j}{n\mathcal{F}} = \frac{D_O(a_O^b - a_O^P)}{\delta} + \frac{jz_O a_O D_O}{\mathcal{F}\sum(z_i)^2 a_i D_i} \qquad [20]$$

The second term is $jt_O/z_O\mathcal{F}$, where t_O is the transport number of O. It will be assumed that the transport number t_O is constant through the diffusion layer. Solving for a_O^P and substituting into equation [6], we obtain

$$\frac{j}{j^0} = \left(\frac{D_O a_O^b}{\delta} + \frac{jt_O}{z_O\mathcal{F}} - \frac{j}{n\mathcal{F}}\right) e^{-\alpha n\mathcal{F}\eta/RT}\left(\frac{D_O a_O^b}{\delta}\right)^{-1}$$

$$- \exp\left[\frac{(1-\alpha)n\mathcal{F}\eta}{RT}\right]$$

or

$$j = \frac{z_O\mathcal{F}a_O^b(1 - e^{n\mathcal{F}\eta/RT})}{(z_O\mathcal{F}a_O^b/j^0)e^{n\mathcal{F}\alpha\eta/RT} - (t_O\delta/D_O) + (z_O\delta/nD_O)} \qquad [21]$$

If the ion is reduced to the neutral, z_O must equal n.

Again we have limiting cases. There will be charge-transfer control of the rate if j^0 is small enough so that the denominator of equation [21] is dominated by the first term. Then

$$j = j^0(1 - e^{n\mathcal{F}\eta/RT}) e^{-\alpha n\mathcal{F}\eta/RT}$$

which is equation [6] with $a_O^P = a_O^b$ (diffusion and migration are fast enough to keep a_O^P equal to a_O^b). This situation occurs for small exchange currents, high concentrations, large diffusion coefficient of the electroactive species, and small diffusion-layer thickness (good stirring).

In the opposite extreme, equation [21] becomes

$$j = (z_O \mathcal{F} a_O^b D_O/\delta)(1 - e^{n\mathcal{F}\eta/RT})\left(\frac{z_O}{n} - t_O\right)^{-1}$$

$$= \frac{z_O \mathcal{F} D_O}{\delta}(a_O^b - a_O^P)\left(\frac{z_O}{n} - t_O\right)^{-1} \quad [22]$$

The second equality follows because, if the exchange current is large compared to j, equation [6] becomes

$$0 = (a_O^P/a_O^b) e^{-\alpha n\mathcal{F}\eta/RT} - e^{(1-\alpha)n\mathcal{F}\eta/RT}$$

which is the Nernst equation corresponding to the concentration cell between P and b: $a_O^P/a_O^b = e^{n\mathcal{F}\eta/RT}$. If migration is unimportant (t_O small compared to z_O/n or unity) the current density of equation [22] becomes $D_O \mathcal{F} n(a_O^b - a_O^P)/\delta$. Equation [22] shows that the current density is increased because of the contribution of migration.

Suppose the electrolyte is $(O^{n+})_m(A^{m-})_n$, where A^{m-} is not electroactive. It was shown in Chapter 1, Section N that the net flow (diffusion plus migration) of A through the diffusion layer is zero, that is,

$$\frac{D_A(a_A^b - a_A^P)}{\delta} - \frac{jt_A}{m\mathcal{F}} = 0 \quad [23]$$

This also follows from equation [18] on writing E as $-j/\kappa$, as above. Local electroneutrality at P requires $m a_A^P = n a_O^P$ and similarly $m a_A^b = n a_O^b$ (approximating concentrations by activities) so equation [23] gives

$$D_A = \frac{jt_A \delta}{n\mathcal{F}}(a_O^b - a_O^P)^{-1}$$

Since $t_O + t_A = 1$ and $t_O/t_A = \Lambda_O/\Lambda_A = (n/m)(D_O/D_A)$,

$$\frac{t_O}{1 - t_O} = \frac{nD_O}{mD_A} = \frac{n^2 D_O \mathcal{F}(a_O^b - a_O^P)}{\delta m j(1 - t_O)}$$

Inserting the expression for t_O into equation [22], one obtains an equation for j which, solved in turn, yields

$$j = \frac{D_O n \mathcal{F} (1 + n/m)(a_O^b - a_O^P)}{\delta} \qquad [24]$$

The factor $1 + n/m$ gives the enhancement of the current density due to transport (compare equation [22] with $t_O = 0$).

C. MULTISTEP REACTIONS

It is common for the rate-determining charge-transfer step to be preceded or followed by a chemical reaction. Equation [15] may then represent the overall reaction, whereas the rate-determining step may be

$$sO + ne^- \rightarrow tR \qquad [25]$$

where O and R do not appear in reaction [15]. The rate of reaction [25] is given by equation [6], but one now has to consider the chemical reactions that create O from the reactants of reaction [15] and which produce the products of reaction [15] from R:

$$\sum_i u_i O_i \rightarrow sO \qquad [26]$$

$$tR \rightarrow \sum_j v_j R_j \qquad [27]$$

We consider the case when both reactions [26] and [27] are at equilibrium, leaving reaction [25] as the rate-determining step. Section E discusses what happens if there is no equilibrium.

The equilibrium conditions may be written in terms of the activities. For reaction [26],

$$a_O^s = K_O \prod a_{O_i}^{u_i} \qquad [28]$$

where K_O is the equilibrium constant. For reaction [27],

$$\prod a_{R_j}^{v_j} = K_R a_R^t \qquad [29]$$

where K_R is the equilibrium constant. If there are several consecutive reactions producing sO from $\{O_i\}$ and producing $\{R_j\}$ from tR, these relations still hold provided that all the reactions are at equilibrium. The activities here refer to bulk electrolyte. Effects of the diffuse layer will be neglected, so that $a_O^P = a_O^b$ and $a_R^P = a_R^b$.

Substitution of the above relations into equation [6], using equation [5] for j^0 and putting $(a_O)^s$ and $(a_R)^t$ for a_O and a_R respectively, we obtain

$$j = n\mathcal{F}\left\{k_r^0 K_O \prod a_{O_i}^{u_i} \exp\left(\frac{-\alpha n\mathcal{F}E'}{RT}\right) - k_o^0(K_R)^{-1} \prod a_{R_j}^{v_j} \exp\left[\frac{(1-\alpha)n\mathcal{F}E'}{RT}\right]\right\} \quad [30]$$

When the magnitude of the overpotential is large enough for one of the exponentials to be neglected compared to the other, the stoichiometric coefficients u_i and v_j may be obtained from the variation of the current density j with activities at constant potential. For example, if the first exponential dominates, we obtain

$$\left(\frac{\partial \ln j}{\partial \ln a_{O_i}}\right)_{E'} = u_i$$

(all other activities being held constant). The transfer coefficient may be found from the slope of a Tafel plot ($\ln j$ vs. η) and the exchange current j^0 may be obtained from the intercept of the two lines which are the asymptotes of the Tafel plot.

The exchange current is given by

$$\frac{j^0}{n\mathcal{F}} = k_r^0 K_O \prod a_{O_i}^{u_i} \exp\left(\frac{-\alpha n\mathcal{F}E_e}{RT}\right)$$

$$= k_o^0(K_R)^{-1} \prod a_{R_j}^{v_j} \exp\left[\frac{(1-\alpha)n\mathcal{F}E_e}{RT}\right] \quad [31]$$

where E_e is the equilibrium value of E'. In terms of the standard potential $E_{1/2}^0$ (for unit activities of O and R) we have

$$E_e = E_{1/2}^0 + \left(\frac{RT}{n\mathcal{F}}\right)\ln\frac{a_O}{a_R}$$

$$= E_{1/2}^0 + \frac{RT}{n\mathcal{F}}\left[\ln K_O \prod a_{O_i}^{u_i} - \ln(K_R)^{-1} \prod a_{R_j}^{v_j}\right]$$

where equations [28] and [29] have been used. Therefore, if the activity of only one reactant is changed, we get

$$\frac{\partial \ln j^0}{\partial \ln a_{O_k}} = \frac{\partial \ln \prod a_{O_i}^{u_i}}{\partial \ln a_{O_k}} - \frac{\alpha n\mathcal{F}}{RT}\frac{\partial E_e}{\partial \ln a_{O_k}}$$

$$= u_k - \alpha u_k \quad [32]$$

C. MULTISTEP REACTIONS

and, if the activity of one product is changed, we get

$$\frac{\partial \ln j^0}{\partial \ln a_{R_k}} = \frac{\partial \ln \prod a_{R_j}^{v_j}}{\partial \ln a_{R_k}} + (1 - \alpha) \frac{n\mathfrak{F}}{RT} \frac{\partial E_e}{\partial \ln a_{R_k}}$$

$$= v_k - (1 - \alpha) v_k \qquad [33]$$

For the single-step reaction (reaction [25]), $(\partial \ln j^0 / \partial \ln a_O) = s(1 - \alpha)$ and $(\partial \ln j^0 / \partial \ln a_R) = t\alpha$.

We now consider a reaction that involves successive charge-transfer steps at the electrode:

$$sO + ne^- \rightarrow uM \qquad [34]$$

$$uM + me^- \rightarrow tR \qquad [35]$$

in which M is an intermediate. Overall,

$$sO + (n + m) e^- \rightarrow tR \qquad [36]$$

Again, it will be assumed that the activities of O and R at the reaction site (P) are the same as in bulk solution and that one need not consider changes in double-layer structure with potential. The overvoltage is hence due to the charge-transfer processes and will depend on the exchange currents and transfer coefficients for processes [34] (j_a^0 and α_a) and [35] (j_b^0 and α_b). The total current density j will be a sum of contributions from the two processes. Since, at steady state, the rates of the processes must be identical, the current density from process [34] is $nj/(n + m)$ and the current density from process [35] is $mj/(n + m)$.

When there is equilibrium for reaction [36], there is equilibrium with respect to both processes [34] and [35]. The overpotentials for both are thus defined relative to the same electrode–solution potential. Therefore

$$\frac{nj}{n + m} = j_a^0 \left\{ \exp\left(\frac{-\alpha_a \mathfrak{F}\eta}{RT}\right) - \exp\left[\frac{(1 - \alpha_a) \mathfrak{F}\eta}{RT}\right] \right\}$$

$$\frac{mj}{n + m} = j_b^0 \left\{ \exp\left(\frac{-\alpha_b \mathfrak{F}\eta}{RT}\right) - \exp\left[\frac{(1 - \alpha_b) \mathfrak{F}\eta}{RT}\right] \right\} \qquad [37]$$

Multiply the first of these two equations by $j_b^0 \exp(-\alpha_b \mathfrak{F}\eta/RT)$ and the second by $j_a^0 \exp[(1 - \alpha_a) \mathfrak{F}\eta/RT]$ and then add. The result is

$$\frac{j}{n + m} \left\{ nj_b^0 \exp\left(\frac{-\alpha_b \mathfrak{F}\eta}{RT}\right) + mj_a^0 \exp\left[\frac{(1 - \alpha_a) \mathfrak{F}\eta}{RT}\right] \right\}$$

$$= j_a^0 j_b^0 \left\{ \exp\left[\frac{-(\alpha_a + \alpha_b) \mathfrak{F}\eta}{RT}\right] - \exp\left[\frac{(2 - \alpha_a - \alpha_b) \mathfrak{F}\eta}{RT}\right] \right\} \qquad [38]$$

For high cathodic overvoltages ($\eta\mathfrak{F}/RT$ large and negative) the first term in each set of curly braces dominates and

$$j_- = (n + m) j_a^0 \exp\left(\frac{-\alpha_a \mathfrak{F}\eta}{RT}\right) \qquad [39]$$

which leads to a Tafel plot. For high anodic overvoltages ($\eta\mathfrak{F}/RT$ large and positive) the second terms dominate and

$$j_+ = (n + m) j_b^0 \exp\left[\frac{(1 - \alpha_b)\mathfrak{F}\eta}{RT}\right] \qquad [40]$$

which is another Tafel plot.

It can be seen, however, that the Tafel lines of equations [39] and [40] refer to different charge-transfer processes. A plot of $\ln j_-$ vs. $-\eta$ will have a slope of $\alpha_a \mathfrak{F}/RT$ and an intercept of $\ln[(n + m) j_a^0]$, whereas a plot of $\ln j_+$ vs. η will have a slope of $(1 - \alpha_b)\mathfrak{F}/RT$ and an intercept of $[(n + m) j_b^0]$. The sum of the slopes is not $\mathfrak{F}/RT$, as would be the case if only one charge-transfer process were involved, and the intercepts are not identical. The fact that the two Tafel lines give different intercepts for $\eta = 0$ is used as evidence of the existence of consecutive charge-transfer steps. For low overpotentials, equation [38] becomes

$$j\{n j_b^0 + m j_a^0\} = (n + m) j_a^0 j_b^0 \left\{\frac{-2\mathfrak{F}\eta}{RT}\right\}$$

so that the charge-transfer resistance is

$$R_c = -\left(\frac{\partial \eta}{\partial j}\right)_{j \to 0} = \frac{RT}{2\mathfrak{F}} \left\{\frac{n}{j_a^0} + \frac{m}{j_b^0}\right\}(m + n)^{-1} \qquad [41]$$

R_c depends on the exchange currents for both processes [34] and [35]; measured values of charge-transfer resistance may be used to check the values for these quantities obtained from Tafel plots.

The magnitude of exchange currents varies widely, which may lead to complications in the interpretation of Tafel plots. Suppose j_b^0/j_a^0 is large enough so that in equation [38]

$$n j_b^0 \exp\left(\frac{-\alpha_b \mathfrak{F}\eta}{RT}\right) \gg m j_a^0 \exp\left[\frac{(1 - \alpha_a)\mathfrak{F}\eta}{RT}\right] \qquad [42]$$

even for values of η which are positive and large enough so that

$$\exp\left[\frac{(2 - \alpha_a - \alpha_b)\mathfrak{F}\eta}{RT}\right] \gg \exp\left[\frac{-(\alpha_a + \alpha_b)\mathfrak{F}\eta}{RT}\right]$$

Then there is a range of overvoltage for which equation [38] becomes

$$j = -(n+m)j_a^0 \exp\left[\frac{(2-\alpha_a-\alpha_b)\mathcal{F}\eta}{RT}\right] n^{-1} \exp\left(\frac{\alpha_b\mathcal{F}\eta}{RT}\right)$$

so that $\ln j$ is linear in η with slope $(2-\alpha_a)\mathcal{F}/RT$. The apparent transfer coefficient, $2-\alpha_a$, may be greater than unity. Of course, for large enough η the inequality [42] will be reversed and equation [40] will be obtained. Thus for $j_b^0 \gg j_a^0$ the anodic Tafel plot will show two linear regions, the cathodic Tafel plot being of the usual form.

Consider an electrochemical reaction in which two electrons are transferred, such as the reduction of Cu^{2+} to Cu. This usually occurs as two one-electron steps, with an intermediate species, that is,

$$A + e^- \to B \qquad [43]$$

$$B + e^- \to C \qquad [44]$$

Associated with each step is transport of a reactant species to the electrode, electron transfer, and transport of product away from the electrode. Suppose that the concentrations of B and C in bulk are negligibly small. Let the concentration of A infinitely far from the electrode be a^∞, and let the concentrations of A, B, and C near the electrode at steady state be a_0, b_0, and c_0, respectively. To simplify the formulas, we assume that A, B, and C have equal diffusion coefficients D, so that the same effective diffusion rate constant governs the transport of all three species. Thus the flux of A to the electrode is

$$j = k_D(a^\infty - a_0) \qquad [45]$$

where $k_D = D/\delta$ and δ is the width of the diffusion layer, while the flux of B away from the electrode is $k_D b_0$ and that of C is $k_D c_0$.

At steady state the rate per unit area at which A is transported to the electrode is equal to the net rate at which it is converted to B, so

$$k_D(a^\infty - a_0) = k_1 a_0 - k_{-1} b_0 \qquad [46]$$

where k_1 and k_{-1} are rate constants per unit area for the forward and reverse of electrode reaction [43]. Similarly the net rate at which B is produced at the electrode equals the rate at which it is transported away, that is,

$$k_1 a_0 - k_{-1} b_0 - k_2 b_0 + k_{-2} c_0 = k_D b_0 \qquad [47]$$

where k_2 and k_{-2} are the rate constants per unit area for the forward and reverse of reaction [44]. The net rate of production of C is equal to the rate at which it diffuses away:

$$k_2 b_0 - k_{-2} c_0 = k_D c_0 \qquad [48]$$

The current density is the net flux of charge to the electrode; if A carries a charge ze we have, using equation [45] for the flux of A,

$$i = \mathcal{F}[zk_D(a^\infty - a_0) - (z-1)k_D b_0 - (z-2)k_D c_0]$$
$$= k_D \mathcal{F}(b_0 + 2c_0) + zk_D \mathcal{F}(a^\infty - a_0 - b_0 - c_0)$$

From equations [46]–[48], it is easy to see that $a^\infty - a_0 = b_0 + c_0$ (transport of A to the electrode = transport of B and C away), so

$$i = k_D \mathcal{F}(b_0 + 2c_0) \qquad [49]$$

and equation [48] can be written

$$k_2 b_0 = (k_D + k_{-2})(a^\infty - a_0 - b_0)$$

Solving this for b_0 and substituting into equation [46] yields

$$a_0 = a^\infty \left[1 + \frac{k_1(k_2 + k_D + k_{-2})}{k_D(k_2 + k_D + k_{-2}) + k_{-1}(k_D + k_{-2})} \right]^{-1}$$

Then one gets b_0 and c_0 in terms of a^∞ and the rate constants. Finally the current density equation (equation [49]) becomes

$$i = \frac{\mathcal{F} k_1 k_D a^\infty (2k_2 + k_D + k_{-2})}{(k_1 + k_D)(k_2 + k_D + k_{-2}) + k_{-1}(k_{-2} + k_D)} \qquad [50]$$

No assumptions have so far been made as to the relative sizes of the rate constants. The electrode reaction rate constants of course depend on the electrode potential. We can now investigate some limiting cases.

If k_2 is small compared to $k_D + k_{-2}$, the second reduction (of B to C) does not take place, because B is removed by diffusion before it can react and/or C is quickly retransformed to B (the thermodynamics of $B + e^- \to C$ are unfavorable). Then the steady-state current density of equation [50] becomes

$$i_1 = \mathcal{F} k_1 k_D a^\infty (k_1 + k_D + k_{-1})^{-1}$$

As long as the electrode potential is such that $k_2 \ll k_D + k_{-2}$, one sees a one-electron reduction wave (Chapter 1, Section Q). The limiting value of i_1 is approached when k_1 is large compared to k_D and k_{-1}:

$$i_{1,L} = \mathcal{F} k_D a^\infty$$

which is proportional to the diffusion rate and corresponds to $a_0 = 0$ (all the A arriving at the electrode is reduced immediately). The half-wave potential occurs when $k_1 = k_D + k_{-1}$. Of course, changing the electrode potential to increase k_1 will increase k_2 as well, so that one may not be able to approach $i_{1,L}$ before current from the second reduction begins to contribute. The current density $i_{1,L}$ provides a useful reference, so one rewrites equation [50] as

$$\frac{i}{i_{1,L}} = \frac{k_1(2k_2 + k_D + k_{-2})}{(k_1 + k_D)(k_2 + k_D + k_{-2}) + k_{-1}(k_{-2} + k_D)} \qquad [51]$$

Now consider the opposite extreme to that just discussed, corresponding to $k_2 \gg k_D + k_{-2}$. Equation [51] becomes

$$\frac{i}{i_{1,L}} = \frac{2k_1}{k_1 + k_D + k_{-1}(k_{-2} + k_D)/k_2} \qquad [52]$$

When k_2 is also large compared to k_{-1}, the molecules of B, formed by reduction of A, undergo a second reduction to C rather than reoxidation to A. Then

$$\frac{i}{i_{1,L}} = \frac{2k_1}{k_1 + k_D}$$

which looks like the equation for a simple reduction wave, but corresponds to a two-electron reduction. In this case, reaction [43] is the rate-determining step and reaction [44] is fast. The maximum current density is $2i_{1,L}$ so that the half-wave potential is the potential for which $k_1 = k_D$.

Now consider what happens when $k_2 \gg k_D + k_{-2}$ (so equation [52] still holds) but k_{-1} is large compared to k_2. We can also expect $k_1 \gg k_D$, so that

$$\frac{i}{i_{1,L}} = \frac{2k_1 k_2}{k_1 k_2 + k_{-1}(k_{-2} + k_D)} \qquad [53]$$

Since we have $k_{-1} \gg k_2 \gg k_D + k_{-2}$, the formation of B is followed either by its oxidation back to A or by a reduction to C, but not by its diffusion away from the electrode. The maximum current density is $2i_{1,L}$, so that we again have a two-electron wave. The half-wave potential is the potential for which $k_1 k_2 / k_{-1} = k_D$, since k_D is likely to be much larger than k_{-2}. In this case, reduction of B (equation [44]) is rate-determining. The reduction of A to B becomes a pre-equilibrium, so that concentration of B is equal to the concentration of A multiplied by the equilibrium constant $K_A = k_1/k_{-1}$.

The difference between the two-electron waves of the preceding two paragraphs is that, when the rate-determining step is reaction [43], the transition state is between A and B, and a transfer coefficient of about $\frac{1}{2}$ may be expected; but when the rate-determining step is reaction [44], the transition state is between B and C, and a transfer coefficient of about $\frac{3}{2}$ is expected, since B differs from A by 1 electron and C differs from A by 2 electrons. Therefore, the rate constant in the second case will vary more rapidly with electrode potential, so that the two-electron reduction wave will be much steeper when the rate-determining step is reaction [44]. The free energy as a function of reaction coordinate, for each case, is shown schematically in Figure 17, with asterisks indicating the transition states. In the former case, there is an appreciable barrier to the first reaction (A to B) and almost none to the second (B to C); in the latter case, the barrier between A and B is small, but that between B and C is large.

What is important are the relative sizes of k_2 and k_{-1}, and these rate constants vary in opposite directions as the electrode potential is changed. Making the potential of the electrode more negative relative to the solution will lower the energy of B relative to the energy of A and lower the energy of C relative to the energy of B, since going from A to B involves moving one electron from electrode to

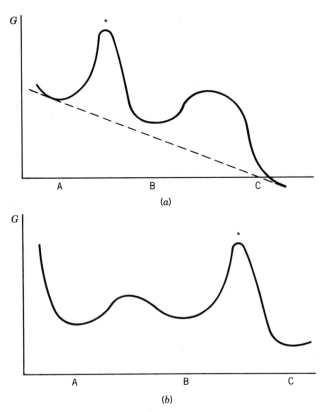

FIGURE 17. Free energy as a function of the reaction coordinate for successive one-electron reductions. In (a) the first reduction is rate-determining; in (b) the second reduction is rate-determining.

solution and going from B to C involves moving one more. On a diagram representing free energy as a function of reaction coordinate R, from A to B to C, one can imagine adding a linear function of R to represent the change in electrode potential, with increasingly negative (positive) slope for more negative (positive) E. A detailed discussion of how the barriers and rates change with potential is given by Albery (1975, pp. 137–144).

Let us denote by E' the electrode potential relative to the standard electrode potential for the A–C system, so $E' = 0$ corresponds to the symmetric free-energy diagram of Figure 18a. For a sufficiently positive value of E', the free-energy diagram may look like Figure 18b, and, for a sufficiently negative value of E', it may look like the diagram of Figure 17a, discussed above. Thus the rate-determining step, which depends on the relative sizes of the rate constants, may change with potential. For $E' > 0$, the barriers for A → B and B → C are large, but the barrier for C → B is small, so that k_1 and k_2 are small and k_{-1} is large. Assuming k_D is smaller than k_{-1}, the forward rate constant for production of C from A is $k_1 k_2 / k_{-1}$ (pre-equilibrium A ↔ B, rate-determining step B → C). As E' becomes

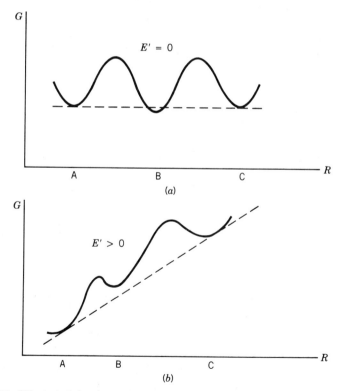

FIGURE 18. Effect of electrode potential on free-energy diagram for A → B → C reduction. (a) At $E' = 0$ the free energies of A and C are equal. (b) For $E' > 0$ the activation energies for the forward reactions (reductions) are increased and those for the reverse reactions (oxidations) are decreased.

less positive, k_1 and k_2 increase, while k_{-1} decreases, so the rate constant for the overall forward reaction increases, but the rate-determining step does not change. At some value of E', k_2 and k_{-1} will be about the same size, as in Figure 18a above. For E' below this value, the rate-determining step will be the first, the rate constant for the overall reaction being just k_1. A typical reduction wave (current as a function of potential) will be seen, the half-wave potential occurring when $k_1 = k_D$.

It is also interesting to consider the reverse reaction (oxidation of C to A). For E' positive, one will have $k_{-1} > k_{-2} > k_D > k_2 > k_1$. As long as $k_{-1} > k_2$, the observed rate constant for the reverse reaction will be k_{-2}: The first oxidation, of C to B (rate constant k_{-2}), will be the rate-determining step, the subsequent oxidation of B to A (rate constant k_{-1}) being fast. Lowering E', one will reach a potential for which $k_{-2} = k_D$: This will be the half-wave potential for the oxidation wave. The rate-determining step will remain C → B + e^- as long as $k_{-1} > k_2$. For E' sufficiently negative that $k_2 > k_{-1}$, the rate-determining step will be the oxidation of B to A, with the oxidation of C to B going to equilibrium. The rate constant for the overall oxidation will be $k_{-2}k_{-1}/(k_2 + k_{-1})$, approaching $k_{-2}k_{-1}/k_2$ for E' becoming large and negative.

D. PARALLEL ELECTRODE PROCESSES

In addition to sequential electrode processes, that is, steps in series, situations involving alternative electrode processes, that is, steps in parallel, are common. With several electrode reactions of the form

$$\sum u_{ij} O_j + n_i e^- \rightarrow \sum v_{ik} R_k \qquad [54]$$

the possibility exists for reaction to occur even at open circuit: One reaction can proceed in the forward direction (as a reduction) and the other can proceed in reverse (as an oxidation). The electron transfer from one solution species to another is then supposed to occur through the electrode, not directly in the solution.

In general, the net current at steady state may be written as a sum of currents for the different processes, each being a cathodic current minus an anodic current:

$$\begin{aligned} j &= \sum_i \left\{ k_{ri}^0 a_O^P \exp\left(\frac{-\alpha_i n_i \mathfrak{F} E}{RT}\right) - k_{oi}^0 a_R^P \exp\left[\frac{(1 - \alpha_i) n_i \mathfrak{F} E}{RT}\right] \right\} \\ &= \sum_i \left\{ k_{ri}' \prod_j a_{O_j}^{u_{ij}} \exp\left(\frac{-\alpha_i n_i \mathfrak{F} E}{RT}\right) \right. \\ &\quad \left. - k_{oi}' \prod_k a_{R_k}^{v_{ik}} \exp\left[\frac{(1 - \alpha_i) n_i \mathfrak{F} E}{RT}\right] \right\} \qquad [55] \end{aligned}$$

In the last member a_O^P and a_R^P for each reaction have been expressed in terms of the activities of $\{O_j\}$ and $\{R_k\}$ by using equilibrium expressions like equations

[28] and [29], so that a constant k' differs from k^0 by a factor of an equilibrium constant; the activities a_{O_j} and a_{R_k} may refer to location P or to bulk electrolyte. When the total current is zero (open circuit) the electrode potential E is called a *mixed potential*. Since there are currents associated with the individual reactions [54], the mixed potential is not an equilibrium potential but a steady-state potential.

For example, if all the n_i are equal to n and all the α_i are equal, the mixed potential E_0 is given by

$$0 = \sum_i \left\{ k'_{ri} \prod_j a_{O_j}^{u_{ij}} - k'_{0i} \prod_k a_{R_k}^{v_{ik}} e^{n\mathfrak{F}E_0/RT} \right\} \quad [56]$$

or

$$E_0 = \frac{RT}{n\mathfrak{F}} \ln \frac{\sum k'_{ri} \prod a_{O_j}^{u_{ij}}}{\sum k'_{oi} \prod a_{R_k}^{v_{ik}}}$$

The Nernst-equation potential for reaction m would be

$$E'_m = \frac{RT}{n\mathfrak{F}} \ln \frac{k'_{rm} \prod a_{O_j}^{u_{mj}}}{k'_{om} \prod a_{R_k}^{u_{mk}}} \quad [57]$$

so that E_0 is a kind of mean of the Nernst potentials for the parallel reactions. If one of these has higher rate constants than the others, E_0 will be closest to the Nernst potential for that reaction. An explicit expression for the mixed potential may also be found if only two terms in equation [55] are important. Thus if k'_{r1} and k'_{o2} are large enough so that other terms may be neglected,

$$k'_{r1} \prod a_{O_j}^{u_{1j}} \exp\left(\frac{-\alpha_1 n_1 \mathfrak{F} E_0}{RT}\right) = k'_{o2} \prod a_{R_k}^{u_{2k}} \exp\left[\frac{(1-\alpha_2) n_2 \mathfrak{F} E_0}{RT}\right]$$

and

$$E_0[n_1\alpha_1 + n_2(1-\alpha_2)] = \frac{RT}{\mathfrak{F}} \ln \frac{k'_{r1} \prod a_{O_j}^{u_{1j}}}{k'_{o2} \prod a_{R_k}^{u_{2k}}}$$

More generally, the mixed potential may be determined from plots of E vs. j (more conveniently, potential vs. $\ln|j|$ for the Tafel-plot regions) for individual reactions, the intersection point giving E_0 for the system.

On the other hand, polarization (potential vs. current) curves can be used to find out, for potentials such that one or the other electron-transfer mechanism dominates, whether the processes are parallel or sequential (Andersen and Eyring 1970, pp. 300–301). Suppose the individual processes give linear Tafel plots, so that (see Figure 19) the line 1–1′ is obtained for process 1 and the line 2–2′ is obtained for process 2. For parallel processes, the faster one (largest k'_1) controls the current,

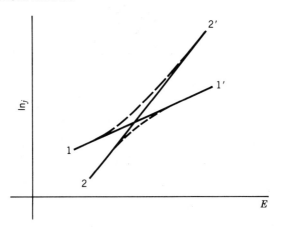

FIGURE 19. Tafel plots for separate reactions, 1–1' and 2–2'. If both can occur, current will follow dashed curve if the reactions are parallel and will follow dotted curve if they are sequential.

so that the overall current will follow 1–1' at lower potentials and 2–2' at higher potential, giving a $\ln j$ vs. E curve that is concave upward (dashed in Figure 19). For consecutive processes, the slowest step is rate-determining and controls the current, and the $\ln j$ vs. E curve will be concave downward (dotted curve in Figure 19), following 2–2' for lower E and 1–1' for higher E.

E. REACTION OVERVOLTAGE

If chemical reactions that precede or follow a charge-transfer step are slow, so that equilibrium is not attained, the rate of the overall reaction will be governed by the chemical reaction rates as well as the rate of the charge-transfer process. Analogously to what happens when diffusion processes are not fast enough to maintain concentrations of O and R at their bulk values, the electrode potential for a given current density will depend on the chemical reaction rates, leading to a *reaction overvoltage*.

We derive the overvoltage for the reaction [15], with the charge-transfer step [25] and the preceding and following reactions [26] and [27]. We assume negligible diffusion overvoltage. Let equations [26] and [27] have forward rate constants k_1^f and k_2^f and reverse rate constants k_1^r and k_2^r; the equilibrium constants used previously are $K_O = k_1^f/k_1^r$ and $K_R = k_2^f/k_2^r$. At steady state,

$$k_1^f \prod a_{O_i}^{u_i} = k_1^r a_O^s + \frac{sj}{n} G \qquad [58]$$

where G is the electrode area divided by the solution volume, so that the last term is the number of moles of O removed from the system per unit volume by the charge-transfer process. Similarly,

$$\frac{tj}{n} G + k_2^f a_R^t = k_2^r \prod a_{R_j}^{v_j} \qquad [59]$$

The current is given by equation [55] of Chapter 1:

$$j = n\mathcal{F} \left[k_r^0 \left(\frac{a_O}{a_{O_e}} \right)^s (a_{O_e})^s e^{-\alpha n\mathcal{F}E/RT} \right.$$
$$\left. - k_0^0 \left(\frac{a_R}{a_{R_e}} \right)^t (a_{R_e})^t e^{(1-\alpha) n\mathcal{F}E/RT} \right]$$

where a_{O_e} and a_{R_e} are the equilibrium values of a_O and a_R, given by equations [28] and [29]. Introducing the exchange current of equation [31], we have

$$j = j^0 \left[\left(\frac{a_O}{a_{O_e}} \right)^s e^{-\alpha n\mathcal{F}\eta/RT} - \left(\frac{a_R}{a_{R_e}} \right)^t e^{(1-\alpha) n\mathcal{F}\eta/RT} \right] \qquad [60]$$

Using equation [58] for a_O and equation [59] for a_R,

$$\left(\frac{a_O}{a_{O_e}} \right)^s = \frac{K_O \prod a_{O_i}^{u_i} - (sj/n)(G/k_1^r)}{K_O \prod a_{O_i}^{u_i}} \qquad [61]$$

$$\left(\frac{a_R}{a_{R_e}} \right)^t = \frac{\prod a_{R_j}^{v_j} - (tj/n)(G/k_2^f) K_R}{\prod a_{R_j}^{v_j}} \qquad [62]$$

It may be seen that if sjG/n is small compared to $k_1^f \prod a_{O_i}^{u_i}$ and tjG/n is small compared to $k_2^r \prod a_{R_j}^{v_j}$, then $a_O/a_{O_e} = a_R/a_{R_e} = 1$, and the current is given by equation [2], with the overvoltage being related to the charge-transfer process.

In the other extreme, the charge-transfer process is fast enough so that equilibrium is attained at the electrode and the electrode potential is given by

$$E = E_{1/2}^0 + \frac{RT}{n\mathcal{F}} \ln \frac{a_R^t}{a_O^s} \qquad [63]$$

but a_R or a_O does not have its equilibrium value (equations [28] and [29]) because of the slowness of equation [26] or [27]. The reaction overvoltage is

$$E - E' = \frac{RT}{n\mathcal{F}} \left(\ln \frac{a_R^t}{a_O^s} - \ln \frac{a_{R_e}^t}{a_{O_e}^s} \right)$$
$$= \frac{RT}{n\mathcal{F}} \left(t \ln \frac{a_R}{a_{R_e}} - s \ln \frac{a_O}{a_{O_e}} \right) \qquad [64]$$

In general, the relationship between current and potential is obtained by substituting equations [61] and [62] in equation [60] and solving for j.

Suppose the slow reaction is the one preceding the charge transfer so $a_R = a_{R_e}$. Then equation [60] becomes

$$j = j^0 \left[\left(K_O \prod a_{O_i}^{u_i} - \frac{sj}{n} \frac{G}{k_1^r} \right) \left(K_O \prod a_{O_i}^{u_i} \right)^{-1} \right.$$

$$\left. \cdot e^{-\alpha n \mathcal{F} \eta / RT} - e^{(1-\alpha) n \mathcal{F} \eta / RT} \right]$$

and, solving for the current density,

$$j = \frac{j^0 (1 - e^{n\mathcal{F}\eta/RT})}{e^{\alpha n \mathcal{F} \eta/RT} + (sj^0/n)(G/k_1^f) \left(\prod a_{O_i}^{u_i} \right)^{-1}} \quad [65]$$

As $\eta \to -\infty$, there is a limiting current given by $nk_1^f (\prod a_{O_i}^{u_i})/sG$. Its value is independent of the exchange current j^0, since it represents the rate at which O is formed by the slow reaction. Double-layer effects on electrode kinetics are discussed by Andersen and Eyring (Andersen and Eyring 1970, Section 3.V), as well as reactions involving specific adsorption (Andersen and Eyring 1970, Section 3.VI).

F. ELECTROCATALYSIS

In Section D we indicated the existence of charge-transfer reactions between solution species which occur very slowly, if at all, in the absence of an electrode, but appreciably fast when electrons are transferred through the electrode, even with no net current flowing. The electrode in such a case is acting as a catalyst, and we would expect that its chemical composition, as well as its state of charge, is important. In fact, this is true for most electrochemical processes: The exchange current for a particular process may change by 10 orders of magnitude with changing electrode material, and different products may be obtained for different electrodes in the same electrolyte. The catalytic activity may be associated with the interaction of the electrode with adsorbed reactants, products, intermediates, or solvent; or it may be explained in terms of changes in the nature or stability of the activated complex. Furthermore, considering electrons from the metal electrode as reactants or products, one knows that their free energy or chemical potential is different for different metals, even at the same metal–electrolyte potential difference.

We first consider the simplest reaction, $O + ne^- \to R$, supposing that O and R are both solution species, with O having a charge ze and R having a charge $(z - n)e$. Interaction of O and R with the metal is assumed unimportant, as is the difference between ϕ^P (potential in the reaction region) and ϕ^{sol}. The free energy of activation for the forward (reduction) reaction is $\tilde{\mu}_{\ddagger}^0 - \tilde{\mu}_O^0 - \tilde{\mu}_e^0$, where the first term is the electrochemical potential of the activated complex. Since the activated complex is intermediate between the reactant O^{z+} and the product $R^{(z-n)+}$, its charge is intermediate, say $(z - \alpha n)\,e$, so that in the activated state a fraction α of the n electronic charges has been transferred from metal to the solution species. The fraction α, it turns out, may be identified with the transfer coefficient (Andersen and Eyring 1970, Section 3.VII.B). The free energy change associated with loss of αn electrons from the metal is $-\alpha n\mu_e^0 + \alpha n \mathcal{F}\phi^M$, where ϕ^M is the inner potential of the metal. With $\phi^{\ddagger}$ as the potential at the location of the activated complex, and the inner potential of the solution ϕ^{sol} taken as zero, we may write

$$\Delta G_f^{0\ddagger} = \mu_{\ddagger}^0 + \mathcal{F}(z - \alpha n)\phi^{\ddagger} - \alpha n \mu_e^0 + \alpha n \mathcal{F}\phi^M - \mu_O^0 \qquad [66]$$

for the free energy of activation.

At the electrode potential corresponding to equilibrium (reversible potential), $\tilde{\mu}_O^0 + n\tilde{\mu}_e^0 = \tilde{\mu}_R^0$ or

$$\mu_O^0 + n\mu_e^0 - \mu_R^0 - n\mathcal{F}\phi^M = 0 \qquad [67]$$

so that, at this potential, the free energy of activation is

$$\Delta G_f^{0\ddagger} = \mu_{\ddagger}^0 + \mathcal{F}(z - \alpha n)\phi^{\ddagger} + \alpha(\mu_O^0 - \mu_R^0) - \mu_O^0$$

If there is no interaction between activated complex and electrode, $\mu_{\ddagger}^0$ is independent of electrode material and $\phi^{\ddagger}$ is ϕ^{sol}, which we have taken as zero. Then the free energy of activation, and hence the rate of reaction, is independent of the nature of the electrode. Note that the reversible potential is different for different electrode materials.

At the potential of zero charge (pzc), the metal–solution potential difference is the difference of two dipole potentials:

$$\phi^M = \chi^{M(sol)} - \chi^{sol(M)} \qquad [68]$$

where $\chi^{i(j)}$ is the potential drop on going from the interior to the exterior of phase i when it is adjacent to phase j. Then, at the pzc (which, like the reversible potential, is different for different electrode materials) the free energy of activation (equation [66]) is

$$\mu_{\ddagger}^0 + \mathcal{F}(z - \alpha n)\phi_{\ddagger}^0 - \alpha n \mu_e^0 - \mu_O^0 + \alpha n \mathcal{F}(\chi^{M(sol)} - \chi^{sol(M)})$$

Again supposing no interaction of the activated complex with the electrode so $\phi^{\ddagger} = \phi^{sol} = 0$, we may introduce the work function $\Phi = \mu_e^0 - \mathcal{F}\chi^{M(vac)}$ to write the free energy of activation as

$$\Delta G_f^{0\ddagger} = \mu_{\ddagger}^0 - \mu_O^0 - \alpha n \Phi + \alpha n \mathcal{F}(-\chi^{M(vac)} + \chi^{M(sol)} - \chi^{sol(M)}) \qquad [69]$$

The effect of the solvent on the surface potential of the metal, $\chi^{M(\text{sol})} - \chi^{M(\text{vac})}$, is relatively small (see Chapter 4, Section B). If the surface potential of the solution on the electrode surface is also small, or if the surface potentials in the last term of equation [69] cancel, the difference of reaction rates at the pzc for different electrode materials should reflect the difference in work functions.

The specific interaction between reactants and the electrode may be by way of the variation of $\mu_{\ddagger}^0$ and $\phi^{\ddagger}$ with electrode material: $\mu_{\ddagger}^0$ may vary just as does the chemical potential of any adsorbed species, and the ratio of $\phi^{\ddagger}$ to ϕ^M may depend on the electrode. If the reaction proceeds by way of an adsorbed intermediate, there are obvious further ways in which the nature of the electrode can affect the rate. An example of this type was considered by Andersen and Eyring (1970, Section 3.VII.C.2) to show the effects.

The reaction was supposed to occur in two steps:

$$O^+ + e^- \rightarrow A \qquad [70]$$

$$A + B \rightarrow P \qquad [71]$$

where A is an adsorbed intermediate and O^+, B, and P are solution species. Double-layer effects are disregarded and A is supposed to adsorb according to the Langmuir isotherm (Chapter 2, Section F, equations [43] and thereafter). Let θ be the fraction of surface covered by A. The rates per unit electrode area of the forward and reverse reactions for process [70] are

$$v_a = \frac{kT}{h} a_{O^+}(1-\theta) \exp\left(\frac{-\Delta G_a^{0\ddagger}}{RT}\right) \exp\left(\frac{-\alpha_a \mathcal{F} \phi^M}{RT}\right)$$

$$v_{-a} = \frac{kT}{h} \theta \exp\left(\frac{-\Delta G_{-a}^{0\ddagger}}{RT}\right) \exp\left[\frac{(1-\alpha_a)\mathcal{F}\phi^M}{RT}\right] \qquad [72]$$

The rates per unit area for the forward and reverse reactions of process [71] are

$$v_b = \frac{kT}{h} \theta a_B \exp\left(\frac{-\Delta G_b^{0\ddagger}}{RT}\right)$$

$$v_{-b} = \frac{kT}{h} a_P(1-\theta) \exp\left(\frac{-\Delta G_{-b}^{0\ddagger}}{RT}\right) \qquad [73]$$

The free energies of activation for process [70] have been separated into chemical and electrical parts; for process [71], all species are supposed to be electrically neutral.

Let ϕ_e be the electrode–solution potential for which process [70] is at equilibrium and let θ_e be the value of the coverage θ at this potential. The equation $v_a = v_{-a}$ or

$$a_{O^+}(1-\theta_e)\exp\left(\frac{-\Delta G_a^{0\ddagger}}{RT}\right)\exp\left(\frac{-\alpha_a \mathscr{F}\theta_e}{RT}\right)$$

$$= \theta_e \exp\frac{-\Delta G_{-a}^{0\ddagger}}{RT}\exp\left[\frac{(1-\alpha_a)\mathscr{F}\theta_e}{RT}\right] \quad [74]$$

may easily be solved for $\exp[\mathscr{F}\theta_e/RT]$. Substituting this into the expression for v_a we have an equation for the exchange current density:

$$j^0 = kT\mathscr{F}h^{-1} a_{O^+}(1-\theta_e)\exp\left(\frac{-\Delta G_a^{0\ddagger}}{RT}\right)$$

$$\times \left\{a_{O^+}(1-\theta_e)\theta_e^{-1}\exp\left[\frac{(\Delta G_a^{0\ddagger} - \Delta G_{-a}^{0\ddagger})}{RT}\right]\right\}^{-(\alpha_a)}$$

$$\ln j^0 = \ln\frac{kT\mathscr{F}}{h} + \alpha_a \ln\theta_e + (1-\alpha_a)\ln\left[a_{O^+}(1-\theta_e)\right]$$

$$-\frac{(1-\alpha_a)\Delta G_a^{0\ddagger}}{RT} - \frac{\alpha_a \Delta G_{-a}^{0\ddagger}}{RT} \quad [75]$$

The standard adsorption free energy of A, ΔG_{ads}^0, is an additive term in the free energy of reaction for reaction [70], and hence it affects the free energies of activation for reaction [70] and its reverse reaction. By the arguments of Section A regarding the effect of raising or lowering one of the potential curves whose intersection marks the activated complex, an increase in ΔG_{ads}^0 by a quantity D will increase $\Delta G_a^{0\ddagger}$ by $\alpha_a D$ and decrease $\Delta G_{-a}^{0\ddagger}$ by $(1-\alpha_a)D$, so the sum of the last two terms in equation [75] is independent of ΔG_{ads}^0. Therefore the effect on the exchange current of the differing interactions between the electrode and A is found only in reaction [71].

The equilibrium condition, $v_b = v_{-b}$, may be written

$$\frac{\theta_e}{1-\theta_e} = \frac{a_P}{a_B}\exp\left[-\frac{\Delta G_{-b}^{0\ddagger} - \Delta G_b^{0\ddagger}}{RT}\right]$$

$$= \frac{a_P}{a_B}\exp\left(\frac{\Delta G_b^0}{RT}\right)$$

where ΔG_b^0 is the standard free-energy change for reaction [71]. The quantity $-\Delta G_{ads}^0$ is an additive term in ΔG_b^0. To single out the role of adsorption, the quantity $\theta_e/(1-\theta_e)$ is written as $K\exp(-\Delta G_{ads}^0/RT)$ (Chapter 2, equation [47]). Then, solving for θ_e and substituting into equation [75], one obtains

$$\ln j^0 = (1 - \alpha_a) \ln a_{O^+} + \alpha_a \ln \left[K \exp \left(\frac{-\Delta G^0_{ads}}{RT} \right) \right]$$

$$- \ln \left[1 + K \exp \left(\frac{-\Delta G^0_{ads}}{RT} \right) \right] + Q \qquad [76]$$

where Q groups together all the terms that do not depend on the free energy of adsorption. If one considers exchange currents for a sequence of different electrode materials, equation [76] shows that $\ln j^0$ will be maximal for $K \exp(-\Delta G^0_{ads}/RT) = \alpha_a/(1 - \alpha_a)$, assuming that α_a depends on the reaction and does not vary from one material to another. For ΔG^0_{ads} large and positive or large and negative, j^0 becomes small. Use of an isotherm other than the Langmuir changes the shape of the $\ln j^0$ vs. ΔG^0_{ads} curve, but not the qualitative behavior. A similar end result also obtains if instead of reactions [70] and [71] one has a two-step reaction in which the first step involves adsorption without charge transfer and the second (desorption) step involves exchange of electrons with the electrode.

G. ELECTRONIC ASPECTS

Any electrode reaction involves transfer of an electron between the electrode and a species in the electrolyte phase. The quantum mechanical aspects of this will be discussed in Chapter 9. Electron transfer is a rapid process compared to atomic movements, so that the arrangement of the atoms does not change during electron transfer (Franck–Condon principle). Conservation of energy requires (unless radiation is involved) that there must be a matching of the energy level of the donor with that of the acceptor. Since molecules in a fluid are constantly undergoing motion and changing their energies, the matching must be to within the thermal energy kT.

In the case of electron transfer between a solid and a molecular species, the chemical nature of the solid is not changed. Because the energy levels of a solid refer to delocalized states, the energy of such a level depends very little on whether it is occupied or not, but this is not the case for a molecular species with localized orbitals. Further, the interaction of a solvated species with the solvent depends strongly on the species' charge. Remembering that electron transfers occur with no movement of nuclei, according to the Franck–Condon principle, we write a reaction for the ionization of the solvent species R,

$$(R)_{solv, R} \rightarrow (O^+)_{solv, R} + e^- \qquad [77]$$

where solv, R implies a solvation sphere appropriate to the species R, and e^- represents an electron at infinity. The energy for this reaction is I^0, the ionization potential of the solvated species R. The neutralization of the charge of O^+ by an electron must be written

G. ELECTRONIC ASPECTS

$$(O^+)_{solv,O} + e^- \to (R)_{solv,O} \qquad [78]$$

and is associated with an energy change $-A^0$ (A^0 is an electron affinity).

Process [78] is not the reverse of process [77]. The rearrangement of solvent around O^+ from a configuration appropriate to R to one appropriate to O^+,

$$(O^+)_{solv,R} \to (O^+)_{solv,O} \qquad [79]$$

involves a free energy change that we denote by $-\lambda_O$; the corresponding rearrangement around R,

$$(R)_{solv,R} \to (R)_{solv,O} \qquad [80]$$

involves a free energy change λ_R. Change in solvation structure means that λ_R and λ_O involve entropy as well as energy changes. The relation between the four free energies, since one has the cycle $(R)_{solv,R} \to (O^+)_{solv,R} \to (O^+)_{solv,O} \to (R)_{solv,O} \to (R)_{solv,R}$, is

$$I^0 - \lambda_O - A^0 - \lambda_R = 0 \qquad [81]$$

The combination $I^0 - \lambda_O$ is the free-energy change for

$$(R)_{solv,R} \to (O^+)_{solv,O} + e^- \qquad [82]$$

It must be noted that the highest occupied energy level of R, from which the electron is removed in reaction [77], depends on the solvent configuration, and the most probable energy, as a function of solvent configurations, is being referred to here. Furthermore, the electron in the actual electrode process is not at rest at infinity, but instead is at rest in the electrode, with an energy equal to the Fermi energy. Thus the electron-transfer reaction is

$$(R)_{solv,R} \to (O^+)_{solv,O} + e^-(\epsilon_F) \qquad [83]$$

and the free-energy change for reaction [83] differs from that for reaction [82] by Φ, which represents the difference between the Fermi energy ϵ_F and the energy of an electron at rest at infinity. The free-energy difference ΔG for reaction [83] changes by $-\mathcal{F}\delta E$ when the electrode-solution potential difference E changes by δE. For reaction [83] to be at equilibrium, the electrode-solution potential difference E must make the free energy of $(O^+)_{solv,O} + e^-(\epsilon_F)$ equal to that of $(R)_{solv,R}$. Thus the electrode-solution potential difference $E^0_{1/2}$, which produces equilibrium, differs from the chemical potential difference of reaction [82] by Φ.

In general, the anodic current density, or the rate for process [83], is proportional to the concentration of oxidized species in solution c_O and to $\int \mathcal{K}(E)\, D_+(E)$

$W_R(E)\,dE$, where $\mathfrak{X}(E)$ is a transfer probability for an electron at an energy E, $D_+(E)$ is the number of unoccupied electronic states per unit energy range in the electrode, and $W_R(E)$ is the probability for finding R in a state corresponding to an energy E. The energy E is actually the energy of an electron state in the metal plus the species O^+ in a solvation sphere corresponding to R, so that $W_R(E)$ is peaked at $E = -I_0$. Similarly, the cathodic current density, corresponding to the reverse of reaction [83], is proportional to the concentration of reduced species c_R and to $\int \mathfrak{X}(E)\,D_-(E)\,W_O(E)\,dE$, where $D_-(E)$ is the number of occupied electronic states per unit energy range. The energy E here is for an electron in the metal plus the species O^+ in a solvation sphere corresponding to O^+, so that W_O is peaked at A_O relative to $(R)_{\text{solv},O}$. The distribution $W_R(E)$ and $W_O(E)$ refer to time-averaged energies for R and O, since the energy of a solvated species fluctuates due to interactions with the solvent.

H. SEMICONDUCTOR ELECTRODES

Gerischer (1970, Sections 5.III.B, C, D) reviews methods for investigating mechanisms for redox reactions on semiconductors and discusses experimental results. We now consider some of the theoretical ideas involved in understanding these results. As mentioned in Sections G and H of Chapter 4, for an intrinsic semiconductor we may suppose that the valence band is filled, except for the levels near its upper band edge, and that the conduction band is empty, except for some population of levels near its lower band edge.

Conservation of energy for electron transfer between a semiconductor electrode and a solution species requires the latter to have an electronic level at an energy within the conduction or valence band. Let A and D^- be acceptor and donor molecules in solution. Writing the equation for the electron-transfer process

$$A + e^- \leftrightarrow D^-$$

implies an electron in a level of the conduction band (normally empty for a semiconductor) and an unfilled level on the acceptor A at the same energy. Alternatively, one writes

$$A \leftrightarrow D^- + h^+$$

to imply that the unfilled level of A is at the energy of a filled level in the valence band, from which an electron may be transferred, leaving a hole. The energy of the acceptor level is the difference in energy between D^- and A.

It is sometimes convenient to divide the cathodic and anodic currents into conduction-band (electron) and valence-band (hole) contributions. One or the other will be important, depending on whether the peaks in W_O and W_R (see end of Section G) occur for energies near one band edge or the other. The electron transfer occurs in a narrow energy range at a band edge, so the integrals over energy reduce to their values at the band edges. Then the net current due to the conduction band can be represented as

$$j_C = k_C^+ \kappa(E_C) \, W_R(E_C) N_C c_R - k_C^- \kappa(E_C) \, W_O(E_C) n_s c_O$$

where N_C is the density of (mostly empty) electron states, and n_s is the density of occupied states or electrons, at the band edge E_C; the net current due to the valence band is

$$j_V = k_V^+ \kappa(E_V) \, W_R(E_V) p_s c_R - k_V^- \kappa(E_V) \, W_O(E_V) N_V c_O$$

where E_V is the valence band edge, N_V is the density of electronic states, and p_s is the density of unoccupied states (positive holes), at E_V. The superscripts on the rate constants refer to anodic ($-$) and cathodic ($+$) currents.

For bulk concentrations c_R^0 and c_O^0, there is an electrode–solution potential that produces equilibrium: j_C and j_V both vanish, with the cathodic and anodic contributions to j_C each becoming equal to j_C^0 and the cathodic and anodic contributions to j_V each becoming j_V^0. Letting n_s^0 and p_s^0 be the equilibrium concentrations of surface electrons and surface holes, we have

$$j_C^0 = k_C^+ \kappa(E_C) \, W_R(E_C) N_C c_R^0 = k_C^- \kappa(E_C) \, W_O(E_C) c_O^0 n_s^0$$

$$j_V^0 = k_V^+ \kappa(E_V) \, W_R(E_V) p_s^0 c_R^0 = k_V^- \kappa(E_V) \, W_O(E_V) N_V c_O^0$$

Now, letting c_R and c_O be concentrations of R and O near the electrode surface, we get

$$j_C = j_C^0 \left(\frac{c_R}{c_R^0} - \frac{c_O n_s}{c_O^0 n_s^0} \right) \qquad [84]$$

$$j_V = j_V^0 \left(\frac{c_R p_s}{c_R^0 p_s^0} - \frac{c_O}{c_O^0} \right) \qquad [85]$$

The value of n_s is proportional to $\exp(-e\phi^s/kT)$, where ϕ^s is the potential at the semiconductor surface relative to the interior, while p_s/p_s^0 is proportional to $\exp(e\phi^s/kT)$.

Let ϕ_0^s be the value of ϕ^s at equilibrium. If the change in the potential difference across the electrolyte's Helmholtz layer is small, then $\phi^s - \phi_0^s$ is the negative of the overvoltage η, so that

$$\frac{n_s}{n_s^0} = \exp\left[\frac{-(\phi^s - \phi_0^s)}{kT} \right] = \exp\left(\frac{-e\eta}{kT} \right)$$

and $p_s/p_s^0 = e^{e\eta/kT}$. Therefore, if the concentrations c_R and c_O at the electrode surface are equal to the concentrations in solution (no concentration polarization), we get

$$j_C = j_C^0(1 - e^{-e\eta/kT})$$
$$j_V = j_V^0(e^{e\eta/kT} - 1)$$

If the electrode current is due to the conduction band, then the rate of the cathodic process, but not the anodic process, is influenced by the applied voltage. If the valence band is responsible for the current, then the anodic, but not the cathodic, process changes with η.

In addition to redox reactions like those just discussed, in which the semiconductor electrode serves only as a source or sink of electrons, one has to consider reactions in which the electrode participates chemically (electrodes of the first, second, or third kinds, Chapter 1, Section D). The interactions between atoms in a semiconductor involve strong covalent bonds, so that the energy barrier to chemical attack is generally higher for a semiconductor than for a metal. On the other hand, the electrons and holes play an important role in decomposition reactions of semiconductors, since they are electronic defects in the bond structure. In an electrochemical oxidation, for instance, electrons associated with bonds involving surface atoms are removed under the influence of an external potential (cell potential), thus breaking bonds and creating species that can react with a constituent of the electrolyte. The chemical reaction at the surface is thus associated with transport of charge, in the form of electrons or holes, through the semiconductor. The chemical reaction is irreversible: An equilibrium distribution of charge carriers is maintained in the interior of the semiconductor but not at the surface.

Gerischer (1970, Section 5.IV.A.1) discusses the electrochemical reaction of A (in the electrode) with X^- (in the solution), in which an A–A bond is broken and an A–X molecular species is formed. The bond is supposed to be in a special (kink) position on the crystal surface, such that removal of two electrons removes an A atom from the chemical bond structure of the solid. The electrons are removed one by one, with removal of the first leading to the formation of a radical intermediate A·, which then reacts with another X^- ligand. The mechanism is thus

$$-A-A + X^- \rightarrow -A-X + A\cdot + e^- \qquad [86]$$
$$-A-A + X^- + h^+ \rightarrow -A-X + A\cdot \qquad [87]$$
$$A\cdot + X^- \rightarrow A-X + e^- \qquad [88]$$
$$A\cdot + X^- + h^+ \rightarrow A-X \qquad [89]$$

where the two reactions [86] and [87] are competitive alternatives, as are the two reactions [88] and [89].

For the rate constant of each reaction, one can write an expression involving a frequency factor and an activation energy factor. The rates of reactions [86] and [87] are proportional to the concentration of X^- and the number of kink sites. The rate of reaction [86] is also proportional to the effective density of states in the conduction band (where the electron goes) N_C, and the rate of reaction [87] is proportional to the surface hole concentration p_s. Since it can be expected that the frequency factors are similar, the preference for reaction [86] and [87] depends on the difference in activation energies and on the ratio N_C/p_s. Similarly, the preference for reaction [88] or [89] depends on the difference in activation energies for these two reactions and on the ratio N_C/p_s. An estimate of these quantities is made by the following arguments.

Since creation of a hole h^+ requires excitation of an electron from the valence band to the conduction band, the energy of the initial state of reaction [87] is higher by the band gap E_g than the energy of the initial state of reaction [86]. One may expect the transition states for the two reactions to have similar energies, so that the energy of activation of reaction [86] exceeds that of reaction [87] by something like E_g, say $\gamma_1 E_g$. The ratio of p_s to N_V, the state density in the valence band, is $\exp[-(\epsilon_F^s - E_V)/kT]$, where ϵ_F^s is the Fermi level in the surface and E_V is the top of the valence band. Since the effective densities of states N_V and N_C are about the same, the ratio of the rate of reaction [87] to that of reaction [86] is approximately $\exp\{[\gamma_1 E_g - (\epsilon_F^s - E_V)]/kT\}$, which is greater than unity if $\epsilon_F^s - E_V$ is less than $\gamma_1 E_g$. The hole reaction is thus more important than the electron reaction if the surface is p-type (Fermi level below the average energy of the band edges), growing in importance for larger band gap since γ_1 will be closer to unity.

Similar arguments are applied to reactions [88] and [89]. In the case of reaction [88] the unpaired electron is initially somewhere in the interband gap, say at E_R, and it must be raised to the conduction band edge to be removed, which requires an energy $E_g - (E_R - E_V)$. For reaction [89], the activation energy is much smaller, so the difference of activation energies for reactions [88] and [89] is $\gamma_2[E_C - E_V - (E_R - E_V)]$, where γ_2 is something less than unity and $E_g = E_C - E_V$ has been used. Since $p_s/N_V = \exp[(\epsilon_F^s - E_V)/kT]$, the ratio of the rate of reaction [89] to the rate of reaction [88] is $(N_V/N_C)\exp\{[\gamma_2(E_C - E_R) - (\epsilon_F^s - E_V)]/kT\}$, and N_V/N_C is of size unity. If E_R is close to the valence band edge, the ratio is similar to that for reactions [86] and [87]. If, however, E_R is close to the middle of the band gap, and $\epsilon_F^s - E_V$ is not too large, the electron process [88] will predominate over the hole process [89]. For larger band gaps, the hole process will again be more likely. When ϵ_F^s is close to E_V and far below the middle of the gap, the electrochemical process will be hole-consuming in both steps.

A similar discussion can be given for compound semiconductors, in which the first step involves reaction of X^- with a heteropolar bond –A–B to form –A–X + B· or –B–X + A·. It is predicted that the more electropositive element of A and B will form a bond to the (nucleophilic) ligand X in this step, with the more electronegative element remaining as a radical. Association of adjacent radicals to form a bond is a possible subsequent reaction. The effect of polarity on the relative importance of the electron and hole mechanisms may be predicted. Gerischer (1970, Section 5.IV.A.2) discusses experimental examples in light of these theoretical arguments.

A reduction may also lead to bond weakening and decomposition of an electrode if the added electron goes into an antibonding orbital of a surface atom. Then reaction with an electrophilic species from the electrolyte removes one partner in the bond and leaves the other behind as a radical. For a heteropolar A–B bond with B more electronegative than A, Gerischer (1970) considers the series of reactions

$$-A-B + e^- + Y^+ \xrightarrow[V_1]{} A\cdot + B-Y \qquad [90]$$

$$-A-B + Y^+ \xrightarrow[V_2]{} A\cdot + B-Y + h^+ \qquad [91]$$

$$A\cdot + e^- + Y^+ \xrightarrow[V_3]{} A-Y \qquad [92]$$

$$A\cdot + Y^+ \xrightarrow[V_4]{} A-Y + h^+ \qquad [93]$$

$$A\cdot + A\cdot \xrightarrow[V_5]{} A-A \qquad [94]$$

where the V_i are reaction rates, each proportional to a product of concentrations of reacting species, a frequency factor, an exponential of an activation energy, and a product of a state density, electron density, or hole density. Reactions [90] and [91] are alternatives, as are reactions [92] and [93]. By arguments similar to those for the oxidation, one shows that

$$\frac{V_1}{V_2} \sim \frac{N_C}{N_V} \exp\left(\frac{\epsilon_F^s - E_C + \gamma_1 E_g}{kT}\right)$$

for the competing reactions [90] and [91]. Since the antibonding orbitals are largely atomic orbitals of the less electronegative element, the energy of the unpaired electron in $A\cdot$ is close to the bottom of the conduction band, so that the hole process [93] becomes unlikely, requiring excitation of an electron from the valence band. The prediction is that, when E_g is large, the reduction process should involve electrons and not holes. The experimental situation is complicated (Gerischer 1970, Section 5.IV.B.2).

Finally, there is the possibility of electrochemical corrosion, in which there is no external current, but simultaneous compensating anodic and cathodic processes on the surface. This is differentiated from direct chemical oxidation by the fact that the rate of an electrochemical corrosion process depends on the state of charge of the surface. In contrast to a metal surface, the field at a semiconductor surface cannot be varied much. The state of charge acts by affecting the surface concentrations of electrons and holes. The surface hole concentration affects the rate of an anodic reaction proceeding in the valence band, while the surface electron concentration affects the rate of a cathodic reaction in the conduction band.

The vanishing of the net current, $j_C^+ - j_C^- + j_V^+ - j_V^- = 0$, may be written as the balancing of the net currents in each energy band

$$j_C = j_C^+ - j_C^- = -j_V = -j_V^+ + j_V^- \qquad [95]$$

or as the equality of anodic and cathodic currents

$$j_C^+ + j_V^+ = j_C^- + j_V^-$$ [96]

The most common situation is that j_C is positive and j_V is negative, meaning electrons and holes are simultaneously injected. Oxidants then pick up electrons from the valence band and pick up holes from (give up electrons to) the conduction band.

9

QUANTUM THEORY OF ELECTRON TRANSFER

The process of electron transfer between electrode and electrolyte species has not been considered in the discussions to this point, except to say it proceeds rapidly, with the nuclei standing still, and to note that there is conservation of energy. Theories of the electron transfer itself must be quantum mechanical in nature. This is the subject of the present chapter.

A. ELECTRONIC TRANSITIONS

Since an electron must change its state or undergo a transition during the charge-transfer process, one must consider solutions to the time-dependent Schrödinger equation:

$$\frac{i\hbar \partial \Phi}{\partial t} = H\Phi \qquad [1]$$

If the time-dependent wave function Φ is expanded in a set of orthogonal and normalized wave functions for stationary states,

$$\Phi = \sum_k c_k(t)\, \phi_k \exp\left(\frac{-iE_k t}{\hbar}\right) \qquad [2]$$

the absolute square of the kth coefficient, $|c_k|^2$, is related to the probability that the system actually is in state k with wave function ϕ_k and energy E_k (Goodisman 1977, Section 13.2). This may be seen by considering the expectation value of H_0, the Hamiltonian operator for which the ϕ_k are eigenfunctions ($H_0 \phi_k = E_k \phi_k$):

$$\langle H_0 \rangle = \frac{\int \Phi^* H_0 \Phi \, d\tau}{\int \Phi^* \Phi \, d\tau}$$

$$= \frac{\sum_l \sum_k c_l^* c_k \int \phi_l^* H_0 \phi_k \, d\tau \exp\left[i(E_l - E_k) t/\hbar\right]}{\sum_l \sum_k c_l^* c_k \int \phi_l^* \phi_k \, d\tau \exp\left[i(E_l - E_k) t/\hbar\right]}$$

Since $H_0 \phi_k = E_k \phi_k$ and the functions are orthonormal ($\int \phi_l^* \phi_k \, d\tau = \delta_{lk}$), the integral in the numerator is $E_k \delta_{lk}$, where $\delta_{lk} = 1$ for $l = k$ and 0 otherwise. Then

$$\langle H_0 \rangle = \frac{\sum_k c_k^* c_k E_k}{\sum_k c_k^* c_k} \qquad [3]$$

which is a weighted mean of the E_k, so that $|c_j|^2 / \Sigma_k |c_k|^2$ is to be interpreted as the probability of being in state j.

We now develop the quantum-mechanical formula for the way the c_k change with time, that is, the transitions between states. These results will be important to the theory of electron transfer, although the electron transfer process also involves other processes (e.g., movement of molecules), which are better described by classical mechanics. The electrochemical reaction rate will depend on state-to-state transition rates of the electrons, on the available initial states for the electron (for example, at a metal surface), and on the available final states for the electron (perhaps on a solution species).

Commonly, one considers the situation in which H is the sum of a time-independent Hamiltonian H_0 and a time-dependent part H_1, such that the ϕ_k are eigenfunctions of H_0 with eigenvalues E_k:

$$H_0 \phi_k = E_k \phi_k \qquad [4]$$

Then, substituting equation [2] into the time-dependent Schrödinger equation (equation [1]), we obtain

$$\sum_k \left(\frac{i\hbar \, dc_k(t)}{dt} + c_k E_k\right) \phi_k \exp\left(\frac{-iE_k t}{\hbar}\right) = \sum_k c_k(E_k + H_1) \phi_k \exp\left(\frac{-iE_k t}{\hbar}\right)$$

Canceling off the terms in $c_k E_k$, we see that transitions between states ($dc_k/dt \neq 0$) are due to the part of the Hamiltonian H_1, called the perturbation. One can multiply the above equation by $\phi_j^* \exp(iE_j t/\hbar)$ and integrate over the coordinates of the ϕ_k, taking advantage of the orthonormality of the ϕ_k, to get

$$i\hbar \frac{dc_j}{dt} = \sum_k c_k \int \phi_j^* H_1 \phi_k \, d\tau \exp\left[\frac{i(E_j - E_k)t}{\hbar}\right] \qquad [5]$$

To investigate transitions from the state m, one seeks solutions to this equation (c_k as a function of time) with the initial condition, $c_k(0) = 0$ for all k except $k = m$. Then $|c_j(t)|^2$ gives the probability that a transition from state m to state j has occurred in time t. The quantity $\omega_{jk} = (E_j - E_k)/\hbar$ is often called a *transition frequency* and the integral $\int \phi_j^* H_1 \phi_k \, d\tau$ is often called a *transition matrix element*. The transition matrix element may be time-dependent because of time-dependence in H_1.

When H_1 oscillates in time at a frequency ω, it is effective in causing transitions between states for which the transition frequency is close to ω (Goodisman 1977, Section 13.3). Usually in electrochemistry, the electronic transitions are between states of the same energy, so that the perturbation must have a time-independent or low-frequency component to be effective. Thus one considers a perturbation which is itself constant in time, but has time-dependence by virtue of being "turned on," either suddenly at $t = 0$ or gradually, starting at $t = -\infty$, so it has full value at $t = 0$. We consider equation [5] for the former situation.

If the system was initially in state m, it remains in state m for $t < 0$: $c_j = \delta_{jm}$. For $t > 0$ some c_j ($j \neq m$) will become nonzero, representing transitions. For small t, c_m will still be about 1 and no other c_j will be of appreciable size, so that equation [5] becomes

$$i\hbar \frac{dc_j}{dt} = H_{jm} \exp(i\omega_{jm} t) \qquad [6]$$

where H_{jm} is the (time-independent) transition matrix element between m and j and $\omega_{jm} = (E_j - E_m)/\hbar$. This corresponds to first-order perturbation theory: Only single transitions from m to other states are being considered.

For $t > 0$, equation [6] is easily integrated to

$$c_j(t) = \frac{H_{jm}}{i\hbar} \frac{\exp(i\omega_{jm} t) - 1}{i\omega_{jm}}$$

$$= \frac{-H_{jm}}{\hbar \omega_{jm}} [\exp(i\omega_{jm} t) - 1]$$

The occupation probability of state j is then

$$|c_j(t)|^2 = \left(\frac{H_{jm}}{\hbar \omega_{mj}}\right)^2 (2 - 2\cos \omega_{mj} t) \qquad [7]$$

As a function of ω_{mj}, the probability is maximized at $\omega_{mj} = 0$ (same energy of initial and final states). For small ω_{mj},

$$(\omega_{mj})^{-2}(2 - 2\cos\omega_{mj}t) \to 2(\omega_{mj})^{-2}\left(\frac{\omega_{mj}^2 t^2}{2!} - \frac{\omega_{mj}^4 t^4}{4!}\cdots\right)$$

$$= t^2 - \frac{\omega_{mj}^2 t^4}{2}\cdots$$

showing a decrease as $|\omega_{mj}|$ increases. However, for $\omega_{mj} = 0$ the probability seems to increase quadratically with time t, which would make the *rate* of transition proportional to t. This unphysical result is due to an unrealistic aspect of the above calculation, which is now corrected.

One should consider transitions to a range of final states. Suppose these are characterized by a density in energy space $\rho_j(E)$, so that there are $\rho_j(E)\,dE$ states with energies between E and $E + dE$. Then the probability of a transition to any one of these states is obtained by integrating equation [7] over states or over energy:

$$P(t) = \int dE\,\rho_j(E)\left(\frac{H_{jm}}{\hbar\omega_{mj}}\right)^2 (2 - 2\cos\omega_{mj}t) \qquad [8]$$

The limits of integration are the energy limits for this band of states but, because the quantity $(\omega_{mj})^{-2}(2 - 2\cos\omega_{mj}t)$ is peaked around $\omega_{mj} = 0$, the main contribution to the integral will come from energies near E_m, and the limits of integration may be extended to infinity. The transition matrix element H_{jm} depends on the final state j but it varies much more slowly with energy than does $(\omega_{mj})^{-2}(2 - 2\cos\omega_{mj}t)$, so that it may be approximated by its value at $E = E_m$ or $\omega_{mj} = 0$. The same is true for $\rho_j(E)$, which is approximately $\rho_j(E_m)$ over the range of E for which the integrand is appreciable. Thus, with $\hbar\omega_{mj} = E_m - E_j$, we get

$$P(t) = \rho_j(E_m)\left|H_{jm}(0)\right|^2 \int_{-\infty}^{\infty} d\omega_{mj}\,\hbar^{-1}(\omega_{mj})^{-2}(2 - 2\cos\omega_{mj}t)$$

Since $2 - 2\cos x = 4\sin^2(x/2)$ the integral is

$$4\hbar^{-1}\int_{-\infty}^{\infty} d\omega\,\omega^{-2}\sin^2\frac{\omega t}{2} = 2\hbar^{-1}t\int_{-\infty}^{\infty} du\,u^{-2}\sin^2 u$$

and the last integral is equal to π. Thus

$$P(t) = \frac{2\pi}{\hbar}\rho_j(E_m)\left|H_{jm}(0)\right|^2 t \qquad [9]$$

which is proportional to t, so that the transition probability per unit time is a constant, as it should be.

The expression for $P(t)/t$, which represents the rate of transitions, is known as the Fermi golden rule. It states that the rate is proportional (a) to the density of final states at the energy of the initial state and (b) to the square of the transition matrix element between the initial state and a final state of the same energy.

On comparing equation [9] to [8], one sees that we have really argued that, for times t of interest to us, the function

$$f(\omega_{mj}) = (\hbar\omega_{mj})^{-2} (2 - 2\cos\omega_{mj}t) \qquad [10]$$

varies rapidly with ω_{mj} near $\omega_{mj} = 0$ and is very small for ω_{mj} appreciably different from zero. Using $\hbar\omega_{mj} = E_m - E_j$, we get

$$\int_{-\infty}^{\infty} dE_m f(\omega_{mj}) = \int_{-\infty}^{\infty} dx\, x^{-2} \left(2 - 2\cos\frac{xt}{\hbar}\right) = \frac{2\pi t}{\hbar}$$

This means we may identify $f(\omega_{mj})$ with $2\pi t/\hbar$ times $\delta(E_{mj})$, where $\delta(a)$ is the Dirac δ-function. The δ-function $\delta(a)$ has the properties that $\delta(a) = 0$ for $a \neq 0$ and

$$\int F(a)\,\delta(a)\,da = F(0) \qquad [11]$$

for any function F, as long as the region of integration passes through $a = 0$. Thus equation [7] is

$$|c_j(t)|^2 = |H_{jm}|^2 \frac{2\pi t}{\hbar} \delta(E_{mj}) \qquad [12]$$

and multiplication by $\rho_j(E)$ and integrating over E yields equation [9]. A function with the properties of the δ-function may be represented in other ways than $f(\omega_{mj})$. For instance, it can be shown that

$$\int_{-\infty}^{\infty} \exp\left[\frac{i(E_m - E_j)t}{\hbar}\right] dt = (2\pi\hbar)\,\delta(E_m - E_j) \qquad [13]$$

which is important in Fourier transform theory.

In fact, $f(\omega)$ (equation [10]) is a damped oscillating function, with its largest maximum at $\omega = 0$: $f(0) = t^2/\hbar^2$. The function vanishes at $\omega = \pm\pi/t$, and the second maxima are for $\omega = \pm 8.98/t$, with the value of f being $0.0472 t^2/\hbar^2$. To treat f as a δ-function, the width of the central peak should be small compared to the energy range over which H_{jm} and $\rho_j(E)$ vary appreciably, which requires t to be large. On the other hand, in deriving the Fermi golden rule from equation [5], we have also assumed that multiple transitions may be ignored, that is, $|c_k| \ll 1$ for $k \neq m$ (see equation [6]). According to equation [9], this is true for times τ such that

$$\frac{2\pi}{\hbar} \rho_j(E_m) |H_{jm}(0)|^2 \tau \ll 1 \qquad [14]$$

so τ cannot be too large either.

One can estimate an "electron transfer time" for the metal–solution interface as the distance between the electrode and an acceptor ion (say 5 Å) divided by the velocity of an electron at the Fermi level (Bockris and Khan 1979, Section 4.7). For an electron density $d = 0.03\,(a_0)^3 = 2.02 \times 10^{23}$ cm^{-3}, the Fermi momentum is $\hbar(3\pi^2 d)^{1/3} = 1.92 \times 10^{19}$ g · cm/s and the electron velocity is 2.2×10^8 cm/s. Thus the electron transfer time is 2.3×10^{-16} s $= \tau$.

B. ROLE OF THE SOLVENT

The theory of the rate of electron transfer between an electrode and an ion in solution, or between two ions, as developed by Levich and Dogonadze (Levich 1970), stresses the role of fluctuations in the solvent. These fluctuations lead to the distribution of states characterized by $\rho_j(E_m)$. The solvent is considered as a continuous medium, characterized by its polarization or dipole moment per unit volume (Chapter 2, equations [70] and thereafter). Because of vibrations and reorientations of solvent molecules, the polarization **P** at any point fluctuates in time. Vibrations have a frequency of $\sim 10^{14}$ s^{-1} and hence a characteristic fluctuation time of 10^{-14} s. Rotations of small molecules have a characteristic time of 10^{-12} s; larger molecules (or molecules in a condensed medium) have characteristic times of perhaps an order of magnitude larger. Since kT at room temperature is much less than $h\nu$ for a vibration, vibrations are not likely to be excited, so we focus our attention on the rotations.

The work required to change the electrical conditions in the volume element dV is (Becker and Sauter 1964, Sections 19 and 32) $(4\pi)^{-1}\,\mathbf{E} \cdot d\mathbf{D}$ or, if the dielectric constant ϵ is independent of field, the work required is $(4\pi\epsilon)^{-1}\,\mathbf{D} \cdot d\mathbf{D}$, which integrates to $D^2/8\pi\epsilon$ or $\epsilon E^2/8\pi$. The value of ϵ to use in an electrolyte system depends on whether the change is carried out slowly, in which case one should use the static dielectric constant ϵ_s, or quickly, in which case one should use the optical dielectric constant ϵ_{op}. The latter is typically much smaller than ϵ_s because it includes only electronic and vibrational polarization, and not the effects of reorganization or reorientation of molecular dipoles. For a given value of **D** the orientation polarization is (Chapter 2, equation [71])

$$\mathbf{P}_{\text{or}} = \frac{(\epsilon_s - 1)\,\mathbf{D}}{4\pi\epsilon_s} - \frac{(\epsilon_{\text{op}} - 1)\,\mathbf{D}}{4\pi\epsilon_{\text{op}}} = (4\pi)^{-1}\left(\epsilon_{\text{op}}^{-1} - \epsilon_s^{-1}\right)\mathbf{D}$$

The work per unit volume that must be done to slowly raise the electrical displacement from 0 to **D** (a function of position) and then quickly reduce it to zero (leaving the orientational polarization) would be

$$\frac{D^2}{8\pi}\left[(\epsilon_s)^{-1} - (\epsilon_{op})^{-1}\right] = 2\pi\left[(\epsilon_{op})^{-1} - (\epsilon_s)^{-1}\right]^{-1} (P_{or})^2 \qquad [15]$$

This is the potential energy per unit volume associated with a local polarization fluctuation, to which one adds a kinetic energy term, $2\pi\left[(\epsilon_{op})^{-1} - (\epsilon_s)^{-1}\right]^{-1} (\omega_0)^{-1} (dP_{or}/dt)^2$. This assumes that the fluctuations in orientational polarization correspond to harmonic oscillators of frequency ω_0.

The process of electron transfer between ions in solution is discussed first. Since the solvent molecules and ions move slowly compared to the electrons, one can invoke the arguments of the Born–Oppenheimer approximation, used to separate electronic from nuclear motions in discussing molecules (Goodisman 1973, Section I.B): The motion of the fast particles (electrons) may be considered as if the heavy particles were stationary, so that one has electronic eigenfunctions in the potential energy field set up by the heavy particles. The corresponding eigenvalues of energy depend on the positions of the heavy particles. Then the eigenvalue of electronic energy (for a particular electronic state), as a function of the heavy-particle positions, acts as a potential energy when the motion of the heavy particles is considered.

Suppose the electron is on ion A and a transfer to ion B is contemplated. The orientation of the solvent dipoles around the ions is appropriate to the charges with the electron on A. If the electron transferred to a state localized on ion B with no change in the solvent orientation, its energy would be higher. As a function of position, the potential felt by the electron, which one would use in the electronic Schrödinger equation, might be as in Figure 20a, having minima at the ion positions and a maximum between them. The electronic energy eigenvalues are shown for a state localized on A (solid line) and for a state localized on B (broken line). In Figure 20b the electronic eigenvalues of energy (electron terms), which are energies of the system, are shown schematically as functions of the solvent polarization; the broken curve is for an electron state localized on B.

Because the energies of the two electronic states are different when the solvent dipoles are oriented for the electron on A, the electronic transition from a state localized on A to one localized on B is improbable. If the electron were on B, and the solvent were polarized appropriately to this situation, the electronic potential energy as a function of position would look like Figure 20c, and the energy of the electronic state localized on B (solid horizontal line) would be lower than that for the state localized on A (broken line). There are some polarization states of the solvent for which the electronic potential energy resembles Figure 20d, and for which the energies of the electronic eigenstates localized on A and B are about equal. When the solvent polarization is in such a state, the electronic transition can occur. This state, which corresponds to the intersection of the curves in Figure 20b, is supposed to be attained by a thermal fluctuation of solvent orientations.

A quantum mechanical treatment of the electronic and solvent motions starts with a Hamiltonian for the electron in the presence of the ions and the solvent polarized by the ions. If ion A is taken as the origin of coordinates, and an ion B

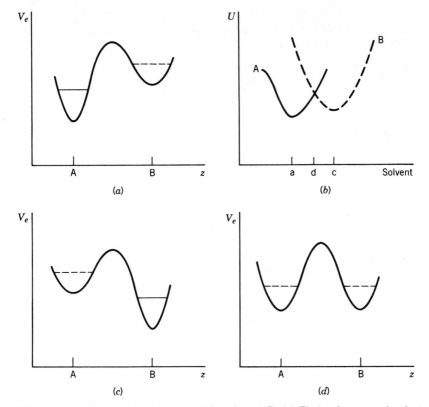

FIGURE 20. Electron transfer from ion at A to ion at B. (a) Electronic energy levels and potential for electrons as a function of position, with solvent polarized for electron on A. (b) Electron terms (energies) as a function of solvent coordinates: Intersection corresponds to (d). (c) Electronic energy levels and potential energy for electrons with solvent polarized for electron on B. (d) Electronic energy levels and potential energy with solvent polarized for the transition state.

is at a position relative to A given by **R**, the interaction potential between an electron and the two ions is

$$V_{ei} = -\left(\frac{z_A e^2}{r} + \frac{z_B e^2}{|\mathbf{r} - \mathbf{R}|}\right)$$

where $z_A e$ and $z_B e$ are the charges on the ions. The Hamiltonian for the electron, were there no solvent, would be this potential energy plus the electronic kinetic energy operator T_e. The solvent contributes polarization terms like those discussed above, due to fluctuations. These are treated as normal vibrational coordinates q_k, with angular frequency ω_0. As usual in translating classical mechanical kinetic energy into a quantum mechanical operator, a kinetic energy term $(m/2)(d\mathbf{q}_k/dt)^2$ is replaced by $(-\hbar^2/2m)$ multiplied by the second derivative with respect to $\mathbf{q}_k$, so that the part of the Hamiltonian representing solvent is

$$H_{\text{sol}} = \frac{\hbar\omega_0}{2} \sum_k \left[(\mathbf{q}_k)^2 - \frac{\partial^2}{\partial(\mathbf{q}_k)^2} \right]$$

The interaction of the ions with the solvent is represented by a polarization term like those of equation [15] and the preceding equations:

$$H_{is} = \int \left[\int_0^{D_i} \mathbf{P} \cdot d\mathbf{D} \right] dV = \int \left[(4\pi)^{-1} \left[1 - (\epsilon_{\text{op}})^{-1} \right] \int_0^{D_i} \mathbf{D} \cdot d\mathbf{D} \right] dV$$

$$= (\epsilon_{\text{op}} - 1)(8\pi\epsilon_{\text{op}})^{-1} \int (D_i)^2 \, dV$$

where $\mathbf{D}_i$ is the electric displacement due to the ions. A similar term represents the interaction of the electron with the solvent; $U_{es} = \int \mathbf{D}_e \cdot \mathbf{P} \, dV$, where $\mathbf{D}_e$ is the electrical displacement due to the electron.

With the ions fixed in position, one now could consider eigenfunctions for the complete Hamiltonian

$$H_{\text{total}} = T_e + V_{ei} + H_{\text{sol}} + H_{is} + U_{es} \qquad [16]$$

such that $H_{\text{total}} \Phi_t(\mathbf{r}, \mathbf{q}_k; \mathbf{R}) = E_t \Phi_t(\mathbf{r}, \mathbf{q}_k; \mathbf{R})$. The Φ_t are functions of the electronic coordinates $\mathbf{r}$ and the solvent coordinates $\mathbf{q}_k$, kinetic energy operators for which appear in H_{total}. They depend parametrically on the ion positions, represented by $\mathbf{R}$, since these enter V_{ei} and H_{is}.

Since the solvent molecules move slowly compared to the electrons, one seeks eigenfunctions for the electrons in the potential energy field of the ions and solvent. The eigenvalue of electronic energy as a function of the heavy-particle (solvent) positions is used in the Hamiltonian for the heavy particles. Then one solves a Schrödinger equation for their wave functions. For each choice of electronic state, there will be a different set of heavy-particle eigenfunctions and eigenvalues. Thus each eigenfunction Φ_t of equation [16] is approximated as a product of a solvent wave function $\chi_j(\mathbf{q}_k; \mathbf{R})$ and an electronic wave function $\phi_m(\mathbf{r}; \mathbf{q}_k, \mathbf{R})$, which is an eigenfunction of H_{total} minus the solvent kinetic energy operator.

The electronic Schrödinger equation, since H_{sol} and H_{is} do not involve electronic coordinates, is

$$(T_e + V_{ei} + U_{es}) \phi_m(\mathbf{r}; \mathbf{q}, \mathbf{R}) = E_m^e(\mathbf{q}, \mathbf{R}) \phi_m(\mathbf{r}; \mathbf{q}, \mathbf{R})$$

Note that ϕ_m is a function of electronic coordinates and depends parametrically on the solvent coordinates and the ion coordinates because they enter the potential energy operators V_{ei} and U_{es}. Note also that there is one set of χ_j (with corresponding energies) for the initial electronic state and another set for the final electronic state. Among the solutions to the electronic Schrödinger equation will be functions corresponding to localization of the electron on ion A; that of lowest energy, $\phi_i(\mathbf{r}; \mathbf{q}, \mathbf{R})$, will be the electronic state before transition. For a state of this kind, V_{ei} might as well be just $-z_A e^2/r$, since $(z_B e^2/|\mathbf{r} - \mathbf{R}|) \phi_i$ is small because there is no region of space for which ϕ_i and $|r - R|^{-1}$ are both appreciable in size. Similarly, there will be solutions $\phi_f(\mathbf{r}; \mathbf{q}, \mathbf{R})$ corresponding to localization of the electron on ion B. These are possible final states for the electronic transition.

The Born–Oppenheimer wave function

$$\Phi_{BO} = \phi_i(\mathbf{r}; \mathbf{q}, \mathbf{R}) \chi_j(\mathbf{q}; \mathbf{R})$$

is not an eigenfunction of H_{total}. In order to have $H_{\text{total}}\Phi_{BO}$ equal a multiple of Φ_{BO} (definition of an eigenfunction), one would have to neglect terms like $\partial^2 \phi_i / \partial (\mathbf{q}_k)^2$, which arise when H_{sol} operates on Φ_{BO}. These terms correspond to a perturbation H_1 (Section A, equation [5]) which leads to electronic transitions. We now must calculate transition matrix elements between two functions of the form $\chi_j(\mathbf{q}_k; \mathbf{R}) \phi_m(\mathbf{r}; \mathbf{q}_k, \mathbf{R})$. One may argue that χ_j is localized and also that the dependence of ϕ_m and H_1 on $\mathbf{q}_k$ is weak, so that ϕ_m and H_1 may be evaluated for some average value of $\mathbf{q}_k$, say $\mathbf{q}_0$ and

$$\int [\chi_f(\mathbf{q}; \mathbf{R}) \phi_f(\mathbf{r}; \mathbf{q}, \mathbf{R})]^* H_1(\mathbf{r}, \mathbf{q}, \mathbf{R})$$
$$\cdot \chi_i(\mathbf{q}; \mathbf{R}) \phi_i(\mathbf{r}; \mathbf{q}, \mathbf{R}) \, d\mathbf{r} \, d\mathbf{q}$$
$$= \int \phi_f(\mathbf{r}; \mathbf{q}_0)^* H_1(\mathbf{r}, \mathbf{q}_0) \phi_i(\mathbf{r}; \mathbf{q}_0) \, d\mathbf{r}$$
$$\cdot \int \chi_f(\mathbf{q}; \mathbf{R})^* \chi_i(\mathbf{q}; \mathbf{R}) \, d\mathbf{q} \quad [17]$$

where the initial and final states for the solvent are also labeled by i and f. The absolute square of equation [17] may have to be averaged over thermal populations of low-lying states χ_i, but no such averaging is necessary for the electronic states ϕ_i and ϕ_f, since excited electronic states are too high in energy to be of interest.

Using equation [17] in the expression for the electronic transition probability (equation [9]), we have

$$P(t) = \frac{2\pi t}{\hbar} \left| \int \phi_f^* H_1 \phi_i \, d\mathbf{r} \right|^2 \sum_f \left\langle \left| \int \chi_f^* \chi_i \, d\mathbf{q} \right|^2 \delta(E_{fi}) \right\rangle_{Av}$$

The angular brackets indicate an average over possible initial solvent states χ_i which may be populated. Because the populations enter, the last factor is temperature-dependent. The transition from one particular electronic state to another involves many possible initial solvent states, and many possible final solvent states, over which we have summed. Using the representation (equation [13]) for the δ-function,

$$P(t) = \hbar^{-2}t|L|^2 \sum_f \left\langle \left| \int \chi_f^* \chi_i \, d\mathbf{q} \right|^2 \int_{-\infty}^{\infty} \exp\left[\frac{i(E_f - E_i)t}{\hbar}\right] \right\rangle_{Av} \quad [18]$$

where the electronic transition integral has been abbreviated by L. The eigenvalues of energy correspond to the solvent wave functions χ_f and χ_i, where the wave functions χ_f belong to the final electronic state and are eigenfunctions of a solvent Hamiltonian H_f, whereas the wave functions χ_i correspond to the initial electronic state and are eigenfunctions of a Hamiltonian H_i.

If χ_i is an eigenfunction of H_i with eigenvalue E_i, and F is some function, then χ_i is an eigenfunction of $F(H_i)$ with eigenvalue $F(E_i)$. Thus $P(t)$ can be rewritten as follows:

$$P(t) = \hbar^{-2}t|L|^2 \int_{-\infty}^{\infty} dt \left\langle \sum_f \left[\int \chi_i^* \chi_f \, d\mathbf{q} \, e^{iE_f t/\hbar}\right] \right.$$
$$\left. \cdot \left[\int \chi_f^* \chi_i \, d\mathbf{q}' \, e^{-iE_i t/\hbar}\right] \right\rangle_{Av}$$

$$= \hbar^{-2}t|L|^2 \int_{-\infty}^{\infty} dt \sum_f \left\langle \left[\int \chi_i^* e^{iH_f t/\hbar} \chi_f \, d\mathbf{q}\right] \right.$$
$$\left. \cdot \left[\int \chi_f^* e^{-iH_i t/\hbar} \chi_i \, d\mathbf{q}'\right] \right\rangle_{Av}$$

The χ_f are supposed to constitute a complete set of functions so that they satisfy the closure relation $\sum_f \chi_f(\mathbf{q}) \chi_f^*(\mathbf{q}') = \delta(\mathbf{q} - \mathbf{q}')$ (Merzbacher 1961, Section 8.4), and

$$P(t) = \hbar^{-2}t|L|^2 \int_{-\infty}^{\infty} dt \left\langle \int \chi_i^* \exp\left[\frac{i(H_f - H_i)t}{\hbar}\right] \chi_i \, d\mathbf{q} \right\rangle_{Av} \quad [19]$$

which expresses the probability as an integral over the initial-state solvent wave functions. These are supposed to be harmonic-oscillator eigenfunctions of frequency ω_0. H_i and H_f are both harmonic-oscillator Hamiltonians (plus a term for the energy of the ions, which is different before and after electron transfer), but H_f is appropriate to the final electronic state, that is, the equilibrium values of the solvent coordinates are different from those of H_i. The thermal averaging over initial states χ_i may be carried out explicitly, assuming thermal equilibrium, to give a transition probability per unit time

$$W_{if} = \frac{P(t)}{t} = \hbar^{-2}|L|^2 \int_{-\infty}^{\infty} dt \sum_j Z^{-1} e^{-E_j/kT}$$

$$\int \chi_j^* \exp\left[\frac{i(H_f - H_i)t}{\hbar}\right] \chi_j \, d\mathbf{q} \qquad [20]$$

where $H_i \chi_i = E_i \chi_i$ and $Z = \sum_j e^{-E_j/kT}$ is the partition function. All the χ_i, which describe the solvent, are eigenfunctions of the Hamiltonian H_i, for which the electron is in its initial state.

Given the harmonic-oscillator form for H_i and H_f, the E_j are equally spaced energy levels and the χ_i (Goodisman 1977) are well-known functions of the coordinates (which here represent solvent polarization). Then the thermal averaging and integration in W_{if} may be carried out to give an explicit but complex expression for the transition probability. A treatment of the problem using a linear response formalism has been developed and, with certain approximations, gives results identical to the formalism of Levich et al. (Bockris and Khan 1979, Section 7.5.3). Bockris and Khan (1979, Section 7.3) discuss improvements in the model. There have been attempts to introduce higher-frequency motions (bond stretching) as well as the low-frequency solvent librations that are responsible for the changes in polarization. This takes into account inner-sphere degrees of freedom, which some authors feel are of primary importance (Schmickler 1976).

For a reduction reaction occurring on a metal electrode, the current from solution to metal results from electronic transitions from metal states with energies ϵ to states on ions at distances x from the electrode. The probability of a transition W_{sm} is given by a formula such as equation [19] or [20]; it depends on x and on ϵ. As discussed in Chapter 8, Section G, one has to integrate over electronic energies in the metal as well as over positions of the ion. The reduction or cathodic current density is thus

$$j_c = \int c_O(x) \, dx \int W_{sm}(x, \epsilon) D_-(\epsilon) \, d\epsilon$$

where $D_-(\epsilon)$, the number of occupied metal states per unit energy range at energy ϵ, is a density-of-states factor multiplied by an occupation probability factor. The concentration $c_O(x)$ is supposed to be given by the equilibrium distribution function for ions in the double layer, that is, the flow of current is not large enough to disturb the equilibrium distribution. A similar expression holds for the anodic current, involving the concentration of reduced species $c_R(x)$ and the number of unoccupied electron states per unit energy range on the metal, $D_+(\epsilon)$. Because equilibrium is assumed, the ratio between $c_O(x)$ and $c_R(x)$ may be calculated using Boltzmann distributions for each species:

$$\frac{c_O(x)}{c_R(x)} = \frac{c_O(\infty)}{c_R(\infty)} \exp\left[-(z_O - z_R) e\phi(x)/kT\right]$$

where ∞ refers to bulk solution and the electrical potential ϕ is taken as zero far from the electrode; for a one-electron reduction, $z_O - z_R = 1$. The ratio of $D_-(\epsilon)$ to $D_+(\epsilon)$ is equivalent to the ratio of $f(\epsilon)$ to $1 - f(\epsilon)$ (Chapter 4, equation [46]),

$$\frac{D_-(\epsilon)}{D_+(\epsilon)} = \exp\left(\frac{\epsilon - \mu}{kT}\right)$$

where μ, the chemical potential, is the Fermi energy ϵ_F.

Using these results, and introducing the overpotential η as the difference between ϕ^M and its equilibrium value, Levich (1970, Section 12.VIII) derived the current density

$$j = ex_1\left(\frac{\pi}{\hbar^2 kTE_s}\right)^{1/2} |L|^2 c_R(x_1)\left[1 - \exp\left(\frac{-e\eta}{kT}\right)\right]\exp\left(\frac{-\phi_1}{kT}\right)$$

$$\times \exp\left\{-\left[E_s + e\eta - e\phi_1 + kT\ln\frac{c_R(\infty)}{c_O(\infty)}\right]^2 (4E_s kT)^{-1}\right\}$$

where x_1 is the outer Helmholtz plane, the distance of closest approach for ions to the electrode; ϕ_1 is the electrical potential at x_1; and E_s is the energy of reorganization or repolarization. E_s is the crucial parameter of the theory. In the solvent Hamiltonian H_{sol} (see above), the equilibrium position for the solvent coordinates is different for the ion–electrode system before and after electron transfer; E_s is the difference in solvent energies when the solvent oscillators are at the equilibrium positions for the two situations (1 and 2):

$$E_s = \frac{\hbar\omega_0}{2}\sum (q_k^{o,1} - q_k^{o,2})^2$$

This corresponds to a reorganization energy such as that given in equation [15].

By writing the last factor in j as a difference of exponentials, Levich is able to write j in the Tafel form, that is, as an exchange current density multiplied by sinh $(e\eta/2kT)$. Theoretical current–potential curves were generated for the case of reduction of ferricyanide on a mercury electrode. Agreement to within 10% accuracy was found between theory and experiment, with $E_s \sim 0.65$ eV.

There has been criticism of the formalism and of the physical basis of continuum theories by various authors, as reviewed by Bockris and Khan (1979, Sections 7.3, 7.6). There has been relatively little comparison of the predictions of continuum theories with experiment. It has been stated (Bockris and Khan 1979, Khan and Bockris 1983) that, for electron transfer between a metal and an ion in solution, they do not correctly predict Tafel's law, which states that the logarithm of electrode current should be linear with potential.

C. QUANTUM MECHANICAL TUNNELING

The electronic transitions of interest for the electron-transfer process involve the movement of an electron from one region of space to another. A particle obeying the laws of quantum mechanics can go from one region of space to another even though the two regions are separated by a region in which the particle has no possibility of existing. Thus, in the time-dependent perturbation theory of Section A, the transition probability is shown to be proportional to $|\int \psi_f^* H_1 \psi_i \, d\tau|^2$, where ψ_f and ϕ_i are final (after transition) and initial (before transition) wave functions for the electron, and H_1 is the perturbation. No reference is made to the process by which the transition occurs, that is, to any intermediate states through which the electron might pass.

If H_1 is time-independent, then ψ_f and ψ_i must have the same energy. Suppose the potential energy of the electron is low in regions I and III of space and high in region II (as in Figure 21). The wave function ψ_i may have appreciable values only in region I, and ψ_f may have appreciable values only in region III. It may well be that, in region II, the potential energy of the electron is so high that its kinetic energy (total energy minus potential energy) is negative, so that the electron cannot exist there. The transition from ψ_i to ψ_f, or from region I to region III, occurs nevertheless, as long as $\int \psi_f^* H_1 \psi_i \, d\tau$ is not zero. One says that "tunneling" has occurred from region I to region III.

An explicit result for tunneling probability may be obtained from steady-state wave functions (instead of from time-dependent perturbation theory) in the simple situation of an electron moving in a one-dimensional potential given by

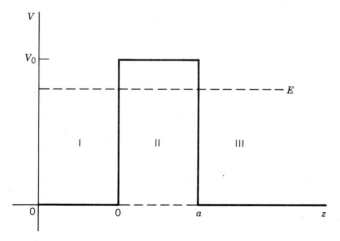

FIGURE 21. Square potential barrier. A particle of energy E can tunnel from region I to region III. Region II is forbidden according to classical mechanics.

9. QUANTUM THEORY OF ELECTRON TRANSFER

$$V(z) = 0, \quad z < 0 \text{ (region I)}$$
$$V(z) = V_0, \quad 0 \leq z < a \text{ (region II)}$$
$$V(z) = 0, \quad z > a \text{ (region III)} \quad [21]$$

(Figure 21). This is referred to as a one-dimensional square potential barrier.

The Schrödinger equation

$$-\frac{\hbar^2}{2m}\frac{d^2\psi}{dz^2} + V\psi = E\psi$$

is easily solved in each of the three regions of equation [21]. In region I, where $V = 0$, the solution is

$$\psi_\text{I} = Ae^{ikz} + Be^{-ikz} \quad [22]$$

where A and B are, at the present time, arbitrary constants and $E = \hbar^2 k^2/2m$ ($\hbar k$ is the momentum). Similarly, in region III the solution is

$$\psi_\text{III} = Fe^{ikz} + Ge^{-ikz} \quad [23]$$

where F and G are now arbitrary, and in region II we have

$$\psi_\text{II} = Ce^{ik'z} + De^{-ik'z} \quad [24]$$

where C and D are arbitrary and

$$\frac{\hbar^2(k')^2}{2m} + V_0 = E$$

The solutions [22] and [23] correspond to plane waves with wavelength $2\pi/k$, or particle beams moving from left to right (e^{ikz}) or from right to left (e^{-ikz}).

The solution [24] corresponds to plane waves with wavelength $2\pi/k'$ if k'^2 is positive, that is, for $E > V_0$. If, however, k'^2 is negative and k' is imaginary, $e^{\pm ik'z}$ is a real exponential, either decreasing or increasing with z. It is just this case ($E < V_0$) that interests us. In classical mechanics, a particle with $E < V_0$ would have a negative kinetic energy in region II and thus could not exist there, so that there would be no connection between regions I and III, to the left and right of the barrier. The quantum mechanical case, examined in terms of solutions to the Schrödinger equation, leads to a different conclusion.

To completely specify the solution, we have to find the six constants A–G. The wave function must be continuous and so must its first derivative with z. Imposing these conditions at $z = 0$ and at $z = a$ gives four relations between the six constants, leaving some arbitrariness. We are free to choose, for instance, $G = 0$ (The remaining constant will appear as a normalization constant.) If $G = 0$, the wave function in region III corresponds to particles moving from left to right only (momentum $\hbar k > 0$), whereas the wave function in region I has particles moving in both directions. The situation corresponds to particles impinging on the barrier from the left, with some of them being reflected back to the left and some, perhaps, having passed through the barrier into region III. The constants A, B, C, and D, can now be expressed in terms of F.

For continuity at $z = 0$, $\psi_I(0) = \psi_{II}(0)$, or

$$A + B = C + D \qquad [25]$$

For continuity of the first derivative at $z = 0$, $(d\psi_I/dz)_0 = (d\psi_{II}/dz)_0$ or

$$ikA - ikB = ik'C - ik'D \qquad [26]$$

Continuity at $z = a$, $\psi_{II}(a) = \psi_{III}(a)$, and continuity of the first derivative at $z = a$, $(d\psi_{II}/dz)_a = (d\psi_{III}/dz)_a$, give

$$Ce^{ik'a} + De^{-ik'a} = Fe^{ika} \qquad [27]$$

$$ik'(Ce^{ik'a} - De^{-ik'a}) = ikFe^{ika} \qquad [28]$$

Equations [27] and [28] allow one to find C and D in terms of F:

$$D = \frac{1}{2}\left(1 - \frac{k}{k'}\right) Fe^{i(k+k')a}$$

$$C = \frac{1}{2}\left(1 + \frac{k}{k'}\right) Fe^{i(k-k')a}$$

Substituting into equation [25] we get B in terms of A, and substituting for B, C, and D in equation [26] we have

$$2kA - \frac{1}{2}k\left(1 - \frac{k}{k'}\right) Fe^{i(k+k')a} - \frac{1}{2}k\left(1 + \frac{k}{k'}\right) Fe^{i(k-k')a}$$
$$= \frac{1}{2}k'\left(1 + \frac{k}{k'}\right) Fe^{i(k-k')a} - \frac{1}{2}k'\left(1 - \frac{k}{k'}\right) Fe^{i(k+k')a}$$

which may be solved for A in terms of F:

$$\frac{4kk'A}{F} = (k' + k)^2 e^{i(k-k')a} - (k' - k)^2 e^{i(k+k')a} \qquad [29]$$

Then we obtain

$$\frac{4kk'B}{F} = (k^2 - k'^2) e^{i(k-k')a} + (k'^2 - k^2) e^{i(k+k')a} \qquad [30]$$

Since we are interested in the case where $E < V_0$ and k' is imaginary (though k is always real), we write $k' = ik''$ and equations [29] and [30] become

$$\frac{4ikk''A}{F} = (ik'' + k)^2 e^{(ik+k'')a} - (ik'' - k)^2 e^{(ik-k'')a}$$
$$\frac{4ikk''B}{F} = (k^2 + k''^2) e^{(ik+k'')a} - (k^2 + k''^2) e^{(ik-k'')a} \qquad [31]$$

Now the flow of particles from left to right in region III, where the wave function is Fe^{ikz}, is $(k\hbar/m)|F|^2$, since $k\hbar/m$ is the velocity (momentum divided by mass). The flow of particles from left to right in region I, where there are particles moving in both directions, is $(k\hbar/m)|A|^2$. The ratio of the flow of particles in region III to the flow toward the barrier (left to right) in region I is the transmission coefficient:

$$T = \frac{|F|^2}{|A|^2} \qquad [32]$$

Before calculating T, we may note that $(k\hbar/m)|B|^2$ is supposed to be the flow of particles from right to left in region I, that is, of particles reflected from the barrier, so that, for our picture to be consistent, $|A|^2$ should equal $|B|^2 + |F|^2$, which we now verify. Taking absolute squares of equations [31], we obtain

$$\frac{16k^2k''^2(|A|^2 - |B|^2)}{|F|^2} = (k^2 + k''^2)^2 e^{2k''a} - (k''^2 - 2ikk'' - k^2)^2$$
$$- (k''^2 + 2ikk'' - k^2)^2 + (k''^2 + k^2) e^{-2k''a}$$
$$- [(k^2 + k''^2) e^{2k''a} - 2(k^2 + k''^2)^2$$
$$+ (k^2 + k''^2)^2 e^{-2k''a}]$$
$$= -2k''^4 + 12k^2k''^2 - 2k^4 + 2(k^4 + 2k^2k''^2 + k''^4)$$

Therefore $|A|^2 - |B|^2 = |F|^2$, so that the incident intensity of particle flow minus the reflected intensity equals the transmitted intensity, confirming our interpretation of the wave function.

Substituting equations [31] into equation [32], we get the transmission coefficient,

$$T = \frac{16k^2k''^2}{(k^2 + k''^2)^2 e^{2k''a} - (k''^2 + 2ikk'' - k^2)^2 - \text{CC} + (k^2 + k''^2)^2 e^{-2k''a}}$$

where CC stands for the complex conjugate of the term preceding. Since $e^{2k''a} + e^{-2k''a} = (e^{k''a} - e^{-k''a})^2 + 2$ and $e^{k''a} - e^{-k''a} = 2\sinh(k''a)$, we may write this as

$$T = \frac{16k^2 k''^2}{(k''^2 + k^2)^2 [4\sinh^2(k''a) + 2] - 2[k''^4 - 4k^2 k''^2 + k^4 - 2k^2 k''^2]}$$

$$= (4k^2 k''^2)[(k^2 + k''^2)^2 \sinh^2(k''a) + 4k^2 k''^2]^{-1}$$

for k'' real or k' imaginary ($E < V_0$). The transmission coefficient approaches 0 exponentially for large values of $k''a$:

$$T \to 4k^2 k''^2 (k^2 + k''^2)^2 \left(\frac{e^{2k''a}}{4}\right)^{-1}, \quad k''a \to \infty \qquad [33]$$

This corresponds to a wide barrier (large a) and/or an energy much below the top of the barrier (large $k''^2 = 2m(V_0 - E)/\hbar^2$). If m is the electron mass, a is in angstroms, and $V_0 - E$ is in electronvolts, then $k''a = 0.35a(V_0 - E)^{1/2}$, so $k''a$ will be large compared to unity when the width is several angstroms and $V_0 - E$ several electronvolts.

D. WENTZEL-KRAMERS-BRILLOUIN TUNNELING FORMULA

Although one can solve the barrier problem explicitly for barriers other than the rectangular one considered above, it is useful to consider a general but approximate treatment for barriers of different shape. This Wentzel-Kramers-Brillouin (WKB) approach is valid when the potential varies gradually with z, so it is not applicable to the rectangular barrier, for which V changes abruptly at $z = 0$ and $z = a$. Of course, the abrupt changes are themselves unphysical. The Schrödinger equation may be written as

$$\frac{d^2\psi}{dz^2} - \frac{2m(V-E)}{\hbar^2}\psi = \frac{d^2\psi}{dz^2} - \mathcal{K}^2\psi = 0 \qquad [34]$$

where $\mathcal{K}$ is a real function of z if $V > E$ (inside a barrier). If $V = V_0$ (a constant), then $\mathcal{K} = k''$ of the treatment of Section C. If $V > E$, then $\mathcal{K}$ is imaginary; it is equal to i times k of the above treatment for $V = 0$.

We seek a solution to equation [34] of the form $A(z) e^{iS(z)}$, in analogy to the solutions [22]-[24] of the rectangular-barrier problem, where A and S are now functions of z instead of being constants. $S(z)$ corresponds to $\hbar^{-1}$ multiplied by the local momentum. Substituting into equation [34] and dividing through by $e^{iS(z)}$, we have

$$A''(z) + 2iA'(z)S'(z) + iA(z)S''(z) - A(z)S''(z)^2$$
$$- A(z)[S'(z)]^2 - \mathcal{X}^2(z)A(z) = 0$$

where the primes indicate derivatives with respect to z. The real and imaginary parts of the lefthand side of the equation are separately set equal to zero:

$$A''(z) - A(z)S''(z) - A(z)[S'(z)]^2 - \mathcal{X}^2(z)A(z) = 0 \qquad [35]$$

$$2A'(z)S'(z) + A(z)S''(z) = 0 \qquad [36]$$

Since $(\ln f)' = f'/f$ for any function f equation [36] may be written as

$$2[\ln A(z)]' + [\ln S'(z)]' = 0$$

which integrates to $\ln A^2 + \ln S' =$ constant, that is,

$$A(z) = C_1[S'(z)]^{-1/2}$$

where C_1 is a constant. Substituting this into equation [35] we have

$$\frac{3}{4}C_1(S')^{-5/2}(S'')^2 - \frac{1}{2}C_1(S')^{-3/2}S'''$$
$$- C_1(S')^{3/2} - C_1\mathcal{X}^2(S')^{-1/2} = 0 \qquad [37]$$

The constant C_1 cancels off and cannot be determined, as should have been anticipated; it is related to the normalizing constant on the wave function.

The treatment up until now is exact and has merely taken us from the unknown function $\psi(z)$ to the unknown function $S(z)$. Now an approximation is made: Because $V(z)$ and $\mathcal{X}^2(z)$ are slowly varying with z, it is assumed that $S(z)$ varies slowly enough so that one can neglect its higher derivatives. Then equation [37] can be approximated as

$$[S'(z)]^{3/2} + \mathcal{X}^2(z)[S'(z)]^{-1/2} = 0$$

which is easily solved to give $S'(z) = \pm i\mathcal{X}(z)$ or

$$S(z) = \pm i \int \mathcal{X}(z)\, dz \qquad [38]$$

The corresponding approximate (WKB) solution to the Schrödinger equation is

$$\psi(z) = A(z)e^{iS(z)} = C_1[\mp i\mathcal{X}(z)]^{-1/2} \exp\left[\pm \int \mathcal{X}(z)\, dz\right]$$
$$= C_2\mathcal{X}(z)^{-1/2} \exp\left[\pm \int \mathcal{X}(z)\, dz\right] \qquad [39]$$

where C_2 is a new constant. The terms neglected in equation [37] mean that the approximate solution is valid if the potential is slowly varying with z. Specifically, it can be shown that the WKB approximation is valid if the change in the classical momentum, $p = [2m(E - V)]^{1/2}$, over one wavelength is small compared to the momentum itself. The wavelength is given by the De Broglie relation, $\lambda = \hbar/p = 1/k$, so the validity condition is that k'/k^2 be small.

We are considering a smoothly varying potential $V(z)$ with a single maximum (at which its value is V_0) such that $V(z) \to 0$ for $z \to \pm\infty$; we are also considering an electron with energy E which is less than V_0 (see Figure 22). On the V vs. z plot, a horizontal line at height E intersects $V(z)$ at two points which we will take to be at $z = 0$ and $z = a$. The potential thus divides the z axis into three regions. In region I, for $z < 0$, $E > V$ and $\mathcal{K}$ is imaginary; in region II, for $0 < z < a$, $E < V$ and $\mathcal{K}$ is real; in region III, for $z > a$, $E > V$ again and $\mathcal{K}$ is imaginary. In a region where $V < E$, we write $\mathcal{K}(z) = ik(z)$ with k real; the function [39] then oscillates in z like a sine or cosine curve. The problem now is the joining of the solution for a region where $V > E$ to the solution for a region where $V < E$. The value of z separating these regions is referred to as a *classical turning point* because a particle moving according to the laws of classical mechanics would exist only in regions where $E > V$ and would turn around when it came to a point where V became equal to E.

The wave function in region I is written as

$$\psi_I = A[k(z)]^{-1/2} \exp\left[i \int_0^z k(z)\, dz\right]$$
$$+ B[k(z)]^{-1/2} \exp\left[-i \int_0^z k(z)\, dz\right] \qquad [40]$$

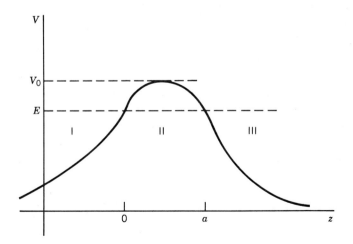

FIGURE 22. Barrier of height V_0, showing turning points (at $z = 0$ and $z = a$) for an energy E.

and the wave function in region II is written as

$$\psi_{II} = C[\mathcal{X}(z)]^{-1/2} \exp\left[\int_0^z \mathcal{X}(z)\, dz\right]$$

$$+ D[\mathcal{X}(z)]^{-1/2} \exp\left[-\int_0^z \mathcal{X}(z)\, dz\right] \qquad [41]$$

Finally, in region III we have

$$\psi_{III} = F[k(z)]^{-1/2} \exp\left[i\int_a^z k(z)\, dz\right]$$

$$+ G[k(z)]^{-1/2} \exp\left[-i\int_a^z k(z)\, dz\right] \qquad [42]$$

The analogy to equations [22]–[24] for the rectangular barrier may be noted. As we did for the rectangular barrier, we assume $G = 0$, since the term whose coefficient is G represents particles moving from right to left in region III, and we wish to consider particles impinging on the barrier from the left (in region I) and either being reflected back or being transmitted.

In general one shows (Landau and Lifschitz 1958, Section 17) that the particle current density in the z direction is given by

$$i_P = \frac{i\hbar}{2m}\left(\psi\frac{d\psi^*}{dz} - \psi^*\frac{d\psi}{dz}\right) \qquad [43]$$

A real wave function ($\psi = \psi^*$) corresponds to no current density. For the wave function $F[k(z)]^{-1/2} \exp[i\int_a^z k(z)\, dz]$, equation [43] gives

$$i_P = \frac{i\hbar}{2m}|F|^2 [k(z)]^{-1} [-ik(z) - ik(z)]$$

since the imaginary exponentials cancel. The current is real: $i_P = |F|^2 \hbar/m$. It is also independent of k, which means it is the same for all z (conservation of probability density). The transmission coefficient is the ratio of the left-to-right particle current in region III to the left-to-right particle current in region I, that is, $|F|^2/|A|^2$.

We now want to relate the constants in equations [40]–[42] by imposing the continuity of the wave function and its derivative with z at the turning points. Since $V = E$ at a turning point, k and $\mathcal{X}$ are both zero and one sees there is a problem. The WKB approximation requires $|k'(z)|$ to be small compared to $|k^2(z)|$, so it cannot work at a turning point. We can assume that this approximation is applicable everywhere except very close to turning points and then look for a better approximation in these regions (Merzbacher 1961, Chapter 7). One can see what happens better with the substitutions $\mathfrak{X} = ik$ and $\psi(z) = v(z)[k(z)]^{-1/2}$ in the Schrödinger equation (equation [34]), which then becomes

$$0 = v'' - v'\frac{k'}{k} - \frac{1}{2}v\frac{k''}{k} + \frac{3}{4}v\left(\frac{k'}{k}\right)^2 + k^2 v = 0 \qquad [44]$$

This can be simplified by writing $y = \int^z k(z)\,dz$ so that for any function f we have

$$f' = \frac{df}{dz} = \frac{df}{dy}\frac{dy}{dz} = \frac{df}{dy}k(z) \qquad [45]$$

and (primes still indicate derivatives with respect to z)

$$f'' = \frac{d^2 f}{dy^2}[k(z)]^2 + \frac{df}{dy}k'(z) \qquad [46]$$

Using the relations [45] and [46] with v for f, equation [44] reduces to

$$\frac{d^2 v}{dy^2}k^2 - \frac{1}{2}v\frac{k''}{k} + \frac{3}{4}v\left(\frac{k'}{k}\right)^2 + k^2 v = 0$$

Again using equations [45] and [46] but with $f = k$, we get finally

$$\frac{d^2 v}{dy^2}k^2 + \left[\frac{1}{4}\left(\frac{dk}{dy}\right)^2 - \frac{1}{2}k\frac{d^2 k}{dy^2} + k^2\right]v = 0 \qquad [47]$$

If the derivative terms in the square brackets can be neglected compared to k^2, equation [47] reduces to $d^2 v/dy^2 + v = 0$. This equation has the solution $v = e^{\pm iy}$, which, with the definitions of v and y, is just the WKB result.

At a turning point, $k^2 = 0$, so one cannot neglect the derivatives of k compared to k^2, and a better solution is necessary. Near $z = 0$, $V = E + \alpha z$ plus higher terms in a power series in z. Neglecting these terms, $V - E = \alpha z$ with α positive and $k^2 = -2m\alpha z/\hbar^2$. Thus, near this turning point,

$$k(z) = \left(\frac{-2m\alpha z}{\hbar^2}\right)^{1/2} \qquad \text{for } z < 0$$

$$k(z) = \exp\left(\frac{-i\pi}{2}\right)\left(\frac{2m\alpha z}{\hbar^2}\right)^{1/2} \qquad \text{for } z > 0$$

where the square roots are both to be taken as positive and a particular choice, $\exp(-i\pi/2)$, has been made for the phase. Integrating $k(z)$ from $z = 0$, we obtain expressions for $y(z)$ on either side of the turning point. The relations between these and the solutions of equation [47] far from the turning point (the WKB solutions) can then be elucidated, although the mathematics is not easy (Merzbacher 1961, Section 7.5). The solutions to the right of the turning point approach the WKB solutions $\exp(\pm|y|)\mathfrak{X}^{-1/2}$, whereas those to the left approach exponentials in $\pm i|y|$ divided by $k^{1/2}$, or sines and cosines in $|y|$ divided by $k^{1/2}$. In particular, the WKB solution $\exp(-|y|)\mathfrak{X}^{-1/2}$ on the right is shown to go over smoothly into $2\cos(-y-\pi/4)/k^{1/2}$ on the left, whereas $\exp(|y|)\mathfrak{X}^{-1/2}$ goes over smoothly into $\sin(-y-\pi/4)/k^{1/2}$ on the left.

In terms of the original variables, the connection formulas are

$$2k^{-1/2}\cos\left(\int_z^0 k\,dz - \tfrac{1}{4}\pi\right) \leftrightarrow \mathfrak{X}^{-1/2}\exp\left(-\int_0^z \mathfrak{X}\,dz\right) \quad [48a]$$

$$k^{-1/2}\sin\left(\int_z^0 k\,dz - \tfrac{1}{4}\pi\right) \leftrightarrow -\mathfrak{X}^{-1/2}\exp\left(\int_0^z \mathfrak{X}\,dz\right) \quad [48b]$$

The forms on the left are for the region to the left of the turning point, so $\int_z^0 k\,dz > 0$; the forms on the right are for the region to the right of the turning point, so $\int_0^z \mathfrak{X}\,dz > 0$. The second turning point in our problem is at $z = a$. The connection formulas are slightly different because the classically forbidden region is now to the left of the turning point rather than to the right. One gets

$$\mathfrak{X}^{-1/2}\exp\left(-\int_z^a \mathfrak{X}\,dz\right) \leftrightarrow 2k^{1/2}\cos\left(\int_a^z k\,dz - \tfrac{1}{4}\pi\right) \quad [49a]$$

$$-\mathfrak{X}^{-1/2}\exp\left(\int_z^a \mathfrak{X}\,dz\right) \leftrightarrow k^{-1/2}\sin\left(\int_a^z k\,dz - \tfrac{1}{4}\pi\right) \quad [49b]$$

One now has to apply equations [48] and [49] to the solutions [40]–[42]; the results are simpler than one might expect.

Since $G = 0$, the function ψ_{III} may be written, with $x = \int_a^z k\,dz$,

$$\psi_{\text{III}} = Fk^{-1/2}(\cos x + i\sin x)$$

$$= Fk^{-1/2}\left[\cos\left(x - \tfrac{1}{4}\pi\right)\cos\left(\tfrac{1}{4}\pi\right) - \sin\left(x - \tfrac{1}{4}\pi\right)\sin\left(\tfrac{1}{4}\pi\right)\right.$$

$$\left. + i\sin\left(x - \tfrac{1}{4}\pi\right)\cos\left(\tfrac{1}{4}\pi\right) + i\cos\left(x - \tfrac{1}{4}\pi\right)\sin\left(\tfrac{1}{4}\pi\right)\right]$$

$$= 2^{-1/2}(1+i)Fk^{-1/2}\cos\left(x - \tfrac{1}{4}\pi\right) + 2^{-1/2}(-1+i)Fk^{-1/2}\sin\left(x - \tfrac{1}{4}\pi\right)$$

Therefore, according to the formulas [49], the solution in region II is

$$\psi_{\text{II}} = 2^{-3/2}(1+i)F\mathfrak{X}^{-1/2}e^{-w} - 2^{-1/2}(-1+i)F\mathfrak{X}^{-1/2}e^{w} \quad [50]$$

where $w = \int_z^a \mathcal{X}\, dz$. We now proceed to the left, to generate ψ_I from ψ_{II} by using equation [48]. One has

$$w = \int_z^a \mathcal{X}\, dz = \int_0^a \mathcal{X}\, dz - \int_0^z \mathcal{X}\, dz = t - u$$

where $t = \int_0^a \mathcal{X}\, dz$ and $u = \int_0^z \mathcal{X}\, dz$, so that

$$\psi_{II} = 2^{-3/2}(1 + i)\, F\mathcal{X}^{-1/2} e^{-t} e^{u} - 2^{-1/2}(-1 + i)\, F\mathcal{X}^{-1/2} e^{t} e^{-u}$$

According to formula [48] the wave function in region I is then

$$\psi_I = -2^{-3/2}(1 + i)\, Fe^{-t} k^{-1/2} \sin\left(\int_z^0 k\, dz - \tfrac{1}{4}\pi\right)$$

$$- 2^{1/2}(-1 + i)\, Fe^{t} k^{-1/2} \cos\left(\int_z^0 k\, dz - \tfrac{1}{4}\pi\right)$$

$$= -2^{3/2}(1 + i)\, Fe^{-t} k^{-1/2}[2^{-1/2} \sin(-s) - 2^{-1/2} \cos(-s)]$$

$$- 2^{1/2}(-1 + i)\, Fe^{t} k^{-1/2}[2^{-1/2} \cos(-s) + 2^{-1/2} \sin(-s)]$$

where $s = \int_0^z k\, dz$. In terms of exponentials, we have

$$\psi_I = Fk^{-1/2} e^{-is}\left(\frac{i}{4} e^{-t} - ie^{t}\right) + Fk^{-1/2} e^{is}\left(\frac{e^{-t}}{4} + e^{t}\right) \qquad [51]$$

Then, comparing with equation [40], we have $A = F(\tfrac{1}{4} e^{-t} + e^{t})$.

Thus the transmission coefficient is

$$T = \frac{|F|^2}{|A|^2} = \left(\frac{1}{4} e^{-t} + e^{t}\right)^{-2}$$

with $t = \int_0^a \mathcal{X}\, dz$ (Bohm 1951, Chapter 12). In order for this result to be correct, the turning points must be far enough apart so that there is a region between them in which the WKB approximation is valid. This means that t should be large compared to unity, so $e^{t} \gg \tfrac{1}{4} e^{-t}$ and

$$T = e^{-2t} = \exp\left(-2 \int_0^a \mathcal{X}(z)\, dz\right) \qquad [52]$$

$$= \exp\left\{-2\left(\frac{2m}{\hbar^2}\right)^{1/2} \int_0^a [V(z) - E]^{1/2}\, dz\right\}$$

If V is roughly constant with value V_0 across the barrier, then $\int_0^a \mathfrak{K}\, dz = a[2m(V_0 - E)/\hbar^2]^{1/2}$. In the terminology of the rectangular-barrier problem of Section C, T is $e^{-2k''a}$. The exact result for this problem, equation [33], was $e^{-2k''a}$ multiplied by $16[(k/k'') + (k''/k)]^{-2}$, and this factor generally is of order unity. (Note that, because of the discontinuities in the rectangular-barrier potential, the WKB theory breaks down for this problem.)

E. ELECTRON EMISSION

The emission of electrons from a metal can be considered using equation [52]. One knows that conduction electrons encounter a barrier of 5–10 eV at the surface of a metal (electrical potential lower inside than outside), which corresponds to the barrier V_0. The electrons with highest energy are at the Fermi level, and $V_0 - \epsilon_F$ is simply the work function Φ. Now consider a field $\mathcal{E}$ just outside the surface at $z = 0$, which gives an additional potential energy $e\mathcal{E}z$. Since the barrier is rather steep, $V - E$ is essentially Φ up to $z = 0$ and about $\Phi + e\mathcal{E}z$ for $z > 0$. If $\mathcal{E}$ is negative there will be turning points at $z = 0$ and at $z = a = -\Phi/e\mathcal{E}$. Although the actual potential differs from $\Phi + e\mathcal{E}z$ in detail, the transmission coefficient, involving an integral over the barrier region, should not be greatly affected. Similarly, while the steepness of the potential near $z = 0$ makes the WKB approximation invalid near this point, this region makes a relatively small contribution to the transmission coefficient. Thus we have

$$\int_0^a \mathfrak{K}(z)\, dz = \int_0^a \left(\frac{2m}{\hbar^2}\right)^{1/2} (\Phi + e\mathcal{E}z)^{1/2}\, dz$$

$$= \frac{2}{3}\left(\frac{2m}{\hbar^2}\right)^{1/2} \Phi^{3/2}(-e\mathcal{E})^{-1}$$

so that

$$T = \exp\left[-\frac{4}{3}\left(\frac{2m}{\hbar^2}\right)^{1/2} \Phi^{3/2}(-e\mathcal{E})^{-1}\right] \qquad [53]$$

The number of electrons passing through unit surface per unit time, or current density, is calculated by multiplying T by the number of electrons striking the surface per unit time (as in gas kinetic theory).

According to equation [53], T is proportional to $e^{K/\mathcal{E}}$ for $\mathcal{E} < 0$; the formula is not applicable for $\mathcal{E} \to 0$. Thus the current density should increase rapidly with field strength $\mathcal{E}$ for $\mathcal{E} < 0$. For a given $\mathcal{E}$ the current density should be greater for metals with a smaller work function. The actual electric field may, however, differ significantly from the nominal electric field because the surface is not likely to be smooth on a submicroscopic level, and a small change in electric field makes a major change in the value of T.

E. ELECTRON EMISSION

More relevant to the problem of electron transfer from an electrode to a species outside (since there must be levels of energy E outside the metal in this case) is the following potential energy for an electron:

$$\begin{aligned} V &= 0, & z &< 0 \\ V &= -eW, & 0 &\leq z \leq a \\ V &= E, & z &> a \end{aligned} \qquad [54]$$

The energy E is equal to the Fermi energy, the energy of the state from which the electron is emitted. In the absence of other electrical fields, $\int_0^a \mathcal{K}(z)\,dz = (2m/\hbar^2)^{1/2}(-eW-E)^{1/2}a$. Now suppose there is an electric field $\mathcal{E}$ in the region $0 \leq z \leq a$, so $V = -eW + e\mathcal{E}z$ in this region. Then as long as $-eW - E + e\mathcal{E}a > 0$, we have

$$\int_0^a \mathcal{K}(z)\,dz = \left(\frac{2m}{\hbar^2}\right)^{1/2} \int_0^a (-eW - E + e\mathcal{E}z)^{1/2}\,dz$$

$$= -\frac{2}{3e\mathcal{E}}\left(\frac{2m}{\hbar^2}\right)^{1/2}\left[(-eW-E)^{3/2} - (-eW - E + e\mathcal{E}a)^{3/2}\right]$$

Abbreviate $-eW - E$, which is a work function, by Φ. Then, for small $\mathcal{E}$, the current density i is proportional to

$$T = \exp\left\{-\frac{4}{3e\mathcal{E}}\left(\frac{2m}{\hbar^2}\right)^{1/2}\Phi^{3/2}\left[1 - \left(1 + \frac{e\mathcal{E}a}{\Phi}\right)^{3/2}\right]\right\}$$

$$= \exp\left\{\frac{4}{3e\mathcal{E}}\left(\frac{2m}{\hbar^2}\right)^{1/2}\Phi^{1/2}\left[\frac{3}{2}e\mathcal{E}a + \frac{3}{8}(e\mathcal{E}a)^2\,\Phi^{-1} + \cdots\right]\right\}$$

Letting i_0 be the current density when $\mathcal{E} = 0$ we have

$$\frac{i}{i_0} = \exp\left[4\left(\frac{2m}{\hbar^2}\right)^{1/2}\Phi^{1/2}\frac{e\mathcal{E}a^2}{8\Phi} + \cdots\right]$$

If $\mathcal{E}$ is proportional to the difference between the potential drop across an interface, consisting of the metal and another phase, and this potential drop when no current is flowing, $\ln(i/i_0)$ will be proportional to this difference of potential drops. This difference is the overpotential.

For field emission from an electrode covered by a layer of adsorbed atoms, the effective potential combines the electric field, used in equation [53], with a rectangular-barrier potential such as that given in equation [54]. In the absence of the field, and taking the potential inside the metal as zero, one writes (Duke and Alferieff 1967)

$$V = 0, \qquad z < 0$$
$$V = \epsilon_F + \Phi, \qquad 0 \leq z \leq D - W$$
$$V = -V_a, \qquad D - W < z < D + W \qquad [55]$$
$$V = \epsilon_F + \Phi, \qquad D + W \leq z$$

The adsorbed layer contributes a potential well of depth V_a, having a thickness $2W$ and centered at $z = D$. As an alternative to the square-well potential, an attractive δ-function potential was also used by Duke and Alferieff (1967) to represent the adsorbed layer. The electric field gives a potential

$$V_\mathscr{E} = -\mathscr{E}ez, \qquad z \geq 0 \qquad [56]$$

to be added to equation [55]. A more realistic model would include (a) an image potential varying proportionally with z^{-1} in the region $z > 0$ and (b) some representation of the effect of polarization of the adsorbed layer by the field. Neglecting these, one has to consider solutions to the Schrödinger equation with potential function $V + V_\mathscr{E}$. This potential is shown in Figure 23.

The one-electron Schrödinger equation for a region of space in which the potential is linear, that is, $V = a + bz$, may be simplified (Landau and Lifschitz 1958, Section 22) by defining the coordinate y as $[z - (E - a)/b](2mb/\hbar^2)^{1/3}$. The Schrödinger equation then becomes

$$\frac{d^2\psi}{dy^2} = y\psi$$

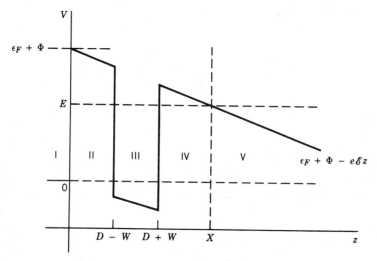

FIGURE 23. Potential for electric-field-induced emission from electrode with covering layer.

which is the Airy equation, whose solutions are known. The solutions to the equation in the various regions of z defined by equation [55] must be connected, using continuity conditions, and the solution for $z > D + W$ must be related to that for $z < 0$ to derive a transmission coefficient. Alternatively, one can use the WKB approximations, although the conditions for their validity may not obtain in the narrow regions of equation [55].

We will now give the treatment according to the WKB theory. We consider an energy below $\epsilon_F + \Phi$, and a field $\mathcal{E}$ which is positive but not too large, so that the electron energy E is less than $e_F + \Phi - e\mathcal{E}(D + W)$. Then there are two barriers, between $z = 0$ and $z = D - W$, and between $z = D + W$ and $z = X$, where $\epsilon_F + \Phi - e\mathcal{E}X = E$. One has to consider five regions of z: Let us denote by I, II, III, IV, and V the regions $z < 0$, $0 \leq z \leq D - W$, $D - W \leq z < D + W$, $D + W \leq z < X$, and $z \geq X$, respectively. The wave function ψ_V, which should be a transmitted wave, may be written in the same form as equation [42] with $G = 0$, that is, $\psi_V = F[k(z)]^{-1/2} \exp[i\int_x^z k_V(z)\,dz]$, where $k_V(z)^2 = (2m/\hbar^2)(E - \epsilon_F - \Phi + e\mathcal{E}z)$. Then the connection formulas [48] and [49] may be used twice, at the classical turning points at $z = 0$, $z = D - W$, $z = D + W$, and $z = X$, to produce ψ_I from ψ_V; ψ_I will consist of an incident and a reflected wave, as in the single-barrier problem.

The wave function ψ_{IV}, in a region where $V > E$, is similar to equation [41] or [50], and ψ_{III}, in a region where $V < E$, is similar to equation [40]. Now, however, the barrier is between $z = D + W$ and $z = X$, whereas in the previous case it was between $z = 0$ and $z = a$. Thus in region III,

$$\psi_{III} = F[k_{III}(z)]^{-1/2}\left[ie^{-is}\left(\frac{e^{-t}}{4} - e^t\right) + e^{is}\left(\frac{e^{-t}}{4} + e^t\right)\right]$$

where $s = \int_{D+W}^{z} k_{III}(z)\,dz$, $(k_{III})^2(z) = (2m/\hbar^2)(E + V_a + e\mathcal{E}z)$, and $t = \int_{D+W}^{x} dz\,(2m/\hbar^2)^{1/2}(\epsilon_F + \Phi - e\mathcal{E}z - E)^{1/2}$. To apply the connection formulas [48] and [49] again, we write

$$s = \int_{D+W}^{z} k_{III}\,dz = \int_{D-W}^{z} k_{III}\,dz - \int_{D-W}^{D+W} k_{III}\,dz = w - u$$

so that the wave function is

$$\psi_{III} = F[k_{III}(z)]^{-1/2}\left[ie^{iu}\left(\frac{e^{-t}}{4} - e^t\right)e^{-iw} + e^{-iu}\left(\frac{e^{-t}}{4} + e^t\right)e^{iw}\right] \quad [57]$$

Because we have already given the connection formula for taking e^{iw} across a barrier to the left, we need to consider only e^{-iw}. We write

$$e^{-iw} = \exp\left[-i\left(w - \frac{1}{4}\pi\right)\right]\exp\left(\frac{-i\pi}{4}\right)$$

$$= \left[\cos\left(w - \frac{1}{4}\pi\right) - i\sin\left(w - \frac{1}{4}\pi\right)\right](1-i)\,2^{-1/2}$$

Then according to equation [49], $[k_{\mathrm{III}}(z)]^{-1/2}e^{-iw}$ becomes, in region II,

$$(1-i)\,2^{-3/2}\,\mathfrak{X}^{-1/2}\exp\left(-\int_z^{D-W}\mathfrak{X}\,dz\right)$$

$$+\,i(1-i)\,2^{-1/2}\,\mathfrak{X}^{-1/2}\exp\left(\int_z^{D-W}\mathfrak{X}\,dz\right)$$

$$=(1-i)\,2^{-3/2}\,\mathfrak{X}^{-1/2}e^{-v}\exp\left(\int_0^z\mathfrak{X}\,dz\right)$$

$$+\,(i+1)\,2^{-1/2}\,\mathfrak{X}^{-1/2}e^v\exp\left(-\int_0^z\mathfrak{X}\,dz\right)$$

where $v = \int_0^{D-W}\mathfrak{X}\,dz$. Using equation [48] and some algebra, this becomes, in region I,

$$(1-i)\,e^{-v}2^{-3/2}(-k^{-1/2})\sin\left(\int_z^0 k\,dz - \frac{1}{4}\pi\right)$$

$$+\,(i+1)\,e^v 2^{1/2}k^{-1/2}\cos\left(\int_z^0 k\,dz - \frac{1}{4}\pi\right)$$

$$= k^{-1/2}\exp\left(-i\int_0^z k\,dz\right)\left(\frac{e^{-v}}{4} + e^v\right)$$

$$+\,k^{-1/2}\exp\left(i\int_0^z k\,dz\right)\left(\frac{-ie^{-v}}{4} + ie^v\right)$$

where $k^2 = (\hbar^2/m)E$ since $V=0$ in region I.
Therefore the result of connecting the wave function ψ_{III} into region I is

$$\psi_{\mathrm{I}} = Fie^{iu}\left(\frac{e^{-t}}{4} - e^t\right)k^{-1/2}\left\{\left[\exp\left(-i\int_0^z k\,dz\right)\left(\frac{1}{4}e^{-v} + e^v\right)\right.\right.$$

$$\left.\left.+\,i\exp\left(i\int_0^z k\,dz\right)\left(-\frac{1}{4}e^{-v} + e^v\right)\right]\right\}$$

E. ELECTRON EMISSION

$$+ Fe^{-iu}\left(\frac{e^{-t}}{4} + e^t\right) k^{-1/2} \times \left\{\left[i \exp\left(-i \int_0^z k\, dz\right)\left(\frac{1}{4} e^{-v} - e^v\right)\right.\right.$$

$$\left.\left. + \exp\left(i \int_0^z k\, dz\right)\left(\frac{1}{4} e^{-v} + e^v\right)\right]\right\} \qquad [58]$$

Here,

$$t = \int_{D+W}^{x} dz \left(\frac{2m}{\hbar^2}\right)^{1/2} (\epsilon_F + \Phi - e\mathcal{E}z - E)^{1/2}$$

$$u = \int_{D-W}^{D+W} dz \left(\frac{2m}{\hbar^2}\right)^{1/2} (E + V_a + e\mathcal{E}z)^{1/2} \qquad [59]$$

$$v = \int_0^{D-W} dz \left(\frac{2m}{\hbar^2}\right)^{1/2} (\epsilon_F + \Phi - e\mathcal{E}z - E)^{1/2}$$

Now the coefficient of the left-to-right wave in ψ_I is

$$-Fe^{iu}(\tfrac{1}{4} e^{-t} - e^t)(-\tfrac{1}{4} e^{-v} + e^v) + Fe^{-iu}(\tfrac{1}{4} e^{-t} + e^t)(\tfrac{1}{4} e^{-v} + e^v) \qquad [60]$$

The transmission coefficient T would be the absolute square of F divided by equation [60], so

$$T^{-1} = \left| e^{iu}\left(\frac{1}{4} e^{-t} - e^t\right)\left(\frac{1}{4} e^{-v} - e^v\right) + e^{-iu}\left(\frac{1}{4} e^{-t} + e^t\right)\left(\frac{1}{4} e^{-v} + e^v\right)\right|^2$$

$$= \left(\frac{1}{4} e^{-t} - e^t\right)^2 \left(\frac{1}{4} e^{-v} - e^v\right)^2 + \left(\frac{1}{4} e^{-t} + e^t\right)^2 \left(\frac{1}{4} e^{-v} + e^v\right)^2$$

$$+ 2 \cos(2u)\left(\frac{e^{-2t}}{16} - e^{2t}\right)\left(\frac{e^{-2v}}{16} + e^{2v}\right)$$

We may anticipate that $e^t \gg e^{-t}$ and $e^v \gg e^{-v}$, in which case

$$T = [2e^{2t+2v} - 2e^{2t+2v} \cos(2u)]^{-1} = \tfrac{1}{2} e^{-(2t+2v)} [1 - \cos(2u)]^{-1} \qquad [61]$$

The exponential factor is familiar, being a product of exponentials such as equation [52]. Its interpretation is simple: There are two barriers through which the particle must pass, so the transmission probability is a product of two probabilities, one for each barrier. For each barrier, increased height or thickness decreases penetration. The other factor is strictly quantum mechanical: $[1 - \cos(2u)]^{-1}$ is a quantum mechanical resonance term. According to equation [61] the transmission coefficient can become infinite if $\cos(2u) = 1$. Because the derivation used the WKB approximation, one can only conclude that T may become very large.

Using equation [59] for the definition of u, the condition for T becoming large reads

$$\left(\frac{2m}{\hbar^2}\right)^{1/2} \frac{2}{3e\mathcal{E}} \left\{ [E + V_a + e\mathcal{E}(D + W)]^{3/2} - [E + V_a + e\mathcal{E}(D - W)]^{3/2} \right\} = n\pi$$

for an integer n. Since one can expect $E + V_a + e\mathcal{E}(D - W)$ to be large compared to $e\mathcal{E}(2W)$, one may write the condition as

$$n\pi = \left(\frac{2m}{\hbar^2}\right)^{1/2} \frac{2}{3e\mathcal{E}} [E + V_a + e\mathcal{E}(D - W)]^{3/2}$$

$$\cdot \left[1 + \frac{3}{2}\left(\frac{2e\mathcal{E}W}{E + V_a + e\mathcal{E}(D - W)}\right) + \cdots - 1 \right]$$

$$= \left(\frac{2m}{\hbar^2}\right)^{1/2} [E + V_a + e\mathcal{E}(D - W)]^{3/2} \, 2W[E + V_a + e\mathcal{E}(D - W)]^{-1}$$

[62]

Now the stationary-state wave functions for a particle in a rectangular well with infinitely high walls, potential energy inside equal to $-[V_a + e\mathcal{E}(D - W)]$, and width $2W$ are (Goodisman 1977, Sections 3.2 and 3.3) $\sin(n\pi z/2W)$, and the corresponding eigenvalues of energy are

$$E_n = n^2 \hbar^2 \left[8m(2W)^2 \right]^{-1} - [V_a + e\mathcal{E}(D - W)]$$

where the terms represent kinetic and potential energy for the wave function $\sin(n\pi z/2W)$. Rearranging this we have

$$n = \left(\frac{8m}{\hbar^2}\right)^{1/2} (2W)[E_n + V_a + e\mathcal{E}(D - W)]^{1/2}$$

which, on multiplication by π, is exactly equation [62]. We conclude that one may find a large transmission coefficient if the energy E of the electron in the metal is equal to a stationary-state energy for an electron in the adsorbed layer, that is, in an adsorbed atom. This resonance enhancement or resonance transmission also appears in the exact treatment of the potential [55] + [56].

The remarkable result is that addition of an adsorbed layer may enhance, as well as diminish, the electron emission rate. This is expected to be true for emission from metal into an electrolytic solution, as well as for emission into vacuum. The resonance tunneling effect can also be seen in a treatment by time-dependent perturbation theory. In this case $-e\mathcal{E}z$ is the perturbation that is suddenly turned on (Gadzuk and Plummer 1973).

A detailed treatment of electron-transfer reactions at metal electrodes covered with oxide films has been given (Schmickler and Ulstrup 1977). The potential is supposed to be constant inside and outside the oxide layer (metal and electrolyte regions) and to be a linear function of position within the layer. At the metal and electrolyte sides of the layer the potential changes discontinuously, with the potential difference between oxide and electrolyte being a linear function of overpotential. The tunneling probability was calculated exactly from the Schrödinger equation (no WKB approximation), and calculated currents were compared to experiment.

F. OTHER APPLICATIONS

The neutralization of a gaseous ion at an electrode can be considered from the point of view of the electron tunneling formulas developed above. If the transfer of an electron from the interior of the metal to a nearby ion occurs without absorption or emission of radiation, there must be a vacant state on the ion with energy equal to that of the electron in the metal. The energy of the empty state is $-I$, where I is the ionization energy of the atom from which the ion is formed, because if an electron is brought from vacuum at an infinite distance to fill the state, the energy of the system will decrease by I. The electrons in the metal have energies up to the Fermi level ϵ_F. Relative to vacuum at infinity, the energy of the Fermi level is $-\Phi$, where Φ is the work function of the metal. The presence of the positively charged ion means that the potential energy of an electron is below its vacuum level outside the metal at points near the ion. Thus there is a barrier, although it is not a function of the single coordinate z.

Because the state of the electron after transfer to the ion has energy $-I$, and because electrons are available in the metal with a range of energies up to $-\Phi$, one must have $-\Phi \geq -I$ or $\Phi \leq I$. However, according to equation [52] the tunneling probability decreases rapidly as the electronic energy decreases relative to the top of the barrier, so one really needs to consider only electrons at the Fermi level, that is, of energy $-\Phi$. An electron of this energy, tunneling out of the metal, might be at an energy for which no stationary state existed, but, if a medium were present to take up some energy as heat, the electron could drop to the energy $-I$, corresponding to the empty state in the ion. In this case also, one must have $-\Phi \geq -I$ and $\Phi \leq I$.

In a medium such as a solution, the energy of the empty electron state in the ion is of course changed, and, as discussed in Section B, will actually become a range of energies, since one must consider the solvated ion, or the ion plus a certain number of rotating and vibrating solvent molecules. One can establish the energy of the empty electron state in a solvated ion, relative to vacuum at infinity, by considering the reaction of a solvated ion with an electron to produce a neutral species. If the electron is considered to move quickly so as not to polarize the solvent (Chapter 6, Section C), the energy of the electron in solution differs from its energy in vacuum only by a small amount due to the surface potential of the solution (~ 0.1 V for water), which we neglect. Because the solvation energy of the neutral species formed is negligible compared to that of the ion, the change in energy on neutralization of the ion is $(-I) - L$, where L is the solvation energy of the ion ($L < 0$). Applying the arguments of the preceding paragraph, $-\Phi$ must equal or exceed $-I - L$, that is,

$$\Phi \leq I + L$$

If there is a potential difference between metal and solution, the energy of the Fermi-level electron in the metal becomes $-\Phi - e\phi^{(M)}$ and the energy of the electron, if it enters the empty state on the ion, would be $-I - L - e\phi^{(S)}$, where $\phi^{(M)}$ and $\phi^{(S)}$ are the inner potentials of metal and solution. Now $-\Phi - e\phi^{(M)}$ must equal or exceed $-I - L - e\phi^{(S)}$, or

$$\Phi + e(\phi^{(M)} - \phi^{(S)}) \leq I + L \quad [63]$$

Some additional contributions still need to be considered (Bockris and Khan 1979, Section 8.5).

First, one notes that, if the neutralization of the ion is rapid, it occurs without change of nuclear conformation (Franck–Condon principle), leaving the solvent molecules around the neutralized ion in a nonequilibrium conformation. The energy of these molecules in this conformation is higher than the energy for the equilibrium conformation by an amount R, where R is a (positive) reorganization energy. L is the energy change for the process, "solvent at equilibrium" + "ion outside" → "solvated ion." The energy change for "solvent not at equilibrium" + "ion outside" → "solvated ion" is $L - R$. Thus L in equation [63] should be replaced by $L - R$. Second, one should consider the difference in the interaction energies of the neutralized species and the ionic species with the metal. Letting A represent the metal–neutral interaction energy minus the metal–ion interaction energy, we must add A to $-I - L$, which was the change in energy on charge neutralization in the preceding paragraph. Now equation [63] reads

$$\Phi + e(\phi^{(M)} - \phi^{(S)}) \leq I + L - R + A$$

A is probably negative and may amount to several electronvolts.

The ion in solution really has to be replaced by an ion–solvent cluster, which is like a complex molecule with many vibration–rotation states, closely spaced in energy. The single electronic level, coupled with the vibration–rotation states, becomes a collection of closely spaced molecular levels. These energy levels will depend on the distance of the species from the metal electrode, since interaction with the electrode affects rotations and vibrations. For the process of electron transfer from a solution species to a metal, there would be a large number of initial states for the electron, which one could imagine having occupation probabilities following, perhaps, a Boltzmann distribution. For transfer from metal to solution species, the energy of the initial state for the electron also has to be replaced by a series or band of energy levels, with associated occupation probabilities. This is because the energy of the transition is actually [energy of solvated neutral species] − [energy of an electron at the Fermi level in a metal + energy of solvated ionic species], or Φ + [energy of solvated neutral species] − [energy of solvated ionic species]. Because both the solvated neutral and ionic species have a series of energy levels, the energy of the transition must be considered as a series of energies. As noted above, these levels also depend on the proximity of the ion to the electrode.

A formal expression for the cathodic current density can be constructed by considering transitions from all occupied electron states in the metal to all possible final states. According to the perturbation theory of transitions (Section A) and the Fermi golden rule, initial and final states should be of the same energy. This means the probability for a particular state-to-state transition can be taken as a transmission coefficient from tunneling theory. The electron transfer may occur to a cation at any distance z from the electrode, so that the current density involves an integral over z as well as an integral over energy E. With e for the electronic charge, we have

$$j = e \iint n(E; \phi^{(M)} - \phi^{(S)}) \, N(E, z) \, P(E, z) \, dE \, dz \qquad [64]$$

Here $n(E; \mathcal{E})$ is the number of electrons at the surface of the metal with energy between E and $E + dE$ when the electrode–solution potential difference is equal to $\mathcal{E}$. This number is the density of states $\rho(E; \mathcal{E})$ multiplied by the number of electrons in the state at energy E, which is the Fermi distribution function $\{1 + \exp[(E + e\mathcal{E} - \epsilon_F)/kT]\}^{-1}$. $N(E, z) \, dE \, dz$ is the number of acceptor states with energy between E and $E + dE$, on ions distant between z and $z + dz$ from the electrode; this factor depends on temperature because energies are relative to energies of states of the solvated ion, whose occupation probability depends on temperature. $P(E, z)$ is the probability of tunneling at energy E to an ion at distance z and may be approximated by the WKB expression, $\exp(-2 \int_a^b \mathcal{X} \, dx)$, where $\mathcal{X}^2 = (2m/\hbar^2)[V(x) - E]$; $V(x)$ is the potential energy of the electron as a function of the electrode–ion coordinate, which may be represented as a rectangular barrier of height U_m and effective length l. Then

$$j = e \iint \rho(E; \mathcal{E}) \left[1 + \exp\left(\frac{E + e\mathcal{E} - \epsilon_F}{kT}\right) \right]^{-1} N(E, z)$$

$$\times \exp\left\{ -2 \left(\frac{2m}{\hbar^2}\right)^{1/2} (U_m - E)^{1/2} l \right\} \qquad [65]$$

The electron transfer from an electrode to a proton in solution, producing a hydrogen atom, constitutes one of the most important electrochemical reactions. Much work, both experimental and theoretical, has been done in studying the hydrogen evolution reaction. Bockris and Khan (1979, Sections 10.1–10.8) give an excellent review of the theoretical work, showing the important ideas and how they developed.

For proton discharge at the electrode, the proton must move from a position in solution, where it is solvated, through a barrier, to a position on the electrode, where it may be chemisorbed. The original position may be the inner or outer Helmholtz plane. With the proton on the electrode, the electron transfer may take place, forming a neutral hydrogen atom, which is then desorbed, probably as H_2. The transfer of the proton from one position to another may occur by tunneling. The proton transfer step, however, is not always rate-determining in the hydrogen evolution reaction. In cases where it is, the reaction rate, as measured by the exchange current density, increases with the magnitude of the heat of adsorption, and the overpotential correspondingly decreases: This seems to be the case for Hg, Pb, and other "soft" metals (with loosely bound, polarizable valence electrons) in acid solution. For transition metals in acid solution, the rate-determining step is the desorption of H atoms as H_2, and increased chemisorptive binding decreases the exchange current density and increases the overpotential because it cuts down on the H_2 desorption rate.

Information about the mechanism of a reaction involving hydrogen can be deduced from separation factors. The separation factor is the ratio of c_H/c_D in products to c_H/c_D in reactants, where c denotes concentration, H is ordinary hydrogen, and D is deuterium. A similar factor for hydrogen and tritium can be defined. Note the mass-dependence in T, equation [52]. The separation factor reflects the relative rates for H-containing and D-containing reactants. For a chemical reaction, it can be expressed as a product of ratios of partition functions, for the hydrogen-containing and deuterium-containing molecules and corresponding transition states, since the partition functions measure the number of accessible internal states (rotation, vibration, etc.). The spacing of these states in energy depends on the masses in the molecules, that is, on whether one has H or D. For the discharge of a solvated proton, the separation factor (Bockris and Khan 1979, Section 10.6) is given as

$$S = \frac{1}{2} \frac{T_H}{T_D} \frac{q_H^\ddagger}{q_D^\ddagger} \frac{q_{HDO(g)}}{q_{H_2O(g)}} K_e$$

where T_H and T_D are the tunneling probabilities for hydrogen and deuterium, $q_H^\ddagger$ and $q_D^\ddagger$ are partition functions for the activated complexes involved in the transfer step, $q_{HDO(g)}$ and $q_{H_2O(g)}$ are partition functions for gaseous HDO and H_2O molecules, and K_e is the equilibrium constant for the isotope exchange reaction of HDO(g) with H_2O(l) to give HDO(l) + H_2O(g).

The tunneling probabilities, and hence their ratio, depend on electric field, as already indicated in equation [53]. The electric field is related to the potential difference across the interface, which is in turn the potential difference at equilibrium plus the overpotential. The relation between the change in the electric field and the overpotential is generally assumed to be linear. Let us consider tunneling of a proton through a single barrier in the presence of a field, that is, from region I to region III in Figure 22. The transmission coefficient according to equation [52] would be e^{-2t}, where, since $V_0 - E$ is now Φ,

$$t = \left(\frac{2m}{\hbar^2}\right)^{1/2} \int_0^a (\Phi - e\mathcal{E}z)^{1/2} \, dz$$

$$= \frac{2}{3}\left(\frac{2m}{\hbar^2}\right)^{1/2} \frac{(\Phi - e\mathcal{E}a)^{3/2} - (\Phi)^{3/2}}{-e\mathcal{E}} \qquad [66]$$

where m is now the mass of the proton. For the deuteron, the quantity t is multiplied by $2^{1/2}$, and the transmission coefficient e^{-2t} is correspondingly decreased.

Tunneling ratios have been calculated for various sizes and shapes of barriers and as a function of overpotential. Particularly for large tunneling distances, the results are sensitive to the barrier shape. Calculation of the partition functions is straightforward once one decides on structures for the species involved, including the activated complex. Details of such calculations are given by Bockris and Khan (1979, Section 10.6). Other theories, both classical and quantum mechanical, for proton transfer are reviewed, compared, and critiqued by Bockris and Khan (1979, Sections 10.7–10.14).

Tunneling is also important in understanding proton transfer. Proton transfer, in a solvent that may ionize by giving up a proton, is a means of rapid charge transfer in solution and hence an important contributor to the transport of charge, that is, to the conductivity. The anomalously high equivalent conductivities of H^+ and OH^- in aqueous solutions, and of such ions as CH_3O^- in methanol and $COOH^-$ in formic acid, have long been known. Free (unhydrated) protons do not exist permanently in aqueous solutions, so that the transfer is thought of as occurring from H_3O^+ to H_2O in water. A series of such transfers is required for charge conduction through the solution, with some rearrangement of solvent molecules after each one. The same interactions as those involved in proton transfer are involved in formation of hydrogen bonds, which in turn are important in determining structure. The subject is reviewed by Erdey-Gru'z and Lengyel (1977), who also discuss theories rejecting proton transfer as a conduction mechanism.

10

QUANTUM KINETICS

A. CURRENT DENSITY

The important rate law is of course the Tafel equation, which is the linearity of the logarithm of the current density j with respect to the overpotential η. Writing $\ln j = \ln j^0 - \alpha\eta\mathcal{F}/RT$, where α is the transfer coefficient, we see the law states that the transfer coefficient is independent of potential. Showing the existence of such a transfer coefficient from a detailed molecular model is still a goal of theoretical electrochemical kinetics. The existence of a transfer coefficient implies a proportionality between free energy of activation and free energy of reaction, which is equivalent to a Brønsted relation between rate constants and equilibrium constants (Krishtalik 1983, Section 3.1). Consider two homologous reactions i and j, with rate constants k_i and k_j and equilibrium constants K_i and K_j. It is commonly found that $k_i/k_j = (K_i/K_j)^\alpha$ or that, if $\ln k_i$ is plotted against $\ln K_i$ for a set of homologous reactions i, one obtains a straight line (Hammett plot). In terms of free energies and activation free energies, $k_i/k_j = (K_i/K_j)^\alpha$ implies a linear free-energy relationship, $\Delta G_i^\ddagger - \Delta G_j^\ddagger = \alpha(\Delta G_i - \Delta G_j)$. For an electrode reaction, the free energies ΔG_i are changed by changing the electrode potential instead of going from one member of a homologous series to another, and j/j^0 is a rate constant. The Tafel law thus states that $\ln k_i$ is linear with respect to electrode potential or to η.

A general expression for cathodic current density, as an integral over energy and distance from electrode, was given as equation [64] of Chapter 9, Section F. For purposes of discussion, one can break up some of the factors in that expression further and write

$$j_c = e \iint [f(E)\,\rho(E)]\,[c(z)\,G(E, z)]\,P(E, z)\,dE\,dz \qquad [1]$$

As before, $P(E, z)$ is the probability for an electronic transition from a state of energy E in the metal electrode to an empty state of energy E on an acceptor ion in solution at a distance z from the electrode. The first factor, the number of electrons per unit area at the metal surface with energy between E and $E + dE$, has been written as a product of the density of states $\rho(E)$ and the occupation probability $f(E)$ according to Fermi statistics; $f(E) = \{1 + \exp[(E - \epsilon_F)/kT]\}^{-1}$. Similarly, the number of acceptor states with energy E on ions at a distance z has been written as a product of the ionic concentration at z, $c(z)$, and the probability of having an (empty) acceptor state at energy E on an ion at distance z, $G(E, z)$. The anodic current density would be given by

$$j_a = e \iint [(1 - f)\rho(E)][c'(z)G'(E, z)]P'(E, z)\, dE\, dz \qquad [2]$$

since we are interested in electronic transitions from ions in solution to empty electronic states in the electrode. In the first factor, $1 - f(E)$ appears, since we want the number of empty states at energy E, and $G'(E, z)$ is the probability of having an electron in a state of energy E on an ion at a distance z from the electrode. The transition probability $P'(E, z)$ now refers to an electron going from an ion state at energy E and distance z to an electrode state at this same energy.

Because of the rapid decrease in $P(E, z)$ and $P'(E, z)$ with z, it may be possible to argue that only ions at their closest distance of approach to the electrode, that is, the outer Helmholtz plane, need to be considered. Then the integrations over distance may be dropped in equations [1] and [2], with $c(z)$ and $c'(z)$ replaced by the concentration of ions at the outer Helmholtz plane. The transition probabilities P and P' have been discussed in Chapter 9; in principle, they are calculable quantum mechanically from tunneling theory or from a more exact model (Khan and Bockris 1983, Section 2.5.2.10). The probabilities G and G' of having acceptor or donor states at energy E on an ion are quite complicated to evaluate in general. They depend on temperature because the solution species are rotating, vibrating, and so on, in the thermal bath constituted by the electrolyte phase. Indeed, the necessity to consider changes in the arrangement of molecules simultaneously with electronic transitions is what makes theoretical calculation of electrode reaction rates difficult.

According to equation [1], if it can be assumed that essentially all the electronic transitions relate to ions at the outer Helmholtz plane and to electrons in the metal at the Fermi level, we have

$$j_c = ePf(\epsilon_F)\rho(\epsilon_F)c(z_H)G(\epsilon_F)$$

The value of $f(\epsilon_F)$, using the Fermi distribution function, is $\frac{1}{2}$. The probability of having an empty electronic state of energy ϵ_F on the acceptor ion means the probability that the acceptor ion O is in a state such that the electronic transition occurs with no change of energy, that is, that the energy of the reduced ion R (after addition of the electron) is equal to the energy of the electron in the metal plus the energy of the ion before reduction. [It should be remembered that the energy of an electronic state in an ion or molecule (if we are talking about an eigenvalue of the electronic Schrödinger equation) depends very much on whether the state is filled or empty.] By the "state" of an ion we mean its position relative to the electrode, the arrangement of the solvent molecules around it, and the values of its internal vibrational or rotational quantum numbers.

The probability $G(\epsilon_F)$ should be proportional to $e^{-h/kT}$, where h is an activation energy, the energy that the ion must have over its ground-state energy to put it into a state for transition. There will be more said about this energy later. We can, though, expect that it is some fraction, say β, of the difference between the energy of the reduced ion in its stable state and that of the ion before addition in *its* stable state. Because the energy difference depends linearly on the electrode–solution potential, we may write

$$h = \beta(H_0 + \mathcal{F}\eta)$$

where H_0 is the energy difference at the equilibrium potential and η is thus the overpotential. Now

$$j_c = CP\rho(\epsilon_F)\, c(z_H)\, \exp\left[\frac{-\beta(H_0 + \mathcal{F}\eta)}{kT}\right]$$

where C is a constant. The dependence of j_c on η is only in the last factor, so that

$$kT\left(\frac{\partial \ln j_c}{\partial \eta}\right)_T = \frac{d[-\beta(H_0 + \mathcal{F}\eta)]}{d\eta} = -\beta\mathcal{F}$$

which is the Tafel law. However, several oversimplifications are involved in getting this expression for j_c.

In writing the probability $G(\epsilon_F)$ as proportional to $e^{-h/kT}$, we have ignored degeneracy: Because of the internal motions of the molecules and ions, as well as the orientations of solvent dipoles around the ions, there are many states of the same energy, and a factor giving the number of states per unit energy range at the excitation energy h should be included in j_c. A change of electrode potential changes h, so the degeneracy factor should depend on η. Also, we are not justified in taking the "symmetry factor" β as independent of η without some model for explaining its origin. The neglect of electrons at energies other than the Fermi energy, and of ions at positions other than the Helmholtz plane, has already been mentioned.

To take into account the degeneracy or density-of-states factor, we could replace $G(\epsilon_F)$ by a sum of contributions of all the empty ion states, each with its population probability according to the Boltzmann distribution law. Let E_O^0 be the ground-state energy of the ion before reduction and let E_R^0 be the ground-state energy of the ion after reduction, so

$$E_R^0 - E_O^0 = H_0 + \mathfrak{F}\eta$$

Let h be an excitation energy from E_O^0 and let $g^O(h)$ be the density of states for this ion, so the probability of finding the ion O at an energy between $E_O^0 + h$ and $E_O^0 + h + dh$ is proportional to $g^O(h)\, e^{-h/kT}\, dh$. The corresponding density of states for the reduced ion is $g^R(h)$. We need the probability that ion O is in a state such that its energy, plus the energy of an electron at the Fermi level of the metal electrode, matches the energy of a state of ion R. Thus for a state of ion O with energy h above its ground state we need to consider states of ion R with energy f above *its* ground state such that $E_R^0 + f = E_O^0 + h + \epsilon_F$. The factor $G(E)$ should be replaced by

$$G'(E) = \frac{\int_0^\infty g^R(E_O^0 + \epsilon_F - E_R^0 + h)\, g^O(h)\, e^{-h/kT}\, dh}{\int_0^\infty g^O(h)\, e^{-h/kT}\, dh}$$

and the overpotential enters since $E_O^0 - E_R^0 = -H_0 - \mathfrak{F}\eta$. The problem now is the form of the density-of-states factors g^O and g^R.

If $g^R(f)$ is sharply peaked at $f = f_0$ the integral in the numerator of $G'(E)$ becomes proportional to

$$g^O(f_0 - \epsilon_F + H_0 + \mathfrak{F}\eta) \exp\left(\frac{f_0 - \epsilon_F + H_0 + \mathfrak{F}\eta}{kT}\right)$$

The Tafel law, $d \ln j_c / d\eta =$ constant, is obtained if g^O is not a sharply varying function of its argument. One can also get it if g^O is sharply peaked at $h = h_0$ and g^R is slowly varying as a function of its argument. For some choices of g^O and g^R, the integral in $G'(E)$ can be conveniently evaluated (which does not imply that these choices are physically reasonable). If $g^O(h) = a \exp[-b(h - h_0)^2]$ and $g^R(f) = c \exp[-d(f - f_0)^2]$ the integral is proportional to

$$\int_0^\infty \exp\left[\frac{-b(h - h_0)^2}{kT} - h\right] \exp\left[-d(E_O^0 + \epsilon_F - E_R^0 + h - f_0)^2\right]$$

and, extending the lower limit to $-\infty$, one gets a result proportional to $\exp[-du + (b + d)^{-1}(bh_0 - (2kT)^{-1} + u)^2]$, where $u = E_O^0 + \epsilon_F - E_R^0 - f_0$. Since $d \ln G'/d\eta = -\mathfrak{F}(d \ln G'/du)$ and u is linear with respect to η, $d \ln G'(E)/d\eta$ is a linear function of η. Thus the Gaussian distribution of states does not in general lead to the Tafel law, which requires $d \ln G'(E)/d\eta$ to be a constant independent of η (Khan and Bockris 1983, Section 2.8). If b or d is large enough, though, $d \ln G'/d\eta$ may be approximately independent of η; this corresponds to sharply peaked g^O or g^R, as described above. Another simple case occurs if $g^O(h)$ is proportional to e^{-ah} and $g^R(f)$ is proportional to e^{-bf}; the integral in $G'(E)$ is proportional to

$$I = \int_0^\infty dh\, e^{-ah} \exp\left[-b(E_O^0 + \epsilon_F - E_R^0)\right] e^{-bh} e^{-h/kT}$$

$$= \exp\left[-b(-H_0 + \epsilon_F - \mathfrak{F}\eta)\right] \int_0^\infty dh\, e^{-h(a+b+1/kT)}$$

and $d \ln I/d\eta = b\mathfrak{F}$. This leads to $d \ln G'(E)/d\eta$ independent of η and hence to the Tafel law. Of course, the expressions for densities of states are formal only, and a model is required to calculate them.

The integrals over excitation energies h are equivalent to sums (or integrals) over electronic states and molecular positions. The current density is a sum of currents due to tunneling events between electrode and molecules in different states. A somewhat different point of view of the electron-transfer process between electrode and solution species is that we are interested in radiationless transitions between two electronic states, which must be at the same energy since there is no radiation to bring in or take out energy. This appears in the derivation of electronic transition probabilities under the influence of a time-independent perturbation (Chapter 9, Section A). The two states between which the transition occurs differ in the wave function or density of at least one electron, which is localized on the solution species in one state and on the metal in the other.

According to the Franck–Condon principle, an electronic transition in a molecule occurs without change in position of the heavy particles (nuclei). The motion of the nuclei can be considered to be much slower than the motion of the electrons (here we use classical mechanical terms: One could say quantum mechanically that the transition frequencies between states of nuclear motion are much smaller than those between electronic states). In addition to the Franck–Condon principle, this implies that the motions of the nuclei cause no electronic transitions. For every configuration of the nuclei, there are a set of electronic states, with wave functions ψ_i and energies E_i, such that each state for nuclear configuration $\mathbf{R}$ can be identified with a state for nearby nuclear configuration $\mathbf{R} + d\mathbf{R}$. We can then speak of a set of electronic energies, each of which is a function of nuclear configuration. Let us denote these by $E_i(\mathbf{R})$, $i = 0, 1, 2, \ldots$, where i is a quantum number or a set of quantum numbers; $i = 0$ is the ground state. The $E_i(\mathbf{R})$ are the molecular electronic terms. To say that no electronic transitions are caused by nuclear motions means that, if the nuclear configuration of a molecule $\mathbf{R}$ changes with time t, the energy of the molecule (which is the electronic energy since we are imagining the nuclei to be successively fixed at each of a series of positions) depends on time according to $E_i(\mathbf{R}(t))$, and the state i does not change. The situation is described by saying that the nuclei move adiabatically.

The nuclear configuration $\mathbf{R}$ for a molecule involves several coordinates, so a plot of the function $E_i(\mathbf{R})$ is an energy surface. If there is only one nuclear coordinate of interest, as for the vibrations of a diatomic molecule, the plot is two-dimensional (see Figure 24); the surfaces $E_i(\mathbf{R})$ become the potential energy curves for the electronic states of a molecule. Note that this picture considers the nuclei according to classical mechanics and the electrons according to quantum mechanics. The adiabatic motion of a system signifies that the system remains on one of its energy surfaces, and the Franck–Condon principle means that an electronic transition (caused, perhaps, by radiation) is vertical, from one surface to another at fixed $\mathbf{R}$ (see Figure 24 for the case of a single nuclear coordinate). An electronic transition between two states of equal energy can occur for a value of $\mathbf{R}$ at which two curves cross.

In considering a chemical reaction we usually isolate a single nuclear motion of interest, referred to as the reaction coordinate, such that the displacement along this coordinate converts reactant molecules into product molecules. It is expected that the electronic energy as a function of this coordinate (which is an electronic term) goes through a maximum, referred to as the *activated complex*, and has minima on either side (see Figure 25). The minima correspond to the reactant species and product species, each of which is a stable molecule, that is, at a local minimum in electronic energy. The maximum corresponds to the activated complex, which is a structure intermediate between those of reactants and products. For reaction to occur, the system must pass through this state; the energy required, starting from the state of reactants, is the activation energy.

The motion of the molecular system, as it undergoes reaction, is adiabatic: There

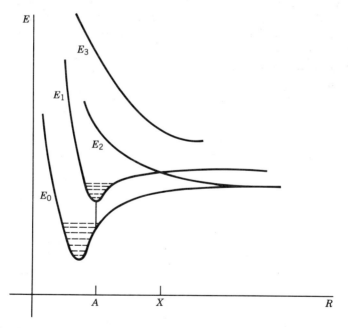

FIGURE 24. Potential energy curves or terms (for a diatomic molecule). Vibrational energy levels are shown as broken lines. A transition from one state to another at fixed R is shown for $R = A$. A radiationless transition can occur where terms cross, at $R = X$.

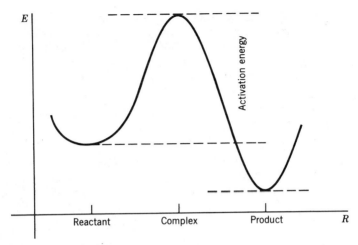

FIGURE 25. Schematic diagram of energy as a function of reaction coordinate, showing position of reactant, product, and activated complex.

is no electronic transition, only a change in electronic energy of the type $E_i(\mathbf{R})$ discussed above. In fact, it is generally assumed that nuclear motions other than motion along the reaction coordinate are fast enough to be treated like the electronic motions, that is, there is a set of states for molecular vibrations and rotations in coordinates other than the reaction coordinate, whose energies may depend on the reaction coordinate, and the motion corresponding to reaction is adiabatic with respect to transitions between these states.

For an electron-transfer process, one has to consider molecular electronic states for different numbers of electrons. The electronic transition is between two states corresponding to different numbers of electrons on the molecule (the missing electron is on the electrode, and its energy must be included in the energy of the molecule). For the transition probability to be appreciable, the states must have the same energy, so we are interested in nuclear configurations $\mathbf{R}$ for which $E_i^{(n)}(\mathbf{R}) = E_j^{(n+1)}(\mathbf{R})$ (superscripts refer to numbers of electrons), that is, crossing points for potential energy curves (point X in Figure 24). If an electronic transition occurs, the system changes its electronic state, so it jumps from one potential curve to another. With subsequent change in $\mathbf{R}$ it continues along the new potential curve. A possible path is shown in Figure 26.

We may consider a chemical reaction that takes place on any of the states i, with energy $E_i(\mathbf{R})$ (i now refers to quantum numbers of fast nuclear, as well as electronic, motions). During the activation process that leads to reaction (change along the reaction coordinate) there are no transitions between states, so the observed rate is a sum of contributions of the different states. If there are high-frequency vibrations, however, they can lead to nonadiabaticity, that is, transitions at points other than crossings of states (Dogonadze and Kuznetsov 1983, Section 1.4).

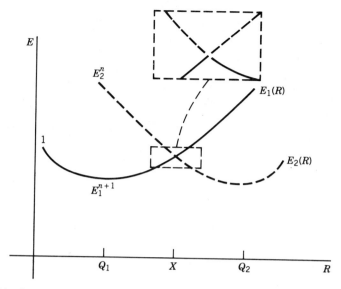

FIGURE 26. Crossing of terms at $R = X$. An electron transfer can occur there. The system would then follow the solid curve (see the inset), jumping from term 1 to term 2 when it reached $R = X$.

Assuming this does not occur, contributions are weighted by the associated occupation probabilities, corresponding to thermal equilibrium, usually expressed by a Boltzmann distribution. If the rate constant for state i is k_i and the concentration of molecules in state i is c_i, the overall rate is $\Sigma \, k_i c_i$. Here, c_i is related to the total concentration of molecules c by the Boltzmann distribution function,

$$\frac{c_i}{c} = q^{-1} e^{-E_i/k_B T} \qquad [3]$$

where q is the partition function for a reactant molecule,

$$q = \sum e^{-E_i/k_B T}$$

and E_i is the energy of a reactant molecule, corresponding to $E_i(\mathbf{R})$, where $\mathbf{R}$ is a configuration corresponding to a reactant; k_B is used for the Boltzmann constant.

B. ACTIVATION PROCESS

According to the absolute reaction rate theory, the rate constant k_i is related to the probability of reaching the activated complex, the maximum in $E_i(\mathbf{R})$ separating the minima corresponding to reactant and product molecules (Figure 25):

$$k_i = k^0 \exp\left(-\frac{E_i^\ddagger - E_i}{k_B T}\right)$$

The energy of the activated complex is $E_i^\ddagger$, and k^0 is a constant independent of i. If an electron transfer at a curve crossing (Figure 26) is involved, $E_i^\ddagger$ is the value of $E_i(\mathbf{R})$ at the crossing point minus its value at the minimum. Now the overall rate is

$$\sum k_i c_i = k^0 \exp\left(-\frac{E_i^\ddagger - E_i}{k_B T}\right) c q^{-1} \exp\left(-\frac{E_i}{k_B T}\right)$$

$$= k^0 q^{-1} \sum \exp\left(\frac{-E_i^\ddagger}{k_B T}\right) c = kc \qquad [4]$$

where k is the overall rate constant. For an electrode reaction, the reaction rate is replaced by a current density, so k is replaced by current density divided by concentration of reacting species.

The sum over states in equation [4] is a partition function, but it is conventional, in a partition function, to take the zero of energy as that of the lowest state. With

$$q^\ddagger = \sum \exp\left(-\frac{E_i^\ddagger - E_0^\ddagger}{k_B T}\right)$$

we may rewrite the rate constant in equation [4] as

$$k = k^0 \frac{q^\ddagger}{q} \exp\left(-\frac{E_0^\ddagger - E_0}{k_B T}\right) \qquad [5]$$

In equation [5] the quantity k^0 is a frequency factor, related to the rate with which molecules move in the direction of the transition state. The energy of activation is $E_0^\ddagger - E_0$. According to statistical thermodynamics (Eyring et al. 1982, Chapter 5), the free energy of N molecules, each with partition function q, is $Nk_B T \ln q$. If all species are to be referred to the same energy zero, a factor of $e^{-E/k_B T}$ must be put into q, so

$$k = k^0 \exp\left(-\frac{\Delta G^\ddagger}{k_B T}\right) \qquad [6]$$

Here $\Delta G^\ddagger$ is a free-energy change per molecule (free energy of activation); $\Delta G^\ddagger = \Delta U^\ddagger - T\Delta S^\ddagger$, where $\Delta U^\ddagger$ is an average of $E_i^\ddagger - E_i$, converted to energy per mole.

If we considered the reverse reaction to the one described by equations [4]–[6], the same electronic states, the same reaction coordinate, and the same transition state or activated complex would be involved. Thus if the reaction were A → B, so that q, E_0, and so on, referred to molecule A, the reverse rate constant would be

$$k' = k^0 \frac{q^\ddagger}{q'} \exp\left(-\frac{E_0^\ddagger - E_0'}{k_B T}\right)$$

where q' and E_0' refer to molecule B, and the equation corresponding to equation [6] would involve $\Delta G^\ddagger$, the free energy difference in going from molecule B to the activated complex. Obviously, k/k' would be $(q'/q) \exp[-(E_0' - E_0)/k_B T]$ or $\exp(-\Delta G/k_B T)$, where ΔG is $G_B - G_A$. All reference to the activated complex disappears from k/k', which is the thermodynamic equilibrium constant. For an electrode reaction, k/k' is the ratio of cathodic to anodic current divided by the ratio of concentrations of oxidized and reduced species and is an equilibrium constant in the context of a Nernst equation, depending on η.

A chemical reaction in solution, or at an electrode, must involve several steps in addition to electron or atom transfer, including the diffusing together of reactants, the rearrangement of solvent shells, and so on. The formalism outlined above is to be applied to the rate-determining step. Then in equation [5], q and E_0 are, respectively, the partition function and the ground-state energy for the reactant in this step. But, by the definition of the rate-determining step, equilibrium is established for every step preceding the rate-determining step. If c_R is the concentration of reactant R for the rate-determining step and c_S is the concentration of reactant for the step preceding, which produces R from S, the equilibrium condition is

$$\frac{c_R}{c_S} = \frac{q}{q_S} \exp\left(-\frac{E_0 - E_0^S}{k_B T}\right) \qquad [7]$$

where q_S and E_0^S are, respectively, the partition function and the ground-state energy for S, so that one can write the rate as

$$kc_R = k^0 \frac{q^\ddagger}{q_S} \exp\left(-\frac{E_0^\ddagger - E_0^S}{k_B T}\right) c_S \qquad [8]$$

The rate is proportional to the concentration of S, and the rate constant involves the properties of S and the activated complex, rather than the properties of R.

For a reaction at an electrode, ΔG will depend on the electrode–solution potential difference E_{es} and hence on the cell potential, because the reaction involves transfer of an electron between electrode and solution. The electrical contribution to the free energy of reaction per mole is $n\mathcal{F}E_{es}$. Since the transition state is intermediate on the reaction coordinate between the initial and final states, so that $\Delta G^\ddagger$ is a fraction of ΔG, we can anticipate that there will be an electrical contribution to $\Delta G^\ddagger$ which is a fraction of $n\mathcal{F}E_{es}$. Then $d \ln k/dE_{es}$ will be a constant, which implies the Tafel law: The current density j is proportional to the reaction rate and hence to k (so $\ln k$ differs from $\ln j$ by a constant), whereas the cell potential differs from E_{es} by a constant.

Before considering electron-transfer reactions at the electrode, we consider homogeneous electron transfer, from molecule A to molecule B, both species being in solution. We use superscripts R and O to denote the reduced and oxidized states, respectively, so that the reactants for the electron-transfer reaction are $A^R + B^O$. The transfer of an electron leads to the products $A^O + B^R$. All species are solvated. In the theories of Marcus, Levich, Dogonadze, and others, the solvation and the way it changes plays a central role. The arrangement of solvent molecules around $A^R + B^O$, assuming the two species to have diffused together, is not appropriate to products $A^O + B^R$. The energy of solvated $A^R + B^O$ is thus appreciably lower than the electronic energy of $A^O + B^R$ would be in the same solvent environment. This means that no electron transfer can take place: A radiationless transition requires electronic states of equal energy in reactant and product. The rearrangement of the solvent shells to produce such a situation becomes the reaction coordinate, which we will denote by Q.

Suppose that, for one value of the reaction coordinate, the solvent arrangement is appropriate to $A^R + B^O$, so that the highest filled electronic level on A is much below the energy of the lowest empty electronic level on B. For another value of Q, the reverse is true; this is the arrangement of solvent appropriate to $A^O + B^R$. For an intermediate value of Q, an arrangement of solvent molecules which is optimum for neither $A^R + B^O$ nor $A^O + B^R$, the electronic energies are equal for reactants and products, and an electronic transition can take place. This is the transition state, not an equilibrium state of the solute–solvent system. Let us now look at the situation from the point of view of the electron which is to be transferred.

For the three values of Q, we can indicate (Figure 27) the potential energy that would enter the *electronic* Schrödinger equation. For Q corresponding to optimum solvation of $A^R + B^O$ (Figure 27a), this potential energy shows a deeper potential well in the vicinity of molecule A and a higher one in the vicinity of molecule B. We refer to this configuration as Q_1. The electronic wave function of lowest energy for $Q = Q_1$ would be localized on A. The electron density and energy for this wave function are shown schematically (broken lines) in the diagram. For $Q = Q_2$, corresponding to the optimum solvation of $A^O + B^R$, the reverse is true and the lowest-energy electronic wave function will be localized on B (Figure 27c). For Q corresponding to the transition state, $Q = Q_T$, the wells are of equal depth. The stationary state electronic wave function of lowest energy will have equal electron density on the two centers (Figure 27b). Alternatively, we can say there are degenerate electronic wave functions, one localized on A and one localized on B, and a transition can take place from one to the other. The formalism of time-independent perturbation theory, or of tunneling, could be applied to calculate the transition probability.

Since motion along the reaction coordinate is adiabatic, a system at $Q = Q_1$, and in the electronic state whose wave function is shown in Figure 27a, would, when solvent rearrangement changes Q to Q_T, be in an electronic state localized

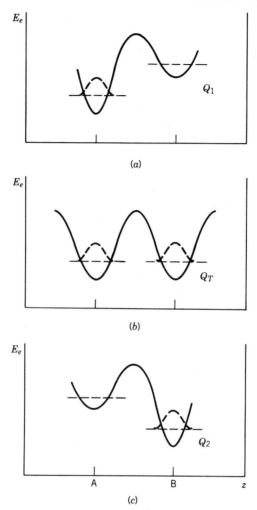

FIGURE 27. Electronic potential energy as a function of position for several values of the solvation coordinate Q. (a) Q corresponding to $A^R + B^O$. (b) Q corresponding to intermediate solvation. (c) Q corresponding to $A^O + B^R$. Electronic wave functions are shown by broken lines.

on A (Figure 28a). This is the state that connects adiabatically with the initial state; its wave function is shown by a broken line in Figure 28a. Electron transfer corresponds to a transition to a state of the same energy, but localized on the other center, as shown in Figure 28b. The change of electronic state from that localized on A (Figure 28a) to that localized on B (Figure 28b) occurs without change in the heavy particle's arrangement (Franck–Condon principle) so we are still at Q_T. A subsequent rearrangement of the solvent shell brings us to the final state (Figure

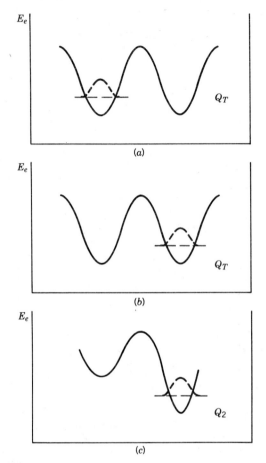

FIGURE 28. Electronic transition. The electronic potential energy as a function of position is shown; the electronic energy and wave function are shown by broken lines. A transition occurs between diagrams (a) and (b) ($Q = Q_T$), then an adiabatic change of Q from Q_T to Q_2 produces the final state (c).

28c); during this rearrangement the reaction coordinate Q changes from Q_T to Q_2. Since the motion is adiabatic, the system remains in the same electronic state, the energy and wave function of which change with Q as shown.

To consider the process from the point of view of the heavy particles, we can indicate the electronic energy (eigenvalue of the electronic Schrödinger equation) as a function of Q for the initial or final electronic state. This is the potential energy that governs heavy-particle displacements (potential surface or term in the case of a molecule). The potential energy for the initial state has a minimum at Q_1, whereas that for the final state has a minimum at Q_2. The two curves cross at $Q = Q_T$, which represents an arrangement of solvent which is equally good (or equally bad) for both electronic states. The situation is the same as in Figure 26; at the crossing point, X or Q_T, the electronic energies of the two states are equal.

An electron transfer between an electrode and a species in the electrolyte can be considered in exactly the same way. The electronic states differ by one electron on a solvent species, and the reaction coordinate could correspond to approach of the solvent species to the electrode or to the outer Helmholtz plane. The energy of each species depends on the distance from the electrode because of interactions between species and electrode, but also because approach of the species involves a rearrangement of the solvent around the species and the solvent on the electrode. The energy of the solvated product species as a function of the electrode–species distance will differ from that of the solvated reactant species. The electronic transition is between a state on the metal (with energy at the Fermi level) and a state on a solvated species. Thus a reduction reaction of A^O would be written $A^O + e^-(\epsilon_F) \to A^R$; one of the curves in Figure 26 would represent the energy of A^R as a function of distance and the other would represent the energy of A^O as a function of distance, plus the energy of the Fermi-level electron in the metal.

Let us suppose the solid curve in Figure 26 is for A^R and the broken curve is for $A^O + e^-$. For several values of the reaction coordinate, the electronic potential energy as a function of distance perpendicular to the metal surface effectively looks like Figure 29. The electronic energy inside the metal may be taken as constant, equal to the Fermi energy ϵ_F independently of position. (In fact ϵ_F is potential plus kinetic energy, but the exclusion principle, which makes the electron's energy ϵ_F rather than the energy at the bottom of the band, may be thought of as a positive potential energy contribution.) At the surface of the metal, the electronic energy passes through a maximum (potential energy barrier), on the other side of which, at the position of A, there is a minimum. There exists an electron state with wave function localized around that position. The energy of this state (including the effect of solvation) is above ϵ_F for $Q = Q_1$, equal to ϵ_F for $Q = Q_T$, and below ϵ_F for $Q = Q_2$. A radiationless transition, in which the electron moves from the metal to the solution species, is possible for $Q = Q_T$. Thus the approach of the oxidized species to the electrode corresponds to decreasing Q along the broken curve of Figure 26, starting at Q_2. When $Q = Q_T$, the electron may be transferred from the metal, converting A^O to A^R and switching the system from the broken curve to the solid curve. Continued change of Q from Q_T to Q_2 produces A^R with its equilibrium (stable) arrangement of solvent. During this adiabatic process, the electronic wave function (dashed curves in Figures 29b and 29c) remains localized on A, with its energy decreasing from ϵ_F to the value corresponding to solvent equilibrium appropriate for A^R.

A change in the electrode–solution potential difference changes the energy of the ion in solution relative to the Fermi energy. It can thus produce changes in the electron potential curve similar to those of Figure 29. The effect of a change in the electrode–solution potential difference (overpotential) on the barrier through which the transferring electron must pass, and hence on the rate of the electron transfer and the current density, is obvious.

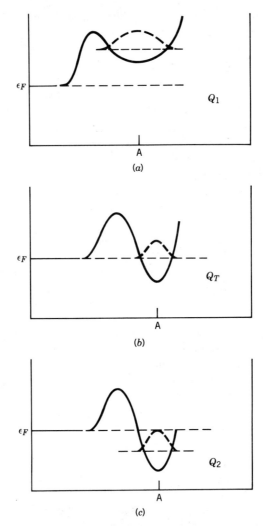

FIGURE 29. Electronic potential energy near electrode surface for various solvent configurations Q. An electronic transition is possible for $Q = Q_T$. Energy levels and wave functions are shown by broken lines.

C. SOLVENT FLUCTUATIONS

In the Levich–Dogonadze theory for an electron-transfer reaction, the fluctuations in the orientation of solvent dipoles around an ion are of central importance. These fluctuations correspond to the change in electrical polarization of the continuum theory (Chapter 9, Section B). It is assumed that the electronic energy of an ion varies quadratically or parabolically as a function of these fluctuations. We can write the energy U as a power series in the coordinate Q representing these fluctuations, with $(dU/dQ)_0 = 0$ if the origin, $Q = 0$, corresponds to the minimum energy, that is, $U = U_0 + (d^2U/dQ^2)_0 (Q^2/2) + \cdots$. Truncating the series after the quadratic term is valid if the fluctuations are small.

Since a number of coordinates are necessary to describe the solvent, we write the electronic energy of an ion pair in the state n as

$$U^n = U_0^n + \sum_k (\tfrac{1}{2} h\nu_k)(x_k - x_{kn})^2 \qquad [9]$$

where the sum is over the solvent coordinates, indexed by k, and the second derivative $[d^2U^n/d(x_k)^2]_0$ has been written as $\tfrac{1}{2} h\nu_k$, with ν_k suggesting a vibrational frequency. The frequencies ν_k have been taken independent of n. The origin, at which U^n has its minimum value, is at $x_k = x_{kn}$, which depends on the state n, because different states have their maximum stability for different solvent configurations. If there is only one solvent coordinate, equation [9] is a parabola; if there are two, it is a paraboloid or parabolic bowl. For another state of the ion pair, say m, we write

$$U^m = U_0^m + \sum_k (\tfrac{1}{2} h\nu_k)(x_k - x_{km})^2 \qquad [10]$$

where the same values of ν_k and the same solvent coordinates have been used for simplicity. The energy of m has its minimum when $x_k = x_{km}$ for each k. Electron transfer, producing state m from state n, can take place for values of the coordinates such that $U^m = U^n$.

The ion pair in the state n was formed by the diffusing together of two ions. The electron transfer transforms an ion pair in the state n into an ion pair in the state m. Subsequent diffusion of the ions away from each other produces two separated ions that differ from those that came together initially. The reaction thus involves three steps. If the electron transfer is the rate-determining step, we consider that equilibrium has been established for the diffusion step that precedes it. We are now interested in the intersection of the surfaces in equations [9] and [10].

Two parabolas intersect in a single point, and two paraboloids intersect in a curve. If there are many solvent coordinates, so that U^m and U^n are surfaces in a many-dimensional space, their intersection may be a complicated entity. However, the probability of solvent fluctuations decreases with their energy, so that we are interested in the intersection of lowest energy. Then things become simpler. We can anticipate that the point (set of values of the x_k) of lowest energy which is common to U^m and U^n lies on a line between the minimum of U^m and the minimum of U^n. This line will in fact become the reaction coordinate. In the space of solvent coordinates, the minimum of U^m is at $x_1 = x_{1m}$, $x_2 = x_{2m}$, and so on, and the minimum of U^n is at $x_1 = x_{1n}$, $x_2 = x_{2n}$, and so on, so the points on the line connecting them are given by $x_1 = x_{1m} + a(x_{1n} - x_{1m})$, $x_2 = x_{2m} + a(x_{2n} - x_{2m})$, and so on, where a is a number between 0 and 1. The point on this line which is common to equations [9] and [10] is given by

$$U_0^m + \sum_k (\tfrac{1}{2}h\nu_k)[a(x_{kn} - x_{km})]^2$$
$$= U_0^n + \sum_k (\tfrac{1}{2}h\nu_k)[(a-1)(x_{kn} - x_{km})]^2 \qquad [11]$$

Rearranging equation [11] we get

$$U_0^m - U_0^n = \left[\sum (\tfrac{1}{2}h\nu_k)(x_{kn} - x_{km})^2\right][(a-1)^2 - a^2]$$

which is easily solved to give

$$a = \frac{U_s + U_0^n - U_0^m}{2U_s} \qquad [12]$$

where $U_s = \sum (\tfrac{1}{2}h\nu_k)(x_{kn} - x_{km})^2$. The interpretation of U_s is that it is the energy of solvent reorganization required to go from the configuration appropriate for ion pair state m to the configuration appropriate for ion pair state n.

If the value of a is used to calculate the x_k in equation [9] or [10], we have the energy

$$U_0^m + \frac{(U_s + U_0^n - U_0^m)^2}{4U_s}$$
$$= \frac{1}{4}[U_s + (U_s)^{-1}(U_0^n - U_0^m)^2] + \frac{1}{2}(U_0^n + U_0^m) = U_M \qquad [13]$$

where U_M is the energy of the transition state at which electron transfer may occur. The energy increase in going from state m to the transition state is

$$U_m^\ddagger = U_M - U_0^m = \tfrac{1}{4}[U_s + (U_s)^{-1}(U_0^n - U_0^m)^2] + \tfrac{1}{2}(U_0^n - U_0^m)$$

With $\Delta U = U_0^n - U_0^m$ (the energy of reaction), the activation energy may be written concisely as

$$U_m^\ddagger = \frac{(U_s + \Delta U)^2}{4U_s} \qquad [14]$$

A similar equation can be derived from the "continuum dielectric" model of the solvent.

In this model, we calculate the amount of work (free-energy change) required to bring the solvent from an equilibrium state of polarization, appropriate to the initial charges on the ions, to a state of polarization, due to orientation of dipoles and vibrations, for which the energy of the ions before charge transfer is equal to

C. SOLVENT FLUCTUATIONS

their energy after charge transfer (Marcus 1977a). This is done by changing the charges on the ions gradually to some intermediate values, with the orientational polarization following the change, and then changing the charges back to their original values with the orientational polarization fixed and only the electronic polarization following (Chapter 6, Section D; Chapter 9, Section B). A more general formulation is also possible (Marcus 1977b).

Consider a gradual change in the charge of ion i from e_i to e'_i, defined by the parameter t, $0 \leq t \leq 1$. When the charge on ion i is $e^t_i = e_i + t(e'_i - e_i)$, the potential in the medium at a point $\mathbf{r}$ is

$$\psi(\mathbf{r}) = \frac{e^t_1}{\epsilon_s r_1} + \frac{e^t_2}{\epsilon_s r_2}$$

where ϵ_s is the static dielectric constant and r_1 (r_2) is the distance from the center of ion 1 (2) to $\mathbf{r}$. Let a_1 and a_2 be the radii of the ions. The potential at a point on the surface of ion 1 due to the medium and to ion 2 is

$$\Phi^t_1 = \frac{e^t_2}{\epsilon_s r_2} + \frac{e^t_1}{a_1} \left[(\epsilon_s)^{-1} - 1 \right]$$

because e^t_1/a_1 is the self-potential of ion 1. The first term, averaged over the surface of the ion, is $e^t_2/\epsilon_s R$, where R is the interionic distance. The total work of charging is

$$W_\mathrm{I} = \int (\Phi^t_1 \, de^t_1 + \Phi^t_2 \, de^t_2) = \int_0^1 [\Phi^t_1(e'_1 - e_1) + \Phi^t_2(e'_2 - e_2)] \, dt$$

which is easily evaluated.

For the reverse process of bringing the charges from e'_i to e_i, the medium is supposed to respond through the electronic part of the dielectric constant only. The work involved, W_II, is like W_I except that ϵ^∞, the high-frequency dielectric constant, replaces ϵ_s. The net work is the free energy of the fluctuation which brings the system to the state of polarization that permits a transition. The result (Marcus 1977) is

$$\Delta G^r = W_\mathrm{I} + W_\mathrm{II} = [(\epsilon^\infty)^{-1} - (\epsilon_s)^{-1}] \left[\frac{(\Delta e_1)^2}{2a_1} \right.$$
$$\left. + \frac{(\Delta e_2)^2}{2a_2} + \frac{\Delta e_1 \Delta e_2}{R} \right] \quad [15]$$

where $\Delta e_i = e'_i - e_i$. The free energy ΔG^P for the fluctuation starting from the product ions instead of the reactant ions has the same form, with the product-ion charges replacing e_1 and e_2.

The charges e'_i are found by minimizing ΔG^r. Of course $\Delta e_1 = -\Delta e_2$ and $\Delta G^r - \Delta G^P = \Delta G$, the free energy of reaction, which is independent of the e'_i. The free-energy barrier to reaction is the work required to bring the reactant ions together plus ΔG^r. On substituting the charges e'_i found from the constrained minimization, one obtains $\Delta G^r = m^2 G_s$ with

$$G_s = (e_1^P - e_1)^2 [(2a_1)^{-1} + (2a_2)^{-1} - R^{-1}] [(\epsilon^\infty)^{-1} - (\epsilon_0)^{-1}]$$

where e_1^P is the charge of ion 1 after electron transfer (in the product ion) and $-(2m+1)G_s$ must be equal to the free energy of reaction ΔG. Thus the activation free energy, except for the small work term for bringing the reactants together, is

$$G^{\ddagger} = \frac{(G_s + \Delta G)^2}{4G_s} \qquad [16]$$

which is quite similar to equation [14], except that free energies replace energies. The activation energy or free energy determines the rate at which solvent reorientation can produce the transition state. This must be combined with the rate of electron transfer when the transition state is reached, to give the rate constant.

The second-order rate constant for homogeneous electron transfer would be written

$$k_2 = K_Q \kappa \nu_s \exp\left(-\frac{G^{\ddagger}}{RT}\right)$$

where κ is the electronic transition probability in the transition state, ν_s is the frequency factor (rate at which solvent orientation fluctuates to the transition state), and K_Q is the equilibrium constant for the diffusion that brings reactants together, producing the ion pair. K_Q gives the ratio of the concentration of ion pairs to the product of the concentrations of the separated reactants. Thus

$$K_Q = \exp\left(-\frac{\Delta G_Q}{RT}\right)$$

where ΔG_Q is the free-energy change in going from separated reactants (say R + O) to reactants in proximity (ion pair). The rate constant is thus $\kappa \nu_s \exp(-\Delta G^{\ddagger}/RT)$, where $\Delta G^{\ddagger} = G^{\ddagger} + \Delta G_Q$. The observed free energy of reaction would be $\Delta G_Q + \Delta G_P + \Delta G$, where ΔG_P is the free energy change for diffusion apart of the product ion pair to separated ions.

Now we consider an electrode reaction, that is, heterogeneous electron transfer, using first the Levich-Dogonadze "solvent vibration coordinate" approach. For the reduction of O to give R, suppose the energy of O as a function of distance from the electrode is

$$U^O = U_0^O + \tfrac{1}{2} h\nu_0 (x - x_0^O)^2$$

C. SOLVENT FLUCTUATIONS

and the energy of R as a function of distance from the electrode is

$$U^R = U_0^R + \tfrac{1}{2}h\nu_R(x - x_0^R)^2$$

(We could also consider multiple solvent fluctuation coordinates for O and R to get an equation resembling equation [10].) The energy U_R includes the energy associated with removal of an electron from the metal, in U_0^R.

We seek the intersection of U^O and U^R, since electron transfer occurs between states of equal energy. Suppose the intersection is at $x = x_0^O + a(x_0^R - x_0^O)$, so that

$$U_0^O + \tfrac{1}{2}h\nu_O(a)^2(x_0^R - x_0^O)^2 = U_0^R + \tfrac{1}{2}h\nu_R(a-1)^2(x_0^R - x_0^O)^2$$

or

$$a^2 b - (a-1)^2 c - \Delta U = 0 \qquad [17]$$

where $\Delta U = U_0^R - U_0^O$, $b = \tfrac{1}{2}h\nu_O(x_0^R - x_0^O)^2$, and $c = \tfrac{1}{2}h\nu_R(x_0^R - x_0^O)^2$. The solution to equation [17] is

$$a = (b-c)^{-1}\left[-c + (bc + b\Delta U - c\Delta U)^{1/2}\right]$$

The activation energy for reduction is the energy at $x = x_0^O + a(x_0^R - x_0^O)$ minus the energy of the oxidized species at its equilibrium value of x, that is,

$$U_0^O + \tfrac{1}{2}h\nu_O a^2(x_0^R - x_0^O)^2 - U_0^O = ba^2$$

The activation energy for the reverse reaction, oxidation of R to O, is $c(a-1)^2$, and the difference of activation energies is $ba^2 - c(a-1)^2 = \Delta U$ (see equation [17]), the energy of reaction.

The activation energy does not resemble equation [16] because U^O and U^R have different frequencies. However, if $d = b - c$ is small, then

$$\left[bc + (b-c)\Delta U\right]^{1/2} = \left[b^2 - bd + d\Delta U\right]^{1/2}$$

$$\simeq b\left[1 - \tfrac{1}{2}b^{-2}d(b - \Delta U) - \tfrac{1}{8}b^{-2}d^2(b - \Delta U)^2\right]$$

so a of equation [17] is $d^{-1}[d - b + b - \tfrac{1}{2}b^{-1}d(b - \Delta U) + \cdots]$ and the activation energy is

$$\Delta U^{\ddagger} = ba^2 \simeq b\left[1 - \tfrac{1}{2}b^{-1}(b - \Delta U)\right]^2 = \tfrac{1}{4}b\left(1 + \frac{\Delta U}{b}\right)^2 \qquad [18]$$

For the "continuum dielectric model" a relation like this can be derived (Marcus 1977a), with free energies replacing energies, as in equations [14] and [16]:

$$\Delta G^{\ddagger} = \frac{1}{4} \Delta G_S \left(1 + \frac{\Delta G^E}{\Delta G_S}\right)^2 \quad [19]$$

Here $\Delta G^{\ddagger}$ is the free energy of activation, ΔG_S is the free energy associated with solvent rearrangement and interaction of O and R with the electrode, and ΔG^E is the free energy change for the electrode reaction, corresponding to ΔU in equation [18].

The rate constant (rate divided by concentration of O) for reduction, proportional to the cathodic current density, is

$$k_E = K_E \kappa_E \nu_E \exp\left(-\frac{\Delta G^{\ddagger}}{RT}\right) \quad [20]$$

where κ_E is the transition probability for electron transfer from the electrode when the transition state has been reached, ν_E is a frequency factor, and K_E is an equilibrium constant for processes preceding the rate-determining step, which bring an ion O from bulk solution to its equilibrium position at x_0^O. If ΔG^0 is the free energy for this process, $K_E = \exp(-\Delta G^0/RT)$. Including the free-energy change ΔG^R for removal of R from electrode to bulk solution, the free-energy change for the reaction $O + e^- \rightarrow R$ is $\Delta G^0 + \Delta G^E + \Delta G^R$.

The process of electron transfer, described by k_E of equation [20], may also be considered as a series of chemical reactions:

$$O \leftrightarrow A^O, \quad A^O \leftrightarrow A^R, \quad A^R \rightarrow R$$

The first is the activation step that forms the activated complex from O (by motion to the electrode, solvent reorganization, etc.) with forward rate constant k_1 and reverse rate constant k_{-1}. The second is the electron-transfer step, with forward and reverse rate constants k_2 and k_{-2}. The third is the movement of reduced species and relaxation of solvent and ion atmosphere to an equilibrium configuration. The rate constants k_1 and k_{-1} are equal to each other and have values of about 10^{13} s^{-1} (Marcus 1977a). The rate constant k_3, for $A^R \rightarrow R$, has about the same value, because the "reaction" is the same kind of fluctuation as in the first step. The constants k_2 and k_{-2} are equal, because k_2/k_{-2} is the equilibrium constant for $A^O \leftrightarrow A^R$, and A^O and A^R differ only by a small entropy change.

Applying the two steady-state conditions on the intermediates, $d[A^R]/dt = d[A^O]/dt = 0$, it is easy to show

$$[A^R] = \frac{k_1 k_2 [O]}{k_{-1} k_{-2} + k_{-1} k_3 + k_2 k_3}$$

so that the rate of the overall reaction is

$$\frac{d[R]}{dt} = k_3[A^R] = k_f[O]$$

with

$$k_f = k_1 k_2 k_3 (k_{-1} k_{-2} + k_{-1} k_3 + k_2 k_3)^{-1} = k_1 \left(2 + \frac{k_3}{k_2}\right)^{-1}$$

In the last term we have used the relations between the rate constants of the preceding paragraph. Now k_f/k_1 represents the probability that an electron transfer occurs during the lifetime of the intermediate A^O, about 10^{-13} s.

D. IMPLICATIONS

To derive the Tafel relation, we consider $(\partial \ln j/\partial E)_{T\mu}$ where E is the electrode-solution potential difference and j is the current density, which is proportional to k_E of equation [20]. Now

$$\left(\frac{\partial \ln j}{\partial E}\right)_{T\mu} = \left(\frac{\partial \ln k_E}{\partial E}\right)_T = -(RT)^{-1}\frac{d\Delta G^\ddagger}{dE}$$

In $\Delta G^\ddagger$, as given by equation [19], only ΔG^E depends on E, so

$$(\partial \ln j/\partial E)_{T\mu} = -(RT)^{-1}\left(\frac{1}{2}\Delta G_S\right)\left(1 + \frac{\Delta G^E}{\Delta G_S}\right)(\Delta G_S)^{-1}\frac{d\Delta G^E}{dE}$$

Now ΔG^E is the free energy difference between oxidized and reduced species near the outer Helmholtz plane. The species differ by a single electron charge so $d(\Delta G^E)/dE = -\mathscr{F}$ (ΔG^E is energy per mole) and

$$\frac{RT}{\mathscr{F}}\left(\frac{\partial \ln j}{\partial E}\right)_{T\mu} = \frac{1}{2}\left(1 + \frac{\Delta G^E}{\Delta G_S}\right) \qquad [21]$$

which is a constant, to the extent that ΔG^E and ΔG_S do not vary much with E. This is the Tafel law.

The transfer coefficient is

$$\alpha = \frac{1}{2}\left(1 + \frac{\Delta G^E}{\Delta G_S}\right)$$

The transfer coefficient is close to $\frac{1}{2}$ when ΔG_S is large compared to ΔG^E, the thermodynamic energy of reaction. The value of ΔG_S, like $b = \frac{1}{2}h\nu_0(x_0^R - x_0^O)^2$ (always positive), reflects the difference between interaction energies of R and O with electrode and solvent, which determines the kinetics according to equations [19] and [16]. Thus $\alpha = \frac{1}{2}$ when a reaction is almost in balance thermodynamically (ΔG^E small) and difficult kinetically (ΔG_S large). Near the equilibrium electrode–solution potential E^0, ΔG^E vanishes and $\alpha = \frac{1}{2}$ according to equation [21]. For fast reactions the theory predicts that α may be much below $\frac{1}{2}$ and may depend on E, since ΔG^E depends on E. Such a dependence has indeed been found for the Cr^{3+}/Cr^{2+} and other systems. If the electrode–solution potential is such that ΔG^E approaches ΔG_S, α approaches 1. Figure 30 may show this more clearly. Here $\Delta G = \Delta G^E$.

According to equation [19], if ΔG is close to ΔG_S, then $\Delta G^{\ddagger}$ is close to ΔG, as shown in Figure 30a. In this case, a change in E by C, which corresponds to displacing the O curve by $\mathscr{F}C$ (to the broken curve), will change ΔG by $\mathscr{F}C$, but

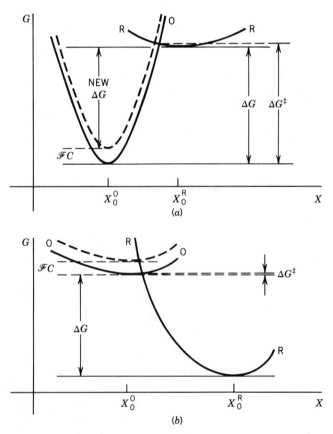

FIGURE 30. Effect of change of electrode–solution potential on ΔG and $\Delta G^{\ddagger}$. (a) $\alpha \simeq 1$. (b) $\alpha \simeq 0$.

the height of the intersection of O and R on the graph hardly changes. Then the change in $\Delta G^{\ddagger}$ is about $\mathfrak{F}C$, corresponding to $\alpha = 1$. In Figure 30b, $\Delta G^{\ddagger}$ is small. Moving the O curve up by $\mathfrak{F}C$ (to the broken curve) hardly changes $\Delta G^{\ddagger}$, because the intersection of the curves is on an almost vertical part of R. This corresponds to $\alpha \cong 0$. However, lowering the O curve by raising E will eventually lead to an increase in α as the intersection of the curves moves to progressively less vertical regions of the R curve.

According to equation [21], the Tafel law (constancy of α with potential) means that $1 + \Delta G^E/\Delta G_S$ is independent of electrode potential. Writing $\Delta G^E = \Delta G^0 - \mathfrak{F}(E - E^0)$ and $E - E^0 = -\eta$ (η = cathodic overpotential), the condition is that $1 + \Delta G^0/\Delta G_S$ should be large compared to $\mathfrak{F}\eta/\Delta G_S$. For many reactions, the Tafel law holds for overpotentials of 0.4 to as much as 1.4 V, while ΔG^0 is of the order of 1 eV. Assuming that we could detect nonconstancy of α to within 5% accuracy over a range of 0.5 V in η, the value of ΔG_S should be 10 eV. ΔG_S, the change in solvent–ion interaction energy on charge transfer, may be estimated from the continuum theories (Section C) as a difference of expressions of the form

$$E^S = \tfrac{1}{2}z^2e^2r^{-1}[(\epsilon^\infty)^{-1} - (\epsilon_0)^{-1}]$$

where r is the ionic radius. With dielectric constants $\epsilon^\infty = 78$ and $\epsilon_0 = 6$, and $r = 1\text{A}$, we have $E^S/z^2 = 1.8 \times 10^{-12}$ erg or 1.1 eV. Thus 10 eV seems rather high; other theories for E^S also suggest values of 1–2 eV, which would make α noticeably a function of η. This and related problems with solvent-reorientation theories have spurred the development of a molecular theory of electrode kinetics (Khan and Bockris 1983). On the other hand, there is some evidence for a dependence of the transfer coefficient on overpotential (Marcus 1977).

Electron transfers, homogeneous or heterogeneous, generally go one electron at a time. If several electrons are involved in the overall process, the existence of an intermediate species is assumed. Thus the two-step process,

$$O^n + e^- \rightarrow O^{n-1}, \qquad O^{n-1} + e^- \rightarrow O^{n-2}$$

involving the intermediate O^{n-1}, is preferred to the direct reaction

$$O^n + 2e^- \rightarrow O^{n-2}$$

This can be understood in terms of the necessity for solvent reorganization. The more solvent reorganization is necessary, the larger the activation energy and the slower the rate. The solvent configuration appropriate to O^{n-1} is expected to be intermediate between those for O^n and O^{n-2}, as shown in Figure 31, where free energy vs. solvent coordinate is indicated for the three species. The activated complexes for the two single-electron steps are at the intersections of curves n and $n - 1$ and of curves $n - 1$ and $n - 2$. The two-electron transfer would have its transition state at the intersection of curves n and $n - 2$ (extension denoted by

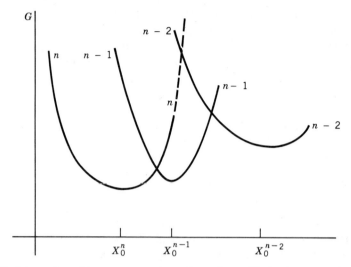

FIGURE 31. Intersection of free-energy surfaces for systems differing by one and two electrons. The activation energy for a direct two-electron transition (broken curve) is much higher than for either one-electron transition.

broken curve). This is likely to be much higher than the other two transition states, and the rate of reaction depends exponentially on the activation energy. Therefore, the rate of the two-electron process would be much smaller than that of the one-electron processes.

Consider now an electrode reaction that involves formation of a product adsorbed on the electrode. For electrodes of different metals, the heat of adsorption differs. If $\Delta\lambda$ is the change in heat of adsorption in going from metal 1 to metal 2, the potential curve for products (say n in Figure 31) on metal 1 is lowered by $\Delta\lambda$ compared to the corresponding curve for metal 2, whereas the curve for reactants (say $n - 1$ in Figure 31) is unaffected. The activation energy is obviously changed. However, a change in overpotential can shift the curve for reactants down, relative to the product curve, by the same amount $\Delta\lambda$. The activation energy, with both curves shifted by the same amount, is obviously the same. Thus, at the same current density (same activation energy) the difference in overpotentials on two metals is equal to the difference in heats of adsorption of the product of the discharge step (Krishtalik 1983, Section 3.2.1).

Since heterogeneous electron transfer ($O + e^- \rightarrow R$ at an electrode) and homogeneous electron transfer ($R' + O \rightarrow O' + R$) are described similarly, a comparison of the processes is of interest. The steps in heterogeneous reduction are: diffusion of the ion O to its equilibrium position near the electrode, change of solvation of O and/or motion of O relative to the electrode surface to create the transition state, electron transfer or electronic transition from a metal state to an ion state to convert O to R, relaxation of the solvation and/or position of R to its equilibrium configuration, and diffusion of R away from the electrode to a position in bulk electrolyte. The oxidation (R → e^- + O) involves these steps in reverse. For the electron transfer that creates O' + R from R' + O, the steps are: diffusion together of separated R' and O ions to form an ion pair, change in solvation around the ion pair to an arrangement for which O' + R and R' + O have the same energy (transition state), transfer of an electron from R' to O to form O' + R at the same energy (denoted in Section C as a transition from state m of the ion pair to state n), relaxation of the solvent configuration to the equilibrium value for O' + R, and diffusion separating the ion pair into ions O' and R.

The energy of solvent reorganization for the homogeneous electron transfer, which we wrote (equations [10] and [14]) as $U_S = \Sigma (\frac{1}{2}h\nu_k)(x_{kn} - x_{km})^2$, is expected to be roughly twice as great as for the heterogeneous electron transfer, since it involves changing the solvation of two ions rather than one. Note that the same positive energy change is involved in changing the solvation around an Fe^{2+} ion to that appropriate to Fe^{3+} as in changing the solvation around Fe^{3+} to that appropriate to Fe^{2+}.

Suppose we consider symmetrical electron-transfer reactions, in which R and R' are chemically identical and O and O' are chemically identical (e.g., Fe^{2+} + Fe^{3+} → Fe^{3+} + Fe^{2+}). The rate of such reactions can be observed by using isotopically labeled reactants. Then (going from energies to free energies) G_S for the homogeneous reaction R' + O → O' + R (see equation [16]) is twice ΔG_S for the heterogeneous reaction O + e^- + R (see equation [19]). Suppose furthermore that ΔG_S is considerably larger than ΔG for the heterogeneous reaction so that

$$k_E = K_E \kappa_E \nu_E \exp\left(-\frac{\Delta G_S}{4RT}\right) \quad [22]$$

Since $\Delta G = 0$ for the homogeneous reaction we have

$$k_2 = K_Q \kappa \nu_S \exp\left(-\frac{G_S}{4RT}\right) = K_Q \kappa \nu_S \exp\left(-\frac{\Delta G_S}{2RT}\right) \quad [23]$$

where K_Q is the equilibrium constant for the ion-pair formation. The frequency factor ν_S is the frequency of solvent dipole reorientation whose value is 10^{11} s^{-1} (bond vibrations have frequencies several orders of magnitude higher). The value of the diffusion equilibrium constant K_Q is between 0.1 dm^3/mol and 1 dm^3/mol. The value of the equilibrium constant K_E, for the equilibrium between ions in bulk solution and ions in a plane near the electrode, is several angstroms. (Note that $K_E \nu_S$ is the number of collisions per second an ion in solution would make with a plane, which is just the average speed in the direction perpendicular to the plane. The thermal velocity $(k_B T/m)^{1/2}$ is about 16 m/s for a molecular mass of 100 g/mol; if $K_E \nu_S \sim$ 16 m/s, $K_E \sim$ 1.6 Å.)

Assuming that ν_E is roughly the same size as ν_S, we get

$$\frac{(k_E)^2}{k_2} \sim \frac{(K_E)^2(\kappa_E)^2\nu_S}{K_Q\kappa} \qquad [24]$$

One can anticipate that κ_E and κ do not vary much from one reaction to another. The values of K_E, K_Q, and ν_S are also relatively unchanged. The large differences in observed rate constants (they vary over some seven orders of magnitude) is to be ascribed to changes in G_S and ΔG_S, which appear in exponentials in the rate constants given in equations [22] and [23], and which are canceled in forming equation [24]. Indeed, one finds that $(k_E)^2/k_2$ is roughly (to an order of magnitude) constant (Albery 1975, p. 120) for a series of reactions for which k_E and k_2 vary over several orders of magnitude.

This suggests that the model at least treats the important features of the reduction processes, that is, that understanding changes in solvation is crucial to understanding electron-transfer processes. The slower reactions (smaller rate constants) are those for which G_S and ΔG_S are large (see equations [22] and [23]), corresponding to more loosely solvated systems requiring larger solvation changes between reactant and transition state. The faster reactions have smaller G_S and ΔG_S and involve less solvent reorganization. The values of G_S and ΔG_S vary in the same way from system to system.

Another comparison carried out by Marcus (Marcus 1977, Albery 1975) is between the rate of a homogeneous reaction of the type

$$A: \quad R' + O \rightarrow O' + R$$

and the rates of the symmetrical homogeneous reactions,

$$B: \quad R^* + O \rightarrow O^* + R$$

and

$$C: \quad R'^* + O' \rightarrow O'^* + R'$$

The asterisks indicate isotopic labeling. Applying equation [16] to reactions A, B, and C, we have

$$\frac{(k_2^A)^2}{k_2^B k_2^C} = \left[\frac{(K_Q^A)^2}{K_Q^B K_Q^C}\right]\left[\frac{(\nu_S^A)^2}{\nu_S^B \nu_S^C}\right] \exp\left(-\frac{2G^{A\ddagger} - G^{B\ddagger} - G^{C\ddagger}}{RT}\right) \qquad [25]$$

where a factor $(\kappa^A)^2/(\kappa^B\kappa^C)$ has been taken as unity because the electronic transition probabilities are all about the same size. A similar argument may be used to replace the first and second factors in equation [25] by unity.

According to equation [15], we have

$$2G^{A\ddagger} - G^{B\ddagger} - G^{C\ddagger} = (4G_S^A)^{-1}(2G_S^{A2} + 4G_S^A \Delta G^A + 2\Delta G^{A2})$$
$$- (4G_S^B)^{-1}(G_S^B)^2 - (4G_S^C)^{-1}(G_S^C)^2$$

since $\Delta G^B = \Delta G^C = 0$. Since $\Delta G^A = -RT \ln K^A$, where K^A is the equilibrium constant for reaction A, equation [25] becomes

$$2 \ln k_2^A - (\ln k_2^B + \ln k_2^C)$$

$$= \ln K^A - \frac{[\tfrac{1}{2}G_S^A + \tfrac{1}{2}(\Delta G^A)^2]/(G_S^A - \tfrac{1}{4}G_S^B - \tfrac{1}{4}G_S^C)}{RT}$$

or

$$\ln k_2^A = \tfrac{1}{2}(\ln k_2^B + \ln k_2^C) + \tfrac{1}{2} \ln K^A$$

$$- \frac{\tfrac{1}{4}[G_S^A + (\Delta G^A)^2/G_S^A]}{RT} + \frac{\tfrac{1}{8}(G_S^B + G_S^C)}{RT}$$

This allows the estimation of the rate constant k_2^A for the unsymmetrical electron-transfer reaction from the equilibrium constant K^A (which can be obtained from emf measurements) and from parameters related to the symmetrical electron-transfer reactions B and C. Agreement between observed and calculated rate constants is reasonably good (Albery 1975, p. 121).

The theory has been used to elucidate the dependence of homogeneous and heterogeneous charge-transfer rate constants on solvent dielectric continuum properties, on changes in bond lengths and angles, and on the number of charges transferred (Marcus 1977). The electron-transfer theory has been extended to reactions in solution which involve proton transfer. There are many similarities, but some differences, as reviewed by Albery (1975, pp. 123–124), Bockris and Khan (1979), and Dogonadze and Kuznetsov (1983). Applications to electrocatalysis should also be noted (Appleby 1983).

It is becoming possible to make detailed quantitative calculations, using accurate theories of electronic structure, for some systems. This allows one to pass from correlations to direct confrontation of experimental results. At the same time, the effect of the simplifying assumptions of the models becomes more evident, and more sophisticated treatments seem to be required at every stage. For example, statistical theories for the fluctuations, dealing with particular molecular motions, have been used instead of the simpler dielectric model; although the equations look similar, the interpretation of ΔG_S is much more complicated.

The detailed calculations, one may hope, will enable decisions to be made as to the relative merits of the different theories. With respect to every aspect of the theories of electrochemical kinetics, "there have been numerous enlightening interactions between theory and experiment, and one can continue to look forward to this fruitful interplay in the future" (Marcus 1977a). However, "there is (still) no agreement upon the nature of the activation process in the fundamental act" (Conway et al. 1983).

REFERENCES

J. W. Albery, *Electrode Kinetics* (Clarendon Press, Oxford, 1975).

J. W. Allen and S. A. Rice, *J. Chem. Phys.* **68,** 5053 (1978).

H. C. Andersen, "Improvements upon the Debye-Hückel Theory of Ionic Solutions," in *Modern Aspects of Electrochemistry*, No. 11, Chapter 1, edited by R. E. White, J. O'M. Bockris, and B. E. Conway (Plenum Press, New York and London, 1975).

T. N. Andersen and H. Eyring, "Principles of Electrode Kinetics," in *Physical Chemistry, An Advanced Treatise*, Vol. IXA, Chapter 3, edited by H. Eyring (Academic Press, New York, San Francisco, and London, 1970).

L. Antropov, *Theoretical Electrochemistry* (Mir Publishers, Moscow, 1972).

A. J. Appleby, "Electrocatalysis," in *Comprehensive Treatise of Electrochemistry*, Vol. 7, edited by B. E. Conway, J. O'M. Bockris, E. Yeager, S. V. M. Khan, and R. White (Plenum Press, New York and London, 1983).

J.-P. Badiali and J. Goodisman, *J. Phys. Chem.* **79,** 223 (1975).

J.-P. Badiali, M.-L. Rosinberg, and J. Goodisman, *J. Electroanal. Chem.* **130,** 31 (1981).

J.-P. Badiali, M.-L. Rosinberg, and J. Goodisman, *J. Electroanal. Chem.* **143,** 73 (1983a).

J.-P. Badiali, M.-L. Rosinberg, and J. Goodisman, *J. Electroanal. Chem.* **150,** 25 (1983b).

C. A. Barlow, Jr., "The Double Layer," in *Physical Chemistry: An Advanced Treatise*, Vol. IXA, edited by H. Eyring (Academic Press, New York and San Francisco, 1970).

R. Becker and F. Sauter, *Electromagnetic Fields and Interactions* Vol. I (Blaisdell, New York and London, 1964).

L. Blum and D. Henderson, *J. Chem. Phys.* **74,** 1902 (1980).

J. O'M. Bockris and S. Argade, *J. Chem. Phys.* **49,** 5133 (1968).

J. O'M. Bockris, M. A. V. Devanathan, and K. Muller, *Proc. Roy. Soc. London* **A274,** 55 (1963).

J. O'M. Bockris and M. A. Habib, *Electrochim. Acta* **22,** 41 (1977).

J. O'M. Bockris and S. U. M. Khan, *Quantum Electrochemistry* (Plenum Press, New York and London, 1979).

J. O'M. Bockris and A. K. N. Reddy, *Modern Electrochemistry* (Plenum Press, New York, 1970).

D. Bohm, *Quantum Theory* (Prentice-Hall, Englewood Cliffs, N.J., 1951).

S. L. Carnie and D. Y. C. Chan, *J. Chem. Phys.* **73**, 2949 (1980).

J. Clavilier and N. Van Laethem-Meuree, "Comparaison des Electrodes Liquides et Solides. Aspects Experimentaux," in *Propriétés Electriques des Interfaces Chargées*, edited by D. Schuhmann (Masson, Paris and New York, 1978).

B. E. Conway, J. O'M. Bockris, E. Yeager, S. U. M. Khan, and R. White, Preface to *Comprehensive Treatise of Electrochemistry*, Vol. 7 (Plenum Press, New York and London, 1983).

C. A. Croxton, *Statistical Mechanics of the Liquid Surface* (John Wiley & Sons, Chichester and New York, 1980).

B. B. Damaskin and A. N. Frumkin, *Electrochim. Acta* **19**, 173 (1974).

N. Davidson, *Statistical Mechanics* (McGraw-Hill, New York and San Francisco, 1962).

P. Delahay, *J. Phys. Chem.* **70**, 647, 2601 (1966); *J. Electrochem. Soc.* **118**, 967 (1966).

M. P. D'Evelyn and S. A. Rice, *J. Chem. Phys.* **78**, 5225 (1983).

R. R. Dogonadze and Yu. A. Chizmadev, *Proc. Acad. Sci. SSSR Phys. Chem. Sect.* **157**, 778 (1964).

R. R. Dogonadze and A. M. Kuznetsov, "Quantum Electrochemical Kinetics: Continuum Theory," in *Comprehensive Treatise of Electrochemistry*, Vol. 7, edited by B. E. Conway, J. O'M. Bockris, E. Yeager, S. U. M. Khan, and R. E. White (Plenum Press, New York and London, 1983).

C. B. Duke and M. E. Alferieff, *J. Chem. Phys.* **45**, 923 (1967).

T. Erdey-Gru'z and S. Lengyel, "Proton Transfer in Solution," in *Modern Aspects of Electrochemistry*, No. 12, Chapter 1, edited by J. O'M Bockris and B. E. Conway (Plenum Press, New York, 1977).

H. Eyring, D. Henderson, B. J. Stover, and E. M. Eyring, *Statistical Mechanics and Dynamics*, Second Edition (John Wiley & Sons, New York and Chichester, 1982).

W. R. Fawcett, *J. Phys. Chem.* **82**, 1385 (1978).

H. S. Frank, "Structural Models," in *Water: A Comprehensive Treatise*, Vol. I, edited by F. Franks (Plenum Press, New York and London, 1973).

A. N. Frumkin, B. B. Damaskin, and O. A. Petrii, *Z. Phys. Chem. (Leipzig)* **256**, 4, 728 (1975).

A. N. Frumkin, O. A. Petrii, and B. B. Damaskin, "Potentials of Zero Charge," in *Comprehensive Treatise of Electrochemistry*, Vol. 1, edited by J. O'M. Bockris, B. E. Conway, and E. Yeager (Plenum Press, New York and London, 1980).

K. Fueki, D.-F. Feng, and L. Kevan, *J. Am. Chem. Soc.* **95**, 1398 (1973).

J. W. Gadzuk and E. W. Plummer, "Field Emission Energy Distribution," *Rev. Mod. Phys.* **45**, 487 (1973).

H. Gerischer, "Semiconductor Electrochemistry," in *Physical Chemistry, an Advanced Treatise*, Vol. IXA, Chapter 3, edited by H. Eyring (Academic Press, New York, San Francisco, and London, 1970).

J. Goodisman, *Diatomic Interaction Potential Theory, Vol. I: Fundamentals* (Academic Press, New York and London, 1973).

J. Goodisman, *Contemporary Quantum Chemistry* (Plenum Press, New York and London, 1977).

J. Goodisman, *Phys. Rev.* **32,** 4835 (1985).

J. Goodisman and S. Amokrane, *J. Phys. Chem.* **86,** 4993 (1982).

D. C. Grahame, *Chem. Rev.* **41,** 441 (1947).

D. C. Grahame, *J. Am. Chem. Soc.* **79,** 2093 (1957).

M. A. Habib, "Solvent Dipoles at the Electrode–Solution Interface," in *Modern Aspects of Electrochemistry*, Vol. 12, Chapter 3, edited by B. E. Conway and J. O'M. Bockris (Plenum Press, New York and London, 1977).

M. A. Habib and J. O'M. Bockris, "Specific Adsorption of Ions," in *Comprehensive Treatise of Electrochemistry*, Vol. I, edited by J. O'M. Bockris, B. E. Conway, and E. Yeager (Plenum Press, New York and London, 1980).

J. W. Halley, S. Pratt, and B. Johnson, *J. Electroanal. Chem.* **150,** 355 (1983).

W. N. Hansen, *J. Electroanal. Chem. Interfacial Chem.* **150,** 133 (1983).

W. A. Harrison, *Pseudopotentials in the Theory of Metals* (W. A. Benjamin, New York, 1966).

W. A. Harrison, *Solid State Theory* (Dover Publications, New York, 1977).

D. Henderson, L. Blum, and M. Lozada-Cassou, *J. Electroanal. Chem.* **150,** 291 (1983).

D. M. Heyes and J. H. R. Clarke, *J. Chem. Soc. Faraday Trans. 2* **75,** 1240 (1979).

D. M. Heyes and J. H. R. Clarke, *J. Chem. Soc. Faraday Trans. 2* **77,** 1089 (1981).

T. L. Hill, *Statistical Mechanics* (McGraw-Hill, New York, Toronto and London, 1956).

H. D. Hurwitz, "Modèles d'Interfaces en Mécanique Statistique," in *Propriétés Electriques des Interfaces Chargées*, edited by D. Schuhmann (Masson, Paris and New York, 1978a).

H. D. Hurwitz, "Isothermes d'Adsorption," in *Propriétés Electriques des Interfaces Chargées*, edited by D. Schuhmann (Masson, Paris and New York, 1978b).

H. D. Hurwitz, "Description des Systèmes a l'Équilibre Mécanique et Electrique," in *Propriétés Electriques des Interfaces Chargée*, edited by D. Schuhmann (Masson, Paris and New York, 1978c).

D. Inman and D. G. Lovering, "Electrochemistry in Molten Salts," in *Comprehensive Treatise of Electrochemistry*, Vol. 7, edited by B. E. Conway, J. O'M. Bockris, E. Yeager, S. U. M. Khan, and R. E. White (Plenum Press, New York and London, 1983).

J. Jortner, *J. Chem. Phys.* **30,** 834 (1959).

J. Jortner, *Rad. Res. Supp.* **4,** 2 (1964).

S. U. M. Khan, "Some Fundamental Aspects of Electrode Processes," in *Modern Aspects of Electrochemistry*, Vol. 15, edited by R. E. White, J. O'M. Bockris, and B. E. Conway (Plenum Press, New York and London, 1983).

S. U. M. Khan and J. O'M. Bockris, "Molecular Aspects of Quantum Electrode Kinetics," in *Comprehensive Treatise of Electrochemistry*, Vol. 7, edited by B. E. Conway, J. O'M. Bockris, E. Yeager, S. U. M. Khan, and R. E. White (Plenum Press, New York and London, 1983).

J. Koryta, J. Dvořák, and V. Boháčková, *Electrochemistry* (Methuen and Co., London, 1970).

L. I. Krishtalik, "Kinetics of Electrochemical Reactions at Metal–Solution Interfaces," in *Comprehensive Treatise of Electrochemistry*, Vol. 7, edited by B. E. Conway, J. O'M.

Bockris, E. Yeager, S. U. M. Khan, and R. E. White (Plenum Press, New York and London, 1983).

L. Kushnick and B. J. Berne, *Mod. Theor. Chem.* **5**, 41 (1977).

C. Lamy, "L'Interface Metal-Solution. Approche Electronique," in *Propriétés Electriques des Interfaces Chargées*, edited by D. Schuhmann (Masson, Paris and New York, 1978).

L. D. Landau and E. M. Lifschitz, *Quantum Mechanics: Non-Relativistic Theory* (Pergamon Press, London and Paris, 1958).

L. D. Landau and E. M. Lifschitz, *Electrodynamics of Continuous Media* (Addison-Wesley, Reading, 1960).

N. D. Lang, "Density-Functional Approach to the Electronic Structure of Metal Surfaces and Metal-Adsorbate Systems," in *Theory of the Inhomogeneous Electron Gas*, edited by S. Lundqvist and N. H. March (Plenum Press, New York, 1981).

N. D. Lang and W. Kohn, *Phys. Rev.* **B1**, 4555 (1970).

V. G. Levich, "Kinetics of Reaction with Charge Transport," in *Physical Chemistry, an Advanced Treatise*, Vol. IXB, Chapter 12, edited by H. Eyring (Academic Press, New York and London, 1970).

G. N. Lewis and M. Randall, *Thermodynamics*, Second Edition, revised by K. S. Pitzer and L. Brewer (McGraw-Hill, New York, Toronto, and London, 1961).

S. H. Liu, *J. Electroanal. Interfac. Chem.* **150**, 305 (1983).

W. L. McMillan and J. E. Mayer, *J. Chem. Phys.* **13**, 276 (1945).

R. A. Marcus, "Theory and Application of Electron Transfers at Electrodes and in Solution," in *Special Topics in Electrochemistry*, Chapter 8, edited by P. A. Rock (Elsevier Scientific, Amsterdam, Oxford, and New York, 1977a).

R. A. Marcus, "On the Theory of Overvoltage for Electrode Processes Possessing Electron Transfer Mechanism. I," in *Special Topics in Electrochemistry*, Chapter 9, edited by P. A. Rock (Elsevier Scientific, Amsterdam, Oxford, and New York, 1977b).

G. A. Martynov, *Russ. J. Phys. Chem.* **37**, 50, 1224 (1963).

G. A. Martynov, *Uspekhi Kolloid Khimiya (Symposium)* (Nauk, Moscow, 1973), p. 86.

E. Merzbacher, *Quantum Mechanics* (John Wiley & Sons, New York, London, and Sydney, 1961).

V. A. Myamlin and Yu. V. Pleskov, *Electrochemistry of Semiconductors* (Plenum Press, New York, 1977).

M. Natori, *J. Phys. Soc. Jpn* **27**, 1309 (1964).

M. Natori and T. Watanabe, *J. Phys. Soc. Jpn* **21**, 573 (1966).

C. W. Outhwaite, L. B. Bhuiyan, and S. Levine, *J. Chem. Soc. Faraday Trans. 2* **76**, 1388 (1980).

R. Parsons, *J. Electroanal. Chem.* **59**, 229 (1975).

R. Parsons, *Electrochim. Acta* **21**, 681 (1976).

R. Parsons, "Thermodynamic Methods for the Study of Interfacial Regions in Electrochemical Systems," in *Comprehensive Treatise of Electrochemistry*, Vol. 1, edited by J. O'M. Bockris, B. E. Conway, and E. Yeager (Plenum Press, New York and London, 1980).

A. K. Pikaev, *The Solvated Electron in Radiation Chemistry* (Israel Program for Scientific Translations, Jerusalem, 1971).

Y. V. Pleskov, "Electric Double Layer on Semiconductor Electrodes," in *Comprehensive Treatise of Electrochemistry*, Vol. 1, edited by J. O'M. Bockris, B. E. Conway, and E. Yeager (Plenum Press, New York and London, 1980).

Yu. V. Pleskov and Z. A. Rotenberg, "Photoemission of Electrons from Metals into Electrolyte Solutions as a Method for Investigating Double-Layer and Electrode Kinetics," in *Advances in Electrochemistry and Electrochemical Engineering*, Volume 11, Chapter 1, edited by H. Gerischer (John Wiley & Sons, New York and Chichester, 1978).

S. Ray, *Chem. Phys. Lett.* **11**, 373 (1971).

R. Reeves, "The Double Layer in the Absence of Specific Adsorption," in *Comprehensive Treatise of Electrochemistry*, Vol. 1, edited by J. O'M. Bockris, B. E. Conway, and E. Yeager (Plenum Press, New York and London, 1980).

F. Reif, *Fundamentals of Statistical and Thermal Physics* (McGraw-Hill, New York and San Francisco, 1965).

P. J. Rossky, *Ann. Rev. Phys. Chem.* **36**, 321 (1985).

A. Sanfeld, *Introduction to the Thermodynamics of Charged and Polarized Layers* (John Wiley & Sons, London and New York, 1968).

W. Schmickler, *Electrochim. Acta* **21**, 161 (1976).

W. Schmickler, *J. Electroanal. Interfac. Chem.* **150**, 19 (1983).

W. Schmickler and J. Ulstrup, *Chem. Phys.* **19**, 217 (1977).

M. J. Sparnaay, "Interface Semiconducteur-Electrolyte," in *Propriétés Electriques des Interfaces Chargées*, Chapter XIII, edited by D. Schuhmann (Masson, Paris and New York, 1978).

F. Stillinger and J. G. Kirkwood, *J. Chem. Phys.* **33**, 1282 (1960).

G. M. Torrie and J. P. Valleau, *J. Chem. Phys.* **73**, 5807 (1980).

G. M. Torrie and J. P. Valleau, *J. Phys. Chem.* **86**, 3251 (1982).

G. M. Torrie, J. P. Valleau, and G. N. Patey, *J. Chem. Phys.* **76**, 4615 (1982).

S. Trasatti, *J. Electroanal. Chem.* **52**, 31 (1974).

S. Trasatti, "The Work Function in Electrochemistry," in *Advances in Electrochemistry and Electrochemical Engineering*, Vol. 10, edited by H. Gerischer and C. W. Tobias (John Wiley & Sons, New York and Chichester, 1977).

S. Trasatti, "Solvent Adsorption and Double-Layer Potential Drop at Electrodes," in *Modern Aspects of Electrochemistry*, No. 13, edited by B. E. Conway and J. O'M. Bockris (Plenum Press, New York and London, 1979).

S. Trasatti, "The Electrode Potential," in *Comprehensive Treatise of Electrochemistry*, Vol. I, edited by J. O'M. Bockris, B. E. Conway, and E. Yeager (Plenum Press, New York and London, 1980).

S. Trasatti, *J. Electroanal. Interfac. Chem.* **150**, 1 (1983).

E. A. Ukshe, N. G. Bukun, D. J. Leikis, and A. N. Frumkin, *Electrochim. Acta* **9**, 437 (1964).

J. P. Valleau and G. M. Torrie, *Mod. Theor. Chem.* **5**, 169 (1977).

J. P. Valleau and G. M. Torrie, *J. Chem. Phys.* **76**, 4623 (1982).

J. P. Valleau and G. M. Torrie, *J. Chem. Phys.* **81**, 6291 (1984).

J. P. Valleau and S. G. Whittington, *Mod. Theor. Chem.* **5,** 137 (1977).

N. Van Laethem-Meuree, "L'Adsorption des Molecules Organiques Neutres aux Interfaces Métal–Solution d'Electrolyte," in *Propriétés Electriques des Interfaces Chargées*, Chapter VI, edited by D. Schuhmann (Masson, Paris and New York, 1978).

F. Vericat, L. Blum, and D. Henderson, *J. Electroanal. Chem.* **150,** 315 (1983); *J. Chem. Phys.* **77,** 5808 (1983).

J. M. Ziman, *Principles of the Theory of Solids* (Cambridge University Press, Cambridge, 1964).

INDEX

Accumulation layer, 169
Activated complex, 36, 268, 289, 341, 344, 352, 356
Activation free energy, 36, 37, 336, 344, 345, 354, 355
Activation overpotential, 35–39, 44
Adiabatic motion, 340–342, 347–348
Adsorption isotherm, 61–62, 70–72, 147, 152, 290
Adsorption of solvent, 201–202, 206–210, 212–216
Amalgam, 23
Asymmetry potential, 25

Band bending, 168, 174–175
Bloch theorem, 177
Boltzmann relation, 95, 97, 106, 137, 167, 207, 343
Born–Green–Yvon–Bogolyubov equation, 93–94
Born–Oppenheimer separation, 306, 309

Calomel electrode, 14
Capacitance, 19, 66, 74, 80, 81, 89, 130, 134, 140, 146, 161, 170–174, 180, 181, 182, 197, 202–205, 206, 218, 222, 224, 233
Cell, electrochemical, 1–5, 8
Charge density, 93
Charge-transfer resistance, 268, 272, 278
Chemisorption, 153, 165, 201, 208, 209. *See also* Specific adsorption
Clausius–Mossotti theory, 110, 113

Cluster theories, 113–114
Compact layer, 133, 136, 144–151, 235, 237
Concentration cell, 20–23, 27
Concentration overpotential, 43–46
Conduction electrons, 153–157, 294, 324
Conductivity, 30, 32, 33, 254, 255
Congruence, 71
Connection formulas, 322, 327
Contact potential, 2, 11, 17, 183–184, 188
Coulomb law, 78, 84, 114
Counter electrode, 28

Daniell cell, 1, 2, 3, 4, 5, 8, 12, 14
Debye–Huckel theory, 26, 32, 114, 139, 226–227
Debye length, 87, 139, 170, 221, 230, 236
Depletion layer, 169, 172, 222
Deposition potential, *see* Discharge potential
Dielectric constant, 74, 108–113, 137, 147, 152, 199, 201, 204, 226, 237, 251, 305, 353, 359, 363
Diffuse layer, 19, 78, 134, 136–144, 170, 235, 237
Diffusion, 43–45, 252–265
Diffusion coefficient, 44, 254, 279
Diffusion layer, 44–45, 47, 255, 261, 270, 274, 279
Diffusion potential, 20, 255, 258
Dipole moment, 81–82, 83, 85, 109, 111, 131
Direct correlation function, 115
Discharge potential, 49–50
Displacement, *see* Electric displacement

371

372 INDEX

Distribution function, 92–94, 96, 197, 229
Double cell, 23–24
Double layer, 3, 18, 19, 50, 63, 84, 85

ecm, see electrocapillary maximum (ecm)
Electric displacement, 83, 105, 244
Electric field, 79
Electrocapillary curve, 66, 196, 217
Electrocapillary maximum (ecm), 19, 66, 68, 203, 219
Electrocatalysis, 288–294, 363
Electrochemical potential, 9–13, 62, 88–89, 97, 102, 116–120, 125, 149, 158, 164, 183, 242–243, 252
Electrode of the first kind, 14, 28
Electrode of the second kind, 14
Electrode of the third kind, 15
Electrode potential, 123–129. See also Half-cell potential
Electrolytic cell, 1, 3
Electromotive force (emf), 2, 4–8, 12
Electron transfer, 14, 234, 301, 305, 311–312, 325, 332, 334, 354–356, 359, 361
Electrostatic units, 78–79
Emersed electrode, 239
emf, see Electromotive force (emf)
Equilibrium constant, 5
Esin–Markov coefficient, 148, 149, 150, 189, 232, 237
Esin–Markov plot, 43
Exchange current density, 39, 41, 267, 271, 276, 278, 288, 291
Extrinsic semiconductor, 167, 168, 175, 222

Faradaic current, 50
Faraday, 3, 33
Faraday's laws, 3
Fermi distribution, 155, 164, 166, 179, 333, 337
Fermi energy, see Fermi level (Fermi energy)
Fermi golden rule, 129, 303, 304, 333
Fermi level (Fermi energy), 116, 120, 121, 129, 156, 157, 158, 166, 240, 269, 293, 331
Fick's laws, 44, 252, 259. See also Diffusion
Field emission, 325–327
Flat band potential, 172, 174
Franck–Condon principle, 292, 332, 340
Frumkin isotherm, 72

Galvanic cell, 1, 3
Galvani potential, 9, 10, 11, 123
Gas electrode, 14

Gibbs adsorption equation, 60, 61, 62, 122
Gibbs dividing surface, 55, 59, 60
Gibbs–Duhem equation, 58, 60
Gibbs–Helmholtz equation, 4
Gibbs–Lippman equation, see Lippman equation
Glass electrode, 24–26
Gouy–Chapman theory, 78, 114, 134, 137–142, 201, 225–226, 229–232. See also Diffuse layer

Half-cell, 6–8
Half-cell potential, 7, 16. See also Electrode potential
Half-wave potential, 52, 281, 284
Hittorf cell, 31, 34
Hole, 166–167, 294–295
Hydrated electron, 239, 240, 243, 247, 249–251
Hydrogen bond, 131, 132, 199
Hydrophilicity, 201, 202, 216

Ideally polarizable electrode, 13, 19, 63, 108, 197
Image force, 234
Inner Helmholtz plane, 136, 145, 151
Inner potential, see Galvani potential
Interface (interfacial) region, 54–56
Intrinsic semiconductor, 167, 168, 171, 174, 222, 294
Inversion layer, 169, 175
Ion atmosphere, 31

Jellium model, 163–165, 232–233, 239

Kronig–Penney model, 176–178

Langevin function, 109, 112, 199, 210, 238, 249
Langmuir isotherm, 71–72, 290
Laplace transform, 262–263
Limiting diffusion current, 46, 50, 270–271
Lippman equation, 64–66, 108, 123, 145, 197
Liquid junction, 1, 2
Liquid–junction potential, 13, 16, 17, 20–22, 255–257
Lorentz field, 110

McMillan–Mayer theory, 227–228
Micropotential, 148–149, 237
Mixed potential, 285
Mobile adsorption, 73
Mobility, 33, 34, 253, 255

INDEX **373**

Molar conductivity, 30, 31, 259
Molecular dynamics, 114–115, 197
Molten electrolyte, 196–198
Monte Carlo method, 114–115, 228, 231, 235
Moving boundary method, 34

Nernst equation, 3, 5, 13, 14, 15, 38, 214, 285, 345
Nernst layer, *see* Diffusion layer
Normal hydrogen electrode (NHE), *see* Standard hydrogen electrode (SHE)
Normal pressure, 99, 100, 101, 102

Ohm's law, 29, 30
Onsager theory, 110, 113, 199, 249
Outer Helmholtz plane, 136, 140, 145, 221, 229, 231, 337
Outer potential, *see* Volta potential
Overpotential (overvoltage), 19, 27–29, 49, 266–268, 270–272, 312, 334, 360
Oxidation–reduction electrode, *see* Redox electrode (oxidation–reduction electrode)

Parallel-plate capacitor, 80, 82, 84, 221
Perturbation theory, 302, 309, 313, 346
Photoemission, 240–242
Poisson equation, 85, 86, 88, 111, 137, 169, 172, 226, 243
Polarizable interface, 18–19, 89, 194, 197. *See also* Ideally polarizable electrode
Polarization:
  electrical, 82, 83, 93, 109, 112, 206, 209, 218, 241, 244, 305, 350, 352–353
  of an electrode, 18, 27
Polarization resistance, 41
Polarographic wave, 51, 52, 53, 280, 284
Polarography, 46, 50–53, 264–265
Potential:
  of mean force, 94–95, 137, 226, 229
  of zero charge (pzc), 66, 126, 172, 184–185, 189, 190, 197, 217, 232, 289
Primitive model, 228–229, 230–231, 235
Proton transfer, 31, 335, 363
pzc, *see* Potential, of zero charge (pzc)

Reaction coordinate, 36, 341, 351
Reaction layer, 47
Reaction overvoltage, 286–288
Real potential, 10, 11, 158
Redox electrode (oxidation–reduction electrode), 15–17, 123, 125, 129

Reduced absolute electrode potential, *see* Vacuum scale electrode potential
Residual current, 50, 51, 53
Resistivity, 30
Reversibility, 13, 17
Reversible electrode, 13–17, 127
Reversible potential, 28, 29, 49, 289

Salt bridge, 20, 21
Schottky barrier, 90, 173
Semiconductor, 165–176, 220–224, 294–299
Separation factor, 334
SHE, *see* Standard hydrogen electrode (SHE)
Solvated electron, *see* Hydrated electron
Solvation, 119, 132, 136, 196, 293, 331, 332, 346
Sommerfeld model, 90, 157, 186–188
Space charge, 165, 168, 169–171, 173, 222
Specific adsorption, 136, 142–143, 145, 151, 153, 219–220
Standard hydrogen electrode (SHE), 8, 14, 17, 126, 127
Stern theory, 147–148
Stokes' law, 254
Surface charge density, 64, 66, 80, 82, 84, 121, 134, 169, 179–180, 216
Surface conductivity, 174
Surface energy, 56
Surface entropy, 56, 60, 62, 131, 136, 198, 199, 213
Surface excess, 60, 62, 67, 68, 69, 121, 140, 174
Surface free energy, 56, 58, 59, 60
Surface potential, 9, 126, 132, 155, 158, 161, 163, 186, 190, 192, 200, 289
Surface pressure, 70
Surface state, 176–182, 223–224
Surface stress, 78
Surface tension, 19, 57–60, 64, 75–77, 100–102, 104, 198, 249
Susceptibility, 83
Symmetry factor, *see* Transfer coefficient

Tafel equation, 39–41, 276, 278, 285–286, 312, 336, 339, 345, 357, 359
Tangential pressure, 99, 101, 102
Thomas–Fermi model, 172–173
Total electrode charge, 122
Transfer coefficient, 37, 38, 41, 267, 269, 276, 282, 289, 357–359
Transference number, *see* Transport number
Transition state, *see* Activated complex

Transmission coefficient, 316, 320, 323–325, 329, 333
Transport number, 21, 33–35, 45, 255–257, 259, 273
Tunneling, 313–317, 326–331, 333, 335, 346

Vacuum scale electrode potential, 124, 125, 128

Viscosity, 254
Volta potential, 9, 10, 127, 183

Water structure, 130–132
WKB approximation, 317–324, 327
Work function, 10, 126, 158, 159, 163, 183–184, 193–195, 240, 324

DATE DUE

PRINTED IN U.S.A.